Low Dimensional Physics and Gauge Principles

Matinyan Festschrift

Low Dimensional Physics and Gauge Principles

Matinyan Festschrift

Editors

V G Gurzadyan
A Klümper
A G Sedrakyan

A I Alikhanian National Laboratory, Armenia & Wuppertal University, Germany

NEW JERSEY • LONDON • SINGAPORE • BEIJING • SHANGHAI • HONG KONG • TAIPEI • CHENNAI

Published by

World Scientific Publishing Co. Pte. Ltd.

5 Toh Tuck Link, Singapore 596224

USA office: 27 Warren Street, Suite 401-402, Hackensack, NJ 07601

UK office: 57 Shelton Street, Covent Garden, London WC2H 9HE

British Library Cataloguing-in-Publication Data
A catalogue record for this book is available from the British Library.

Cover photo: The slopes of Mt. Aragats, photo by V. Gurzadyan.

LOW DIMENSIONAL PHYSICS AND GAUGE PRINCIPLES
Matinyan's Festschrift

ISBN 978-981-4440-33-2

Printed in Singapore.

FOREWORD

We are happy to present this volume dedicated to the 80th birthday of Professor Sergei Matinyan, the founder of the high energy schools in Georgia and Armenia. The volume is based mainly on the contributions to the workshop commenced in Armenia on September 21, 2011, at the spectacular site of Nor Amberd on the slope of Mt. Aragats (4095 m), in a conference center at the cosmic ray station (at 2000 m) of the Alikhanian National Laboratory (former Yerevan Physics Institute).

The workshop continued the long time tradition of conferences organized at this site; in the same hall during an international school on particle physics in April 1961, the speaker, Lev Okun (contributor also to this volume), was suddenly interrupted: the radio announced the first man, Gagarin, in space. To underline the spirit of the place, let us recall a few more episodes. In 1929 two young mathematicians – Alexandroff and Kolmogorov - traveling from Göttingen, climbed this mountain: here they worked on the classical textbook *Topologie* and on a paper on probability theory, respectively. In the 1940s, the physicist-brothers Abram and Artem Alikhanians founded here the first laboratories of the Yerevan Physics Institute, and astronomer Victor Ambartsumian founded the renowned observatory nearby.

The participation of Sergei Matinyan and hence the spirit of the workshop determined its non-standard schedule: having started in Nor Amberd, we then continued in Tbilisi on September 28-29, in Andronikashvili Institute of Physics; below is the welcome speech of J. Chkareuli in Tbilisi.

Ten years ago we compiled the volume *"From Integrable Models to Gauge Theories"* (World Scientific, 2002, ISBN: 9810249276) dedicated to the 70th birthday of Professor Matinyan. The conferences and hence the proceedings are the snapshots of their time, so this one reflects the time of graphene, dark matter and dark energy.

The workshop was funded by Volkswagen Stiftung, as well as by the State Committee of Science of Armenia, and by CREST and NASA Research Centers, NC, for the proceedings. We thank A.L.Kashin and S.Sargsyan for help with the proceedings.

V. G. Gurzadyan
A. G. Sedrakyan

Welcoming Speech

It's my great pleasure to open the Tbilisi part of the Workshop "Low Dimensional Physics and gauge principles" which, as I know, was successfully held in Yerevan at Alikhanian National Laboratory and we proceed now here at Andronikashvili Institute of Physics.

This Institute, among many other things, is a place where Professor Sergei Matinyan started as a theoretical physicist over 50 years ago - our distinguished colleague whose 80th birthday is specially marked by this Workshop.

I well remember that time, though I joined his Particle Theory group here at the Institute later in the mid of 60-ies. We were all very young - Oleg Kancheli, Edward Gedalin, Larisa Laperashvili and some others, all headed by Sergei, and besides, I was his first PhD student. It seems to sound a bit funny when someone of 70, like me, wants to thank another one of 80, like Sergei, for teaching and guidance, however, it is exactly what I would like to do. As is well-known, Richard Feynman used to say with some kind of surprise that, while it takes such a long time for professors to decide who from their students are really good for a scientific job; students in contrast are much quicker regarding to their professors. Indeed, when I met Sergei for the first time - it was when he lectured a Collision Theory Course at Tbilisi State University - I immediately understood two things: first, Elementary Particle Physics is the most fundamental branch of physics and second, Sergei Matinyan is a high-rank professional just in this exciting science who could really teach me something. And I came to him very soon.

Indeed, it was a good time for Particle Physics - the Eightfold Way by Gell-Mann and Ne'eman was just confirmed (with omega-hyperon found) and CP violation was just discovered. And, surely, the gauge revolution of the 70-ies in a form of the celebrated Standard Model was in a plain view. Looking back onto these years, could one surely say that we really got wiser since then? In one sense "how it works?" we had some actual progress with the Standard Model, though in another sense "why it works?" the progress is rather modest. We still have no good answer to the very existence of weak interactions at the low TeV scale rather than at the Planck scale (the so called gauge hierarchy problem), no answers to other even smaller entities in physics, like the quark and lepton masses and, of course, the tiny cosmological constant. And also the famous question "why God invented muons, if electrons already existed?" (quark-lepton family problem) is indeed unanswered yet.

So, possibly with the Standard Model and some of its extensions we are now at another and somewhat deeper layer of the truth, as compared with the 70-ies, however, certainly still too far from its core. Nevertheless, life is long enough to see many things more clearly and, sometimes, even what stands behind them. Hopefully, we will be more certain next time when we will mark the 90th birthday of Professor Sergei Matinyan again here at Andronikashvili Institute of Physics. Thank you!

J. L. Chkareuli
Andronikashvili Institute of Physics, Tbilisi.

International Advisory Committee

B. Altschuler – New York
J. Ambjorn – Copenhagen, Utrecht
A. Belavin – Moscow
J. Chkareuli – Tbilisi
R. Flüme – Bonn
V. Gurzadyan – Yerevan
I. Khriplovich – Novosibirsk
A. Klümper – Wuppertal
L. Lipatov – St. Petersburg
A. Sedrakyan – Yerevan
M. Vysotsky – Moscow

Local Organizing Commitee

H. Khachatryan (chair), T. Ghahramanyan, Sh. Khachatryan,
E. Poghosyan, S. Sargsyan, G. Yegoryan

CONTENTS

This volume is a Festschrift to
Sergei Matinyan

Born on January 8, 1931, in Tbilisi, Georgia.

Education: Tbilisi State University, 1954, with honors; candidate of phys.-math. sciences (PhD), Tbilisi State University, 1958; DrSci, Tbilisi State University, 1966;

Positions: Research Associate, Institute of Theoretical and Experimental Physics, Moscow,, 1956-57; Scientific secretary, senior researcher, 1957-1966; head lab., Institute of Physics, Tbilisi, 1966-1970; Professor, Tbilisi State University, 1959-1970; vice-director, head of theoretical lab., Yerevan Physics Institute, 1970-1992; Professor, Yerevan State University, 1971-1993; Professor, Duke University, North Carolina University, from 1993.

Membership: Academy of Sciences of Armenia, from 1990; Council of Russian Physical Society, 1992-1994; Councils of particle physics and cosmology (1989-1992), cosmic ray physics (1979-1991) and synchrotron radiation (1981-1991) of Academy of Sciences of USSR; Editorial board of *Journal of Nuclear Physics*, Moscow; American Physical Society, from 1987.

Nor Amberd, September, 2011.

A MATRIX MODEL REPRESENTATION OF THE INTEGRABLE XXZ HEISENBERG CHAIN ON RANDOM SURFACES

J. Ambjørn

Niels Bohr Institute, Blegdamsvej 17, 2100, Copenhagen, Denmark,
e-mail: ambjorn@nbi.dk

A. Sedrakyan

Physics Institute, Alikhanian Br. str. 2, Yerevan 36, Armenia
e-mail: sedrak@nbi.dk

We consider integrable models, i.e. models defined by R-matrices, on random Manhattan lattices (RML). The set of random Manhattan lattices is defined as the set dual to the lattice random surfaces embedded on a regular d-dimensional lattice. As an example we formulate a random matrix model where the partition function reproduces annealed average of the XXZ Heisenberg chain over all RML. A technique is presented which reduces the random matrix integration in partition function to an integration over their eigenvalues.

1. Introduction

One of the major goals of non-critical string theory was to describe the non-perturbative physics of non-Abelian gauge fields in two, three and four dimensions. The asymptotic freedom of these theories allowed us to understand the scattering observed at high energies,[1,2] but it also made the long distance, low energy sector of the theories non-perturbative and indicated a non-trivial structure of the vacuum.[3] It became necessary to develop non-perturbative tools which would allow us to study phenomena associated with e.g. confinement. One possibility which attracted a lot of attention was the attempt by Polyakov to reformulate the non-Abelian gauge theories as a string theory. This line of research led Polyakov to his seminal work on non-critical string theory.[4] He showed that the presence of the conformal anomaly forces us to include the conformal factor in the string path integral, and that the action associated with the conformal factor is the Liouville action. In his approach the study of non-critical string theory becomes equivalent to the study of two-dimensional quantum gravity (governed by the Liouville theory) coupled to certain conformal matter fields.

Attempts to understand and define rigorously the quantum Liouville theory triggered the lattice formulation, presenting the two-dimensional random surfaces appearing in the string path integral as a sum over triangulated piecewise linear surfaces.[5,6,8] This sum over "random triangulations" (or "dynamical triangulations" (DT)) could be represented by matrix integrals and in this way certain matrix integrals became almost synonymous to non-critical string theory. Somewhat surprisingly many of the lattice models were exactly solvable and at the same time it was possible to solve the continuum quantum Liouville model via conformal bootstrap, and whenever results of the two non-pertubative methods could be compared, agreement was found. However, the solutions only made sense for matter fields with a central charge $c \leq 1$.[11] Thus the understanding of non-critical string theory in three and four dimensions, where the non-Abelian gauge theories are non-trivial, is still missing. It is thus of great interest to try to study new classes of random surface models which might allow us to penetrate the $c = 1$ barrier. This is one of the main motivations of this paper. We propose to consider a new class of random lattices, the so-called random Manhattan lattices, which we were led to via the 3d Ising model[12,13] and via the study of the Chalker-Coddington network model.[15] The study of the latter model led to the idea that an R-matrix could be associated to a random Manhattan lattice, and we will consider how to couple in general a matter system defined by an R-matrix to a random lattice. By summing over the random lattices (i.e. taking the annealed average) we thus introduce a coupling between the integrable model and two-dimensional quantum gravity.

More precisely we start with an integrable model on a 2d square lattice, assuming we know the R-matrix. We then show that the same R-matrix can be used on a so-called Random Manhattan Lattice (RML) (see Fig. 1), which is a lattice where the links have fixed arrows which indicate the allowed fermion hopping. No hopping is allowed in directions opposite to arrows. The summation over the RMLs can be performed by a certain matrix integral related to the R-matrix. This matrix integral is somewhat different from the the conventional matrix integrals used to describe conformal field theories with $c < 1$ coupled to 2d quantum gravity, and thus there is hope than one can penetrate to $c = 1$ barrier. Below we describe the construction in detail.

2. The model

As mentioned one arrives in a natural way to a RML comes from the study of the 3d Ising model on a regular cubic lattice. The high temperature expansion of the Ising model can be expressed as a sum over random lattice surfaces of the kind shown in Fig. 1, and on these two-dimensional lattice surfaces one constructs a kind of dual lattice by the following procedure: the lattice surface consists of plaquettes. consider the mid-points of the links on the plaquettes as sites of the dual lattice, and consider arrows on the links as shown in Fig. 1.

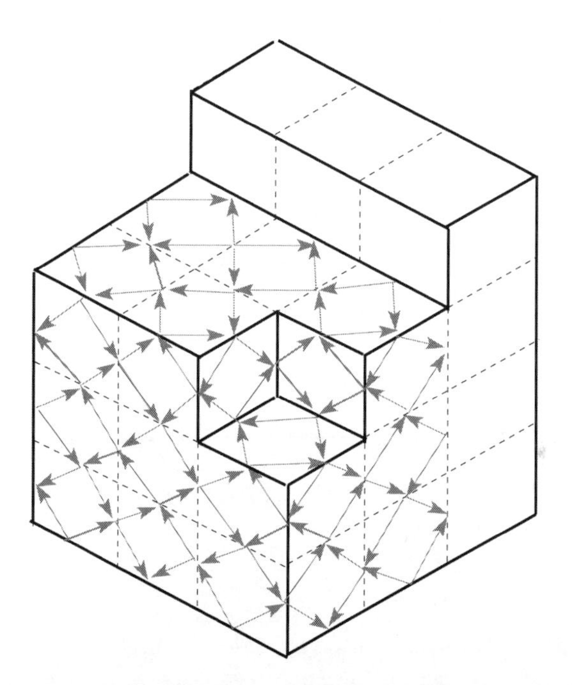

Fig. 1. Dual lattice surface

The allover orientation of arrows on the plaquettes should be such that the flow to neighbouring plaquettes is continuous as illustrated in Fig. 3. This type of dual lattice with arrows will be a finite Manhattan lattice corresponding to a particular plaquette lattice surface on the regular three-dimensional lattice. There is a one to one correspondence between the plaquette surfaces on the regular lattice and the finite Manhattan lattices described above.

A second way of obtaining a RML is by starting from oriented double

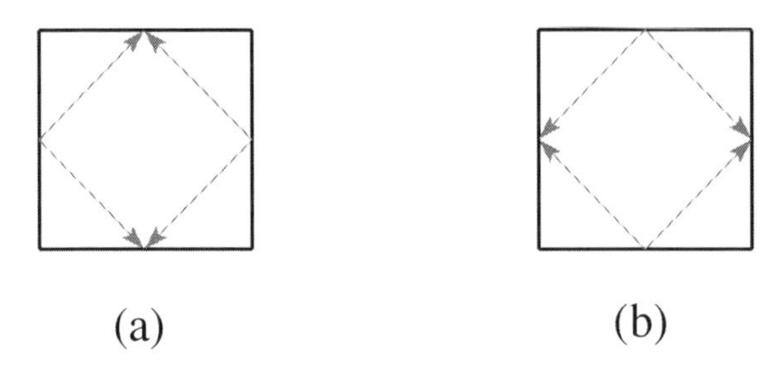

Fig. 2. Assignment of arrows to dual lattice

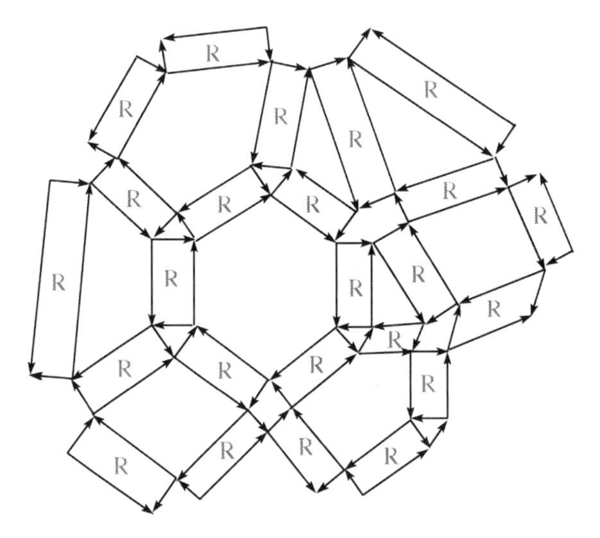

Fig. 3. Random Manhattan lattice

line graphs, like the ones introduced by 't Hooft, and then modify the double line propagator like shown in Fig. 4.

We will now attach an R-matrix of an integrable model to the squares of the RML with the index assignment shown in Fig. 5. Two neighbouring squares will share one of indices, and the same is thus the case for the corresponding R-matrices, and a summation over values of the indices are understood, resulting in a matrix-like multiplication of R-matrices. To a RML Ω we now associate the partition function

$$Z(\Omega) = \prod_{R \in \Omega} \check{R}, \tag{1}$$

where the summation over indices is dictated by the lattice. Our final par-

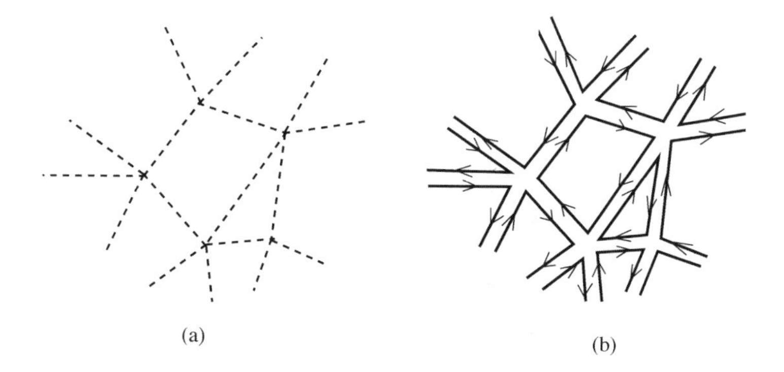

Fig. 4. Manhattan lattices and double line graphs

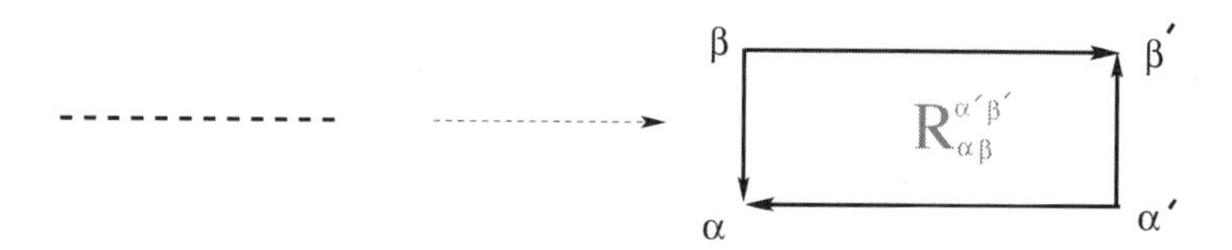

Fig. 5. Index assignment of the R-matrix

tition function is defined by summing over all possible (connected) lattices Ω:

$$Z = \sum_{\Omega} Z(\Omega) \tag{2}$$

The summation over the elements in Ω, i.e. the summation over a certain set of random 2d lattices, is a regularized version of the sum over 2d geometries precisely in the same way as in ordinary DT. It is known the sum over random polygons (triangles, squares, pentagons etc) with positive weights under quite general conditions leads to the correct continuum limit, i.e. the functional integral over 2d geometries, when the link length goes to zero. Thus it is natural to assume that the sum over RML will also represent in correct way the sum over 2d geometries when the link length goes to zero. Under this assumption we have coupled a given two-dimensional model, integrable on a regular lattice, to two-dimensional quantum gravity. In the next Section we will, in order to make the discussion mor eexplicit, consider the XXZ Heisenberg model, where the R-matrix is known.

3. The matrix model

In order to represent the *XXZ* model as a matrix model which at the same time will offer us a topological expansion of the surfaces spanned by the oriented ribbon graphs considered above, we consider the set of $2N \times 2N$ normal matrices. We label the entries of the matrices as $M_{\alpha\beta,ij}$, where α, β takes values 0,1 and i, j takes values $1, \dots, N$. The α, β indices refer to the *XXZ* model, while the i, j indices will be used to monitor the topological expansion. A normal matrix is a matrix with complex entries which can be diagonalized by a unitary transformation, i.e. for a given normal matrix M there exists a decomposition

$$M = UM^{(d)}U^{\dagger} \tag{3}$$

where U is a unitary $2N \times 2N$ matrix and $M^{(d)}$ a diagonal matrix with eigenvalues $m^{(d)}_{\alpha,ii}$ which are complex numbers.

Consider now the action

$$S(M) = M^{*}_{\alpha\beta,ij}\check{R}^{\alpha'\beta'}_{\alpha\beta} M_{\beta'\alpha',ij} - \sum_{n=3}^{\infty} \frac{1}{n} \operatorname{tr}\left(M^{n} + (M^{\dagger})^{n}\right). \tag{4}$$

We denote the sum over traces of M and $M^{\dagger}$ as the potential. The matrix partition function is defined by

$$Z = \int \mathrm{d}M \; \mathrm{e}^{-NS(M)}. \tag{5}$$

When one expands the exponential of the potential terms and carries out the remaining Gaussian integral one will generate all graphs of the kind discussed above, with the R-matrices attached to the graphs as described. The only difference is that the graphs will be ordered topologically such that the surfaces associated with the ribbon graphs appear with a weight N^{χ}, where χ is the Euler characteristics of the surface. If we are only interested in connected surfaces we should use as the partition function

$$F = \log Z. \tag{6}$$

In particular the so-called large N limit, which selects connected surfaces with maximal χ, will sum over to the planar (connected) surfaces generated by F, since these are the connected surfaces with the largest χ.

Explicitly, in the case of the *XXZ* Heisenberg model the R matrix is given by:

$$\begin{aligned}\check{R}^{\alpha'\beta'}_{\alpha\beta} &= \frac{a+c}{2} 1^{\alpha'}_{\alpha} \otimes 1^{\beta'}_{\beta} + \frac{a-c}{2} (\sigma_3)^{\alpha'}_{\alpha} \otimes (\sigma_3)^{\beta'}_{\beta} \\ & \frac{b}{2}\left((\sigma_1)^{\alpha'}_{\alpha} \otimes (\sigma_1)^{\beta'}_{\beta} + (\sigma_2)^{\alpha'}_{\alpha} \otimes (\sigma_2)^{\beta'}_{\beta}\right).\end{aligned} \tag{7}$$

As an abbreviation we will write

$$\check{R}_{\alpha\beta}^{\alpha'\beta'} = \sigma_a \otimes \sigma_a \tilde{I}_a \tag{8}$$

where a summation over index a is understood, $\sigma_0 = 1$, the identity matrix, and σ_a, $a = 1, 2, 3$ are the Pauli matrices.

Our aim is to decompose the integration over the matrix entries of M into their radial part M_d and the angular U-parameters. This decomposition is standard, the Jacobian is the so-called Vandermonde determinant (also in the case of normal matrices,[16]) When we make that decomposition the potential will only depend on the eigenvalues $m_{\alpha,ii}^{(d)}$ and for the measure we have:

$$dM = dU \prod_{\alpha,i} dm_{\alpha,ii}^{(d)}\, dm_{\alpha,ii}^{(d)*} \prod_{\alpha,i\neq\beta,j} \left|(m_{\alpha,ii}^{(d)} - m_{\beta,jj}^{(d)})\right|^2. \tag{9}$$

However, the problem compared to a standard matrix integral is that the matrices U, introduced by the transformation (3), will appear quartic in the action (4). Thus the U-integration does not reduce to an independent factor, decoupled from the rest. Neither is it of the Itzykson-Zuber-Charish-Chandra type.

In order to perform the integral we pass from the transformation (3) which is given in the fundamental representation, to a form where we use the adjoint representation. Let us choose a basis t^A for Lie algebra of the unitary group $U(2N)$ in the fundamental representation. The normal matrix M can also be expended in this basis:

$$M = C_A t^A, \qquad \mathrm{tr}\, t^A t^B = \delta^{AB}, \tag{10}$$

where the last condition just is a convenient normalization. For a given U belonging to the fundamental representation of $U(2N)$ the corresponding matrix in the adjoint representation, $\Lambda(U)$, and the transformation (3) are given by

$$\Lambda(U)_{AB} = \mathrm{tr}\, t^A U t^B U^\dagger, \qquad C_A = \Lambda_{AB} C_B^{(d)}, \tag{11}$$

where $C_B^{(d)}$ denotes the coordinates of $M^{(d)}$ in the decomposition (10). The transformation (3) is now linear in the adjoint matrix Λ and the action (4) will be quadratic in Λ. However, we pay of course a price, namely that the entries of the $(2N)^2 \times (2N)^2$ unitary matrix Λ satisfy more complicated constraints than those satisfied by the entries of the $N \times N$ unitary matrix U. We will deal with the this problem below. First we express the action (4) in terms of the eigenvalues $m_{\alpha,ii}^{(d)}$ and Λ.

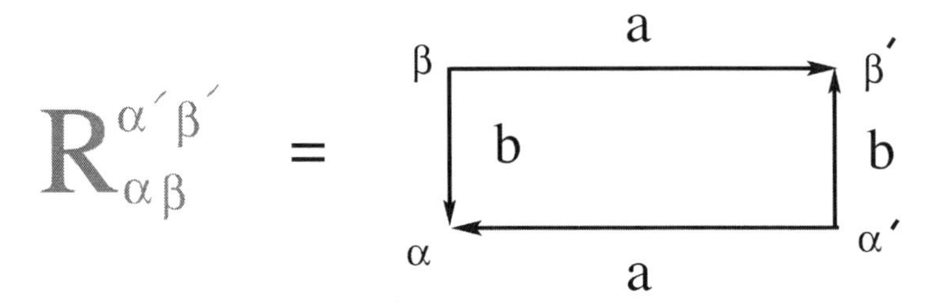

Fig. 6. Index assignment of the R-matrix

For this purpose it is convenient to pass from the representation (5) of the R-matrix to the cross channel (see Fig. 6), which is achieved by:

$$\left(\check{R}^c\right)^{\beta\beta'}_{\alpha\alpha'} = \check{R}^{\alpha'\beta}_{\alpha\beta'}, \tag{12}$$

which amounts to making a particle-hole transformation $0 \leftrightarrow 1$ for the indices α' and β. After some algebra one obtains:

$$\begin{aligned} \left(\check{R}^c\right)^{\beta\beta'}_{\alpha\alpha'} = &\frac{b+c}{2}\sigma_1 \otimes \sigma_1 + \frac{b-c}{2}\sigma_2 \otimes \sigma_2 \\ &\frac{a}{2}\left[1 \otimes 1 + \sigma_3 \otimes \sigma_3\right]. \end{aligned} \tag{13}$$

We can now write

$$\left(\check{R}^c\right)^{\beta\beta'}_{\alpha\alpha'} = \sigma_a \otimes \sigma_a \, I^a, \tag{14}$$

where the summation is over a, $a = 0, 1, 2, 3$, $\sigma_0 \equiv 1$, the identity 2×2 matrix. One has:

$$I_a = \left(\frac{a}{2}, \frac{b+c}{2}, \frac{b-c}{2}, \frac{a}{2}\right) \tag{15}$$

Using the cross channel R-matrix (9) the action (4) becomes:

$$S = M^*_{\alpha\beta,ij} \left(\check{R}^c\right)^{\beta\beta'}_{\alpha\alpha'} \delta^{i'}_i \delta^{j'}_j M_{\beta'\alpha',i'j'} - \mathrm{Re}V(M). \tag{16}$$

Let τ^μ, $\mu = 1, \ldots, N^2$ denote generators of the Lie algebra of $U(N)$, apropriately normalized such that

$$\delta^{i'}_i \delta^{j'}_j = \tau^\mu_{ij} \tau^\mu_{i'j'}. \tag{17}$$

If we insert (12) into the action (11) we obtain

$$S = \mathrm{tr}\,\left(M^\dagger \sigma^a \tau^\mu\right) I^a \mathrm{tr}\,\left(\sigma^a \tau^\mu M\right) \tag{18}$$

where we can view $\sigma^a\tau^\mu$, $a = 1,2,3,4$ and $\mu = 1,\ldots,N^2$ as the $(2N)^2$ generators t^A of the Lie algebra of $U(2N)$. Formulas (10) and (11) with the generators $\sigma^a\tau^\mu$ read

$$m^{(d)}_{a,\mu} = \frac{1}{2}\mathrm{tr}\left(M^{(d)}\sigma^a\tau^\mu\right), \tag{19}$$

$$\Lambda^{a\mu,a'\mu'} = \frac{1}{2}\mathrm{tr}\left(\sigma^a\tau^\mu U\sigma^{a'}\tau^{\mu'}U^\dagger\right). \tag{20}$$

We now want to use (3) and (15) and (16) to express the matrix M in the action (13) in terms of $m^{(d)}_{a,\mu}$ and $\Lambda^{a\mu,a'\mu'}$, and we obtain

$$S = m^{(d)*}_{a'\mu'}\Lambda^{a\mu,a'\mu'}I^a\Lambda^{a\mu,a''\mu''}m^{(d)}_{a''\mu''} - V(m^{(d)}_{a\mu}) \tag{21}$$

If we choose a basis τ^μ such that $\mu = (i,j)$, $i,j = 1,\ldots,N$ and

$$\left(\tau^{ij}\right)_{kl} = \delta_{ik}\delta_{jl} \tag{22}$$

we have from (15), for $\mu = (i,i)$, i.e. $m^{(d)}_{a,\mu} \equiv m^{(d)}_{a,ii}$

$$m^{(d)}_{3,ii} = m^{(d)}_{\alpha=1,ii} - m^{(d)}_{\alpha=2,ii}, \tag{23}$$

$$m^{(d)}_{0,ii} = m^{(d)}_{\alpha=1,ii} + m^{(d)}_{\alpha=2,ii} \tag{24}$$

$$m^{(d)}_{b=1,ii} = m^{(d)}_{b=2,ii} = 0 \tag{25}$$

3.1. *The space of integration*

We now change the integration over the unitary matrices $U(2N)$ in formula (6), which are in the fundamental representation, to the unitary matrices $\Lambda(2N)$ in the adjoined representation. Rather than using the Haar measure expressed in terms of the U-matrices we should express the Haar measure in terms of the Λ-matrices.

Since normal matrices M can be regarded as elements in the algebra $u(2N)$, the action of Λ, defined by the formula (11) on its diagonalized form (23), will form an orbit in the algebra with the basis consisting of all diagonal matrices. Diagonalized elements of M are invariant under the action of the maximal abelian (Cartan) subgroup $\otimes U(1)^{2N}$of $U(2N)$. Therefore the orbits are isomorphic to the factor space $\frac{U(2N)}{U(1)\otimes\cdots\otimes U(1)}$.

Moreover this factor space is isomorphic to a so-called flag-manifold, defined as follows (see[20] and references there): A single flag is a sequence of nested complex subspaces in a complex vector space C_n

$$\{\varnothing\} = C_0 \subset C_{a_1} \subset \cdots \subset C_{a_k} \subset C_n = C^n \tag{26}$$

with complex dimensions $dim_C C_i = i$. For a fixed set of integers $(a_1, a_2 \cdots a_k, n)$ the collection of all flags forms a manifold, which called a flag manifold $F(a_1, a_2, \cdots a_k, n)$. The manifold $F(1, 2, \cdots n)$ is called a full flag manifold, others are partial flag manifolds. Only the full flag manifold $F(1, 2, \cdots 2N)$ is isomorphic to the orbits of the action of the adjoined representation of $U(2N)$ on its algebra

$$F(1, 2, \cdots 2N) = \frac{U(2N)}{\otimes U(1)^{2N}} \tag{27}$$

The set of C_{i-1} hyperplanes in C_i is isomorphic to the set of complex lines in C_i. In differential geometry this set is denoted by $\mathbf{CP}^i$ (and also as Grassmanians $\mathbf{Gr}(1, i)$) and is called a complex projective space. Hence, the complex projective space is a factor space

$$\mathbf{CP}^i = \frac{U(i)}{U(i-1) \otimes U(1)} = \frac{S^{2i-1}}{U(1)} \tag{28}$$

where S^{2i-1} is a real $2i - 1$ dimensional sphere.

According to description presented above the orbit of the action of the adjoined representation of $U(2N)$ on the set of normal matrices M is a sequence of fibre bundles and locally, on suitable open sets, the elements of the flag manifold can be represented as a direct product of projective spaces (the fibers)

$$\frac{U(2N)}{\otimes U(1)^{2N}} \simeq \mathbf{CP}^{2N} \times \mathbf{CP}^{2N-1} \times \cdots \mathbf{CP}^2 \tag{29}$$

In simple words we have the following representation of the orbit: any diagonalized normal matrix in the adjoined representation has a following form

$$\begin{aligned} M^{(d)}_{a\mu} = \Big(& \underbrace{m^{(d)}_{0,NN}, 0, \cdots 0}_{4N-1}, \underbrace{m^{(d)}_{3,NN}, 0, \cdots 0}_{4N-3}, \cdots \underbrace{m^{(d)}_{0,kk}, 0, \cdots 0}_{4k-1}, \underbrace{m^{(d)}_{3,kk}, 0, \cdots 0}_{4k-3}, \\ & \cdots \underbrace{m^{(d)}_{0,11}, 0, 0}_{3}, m^{(d)}_{3,11} \Big) \end{aligned} \tag{30}$$

The action of the adjoined representation Λ on this M transforms it into the elements of $\frac{U(2N)}{\otimes U(1)^{2N}}$ presented in (29) where $\mathbf{CP}^{2k}$ represents image of the part $\underbrace{m^{(d)}_{0,kk}, 0, \cdots 0}_{4k-1}$.

This implies that the measure of our integral over normal matrices M can be decomposed into the product of measures of the base space (the diagonal matrices) and the flag manifold (the fiber)

$$\mathcal{D}\Lambda = \prod_{i=1}^{N} dm_{0,ii}^{(d)} dm_{3,ii}^{(d)} \prod_{k=1}^{2N} \mathcal{D}[\mathbf{CP}^k]$$
$$= \prod_{i=1}^{N} dm_{0,ii}^{(d)} dm_{3,ii}^{(d)} \prod_{k=1}^{N} \mathcal{D}\Big[\frac{S^{4k-1}}{S^1}\Big]\mathcal{D}\Big[\frac{S^{4k-3}}{S^1}\Big] \qquad (31)$$

However, since the diagonal matrix elements $m_{0,ii}^{(d)}$ and $m_{3,ii}^{(d)}$ are complex, our action is invariant over $\otimes U(1)^{2N}$ (one $U(1)$ per marked segment in (30) and we can extend integration measure from (31) to

$$\mathcal{D}\Lambda = \prod_{i=1}^{N} dm_{0,ii}^{(d)} dm_{3,ii}^{(d)} \prod_{k=1}^{N} \mathcal{D}\big[S^{4k-1}\big]\mathcal{D}\big[S^{4k-3}\big] \qquad (32)$$

In other words, we suggest that the action of Λ on the segments $\underbrace{m_{0,kk}^{(d)}, 0, \cdots 0}_{4k-1}$ and $\underbrace{m_{3,kk}^{(d)}, 0, \cdots 0}_{4k-3}$ in (30) form vectors $m_{0,kk}^{(d)} z_{0,k}^r$, $(r = 1 \cdots 4k-1)$ and $m_{3,kk}^{(d)} z_{3,k}^s$, $(s = 1 \cdots 4k-3)$, respectively, where z_k^r belongs to unite spheres S^{4k-1} and S^{4k-3}.

In order to write the measure of integration over the spheres S^{4k-1} and S^{4k-3} we embed them into the Euclidean spaces R^{4k} and R^{4k-2}, respectively, and define

$$\begin{aligned}
\mathcal{D}\big[S^{4k-1}\big] &= \delta\Big(\sum_{s=1}^{4k}[z_{0,k}^s]^2 - 1\Big)\prod_{s=1}^{4k} dz_{0,k}^s \\
&= \int d\lambda_{0,k} \prod_{s=1}^{4k} dz_{0,k}^s e^{i\lambda_{0,k}(\sum_{s=1}^{4k}[z_{0,k}^s]^2 - 1)}, \\
\mathcal{D}\big[S^{4k-3}\big] &= \delta\Big(\sum_{s=1}^{4k-2}[z_{3,k}^s]^2 - 1\Big)\prod_{s=1}^{4k-2} dz_{3,k}^s \\
&= \int d\lambda_{3,k} \prod_{s=1}^{4k-2} dz_{3,k}^s e^{i\lambda_{3,k}(\sum_{s=1}^{4k-2}[z_{3,k}^s]^2 - 1)},
\end{aligned} \qquad (33)$$

where we have introduced Lagrange multipliers $\lambda_{a,k}$, $a = 0, 3$. In order to write this embedded measure we have extended the number of components of the z-vectors by one, compared to the notation used for the action (14).

With this definition of the measure the partition function (5) can be written

$$\int \mathrm{d}Me^{-NS(M)} = \int \prod_{k,a=0,3} \mathrm{d}m^{(d)}_{a,kk}\mathrm{d}\lambda_{a,k}$$
$$\prod_{s=1}^{4k} dz^s_{0,k} \prod_{s=1}^{4k-2} dz^s_{3,k} W(m^{(d)}_{a,kk}) e^{-S(m^{(d)}_{a,kk},\lambda_{a,k},z_{a,k})}, \quad (34)$$

where $W(m^{(d)}_{a,kk}) = \prod_{\alpha,i\neq\beta,j} \left|(m^{(d)}_{\alpha,ii} - m^{(d)}_{\beta,jj})\right|^2$ is Vandermonde determinant and

$$S(m^{(d)}_{a,kk}, \lambda_{a,k}, z_{a,k}) = \sum_{a=0,3,k=1}^{N} \Big[|m^{(d)*}_{a,kk}|^2$$
$$\sum_{b=1}^{s(a,k)} |z^b_{a,k}|^2 I^b - i\lambda_{a,k}\big(\sum_{s=1}^{s(a,k)} [z^s_{a,k}]^2 - 1\big) - V(m^{(d)}_{a,kk})\Big], \quad (35)$$

where $s(0,k) = 4k$ and $s(3,k) = 4k-2$, while according to (10)

$$I = \frac{1}{2}\Big(\underbrace{0, a, b-c, b+c, a\cdots b+c}_{4N}, \underbrace{0, a, a, b-c, b+c, a\cdots a}_{4N-2}, \cdots$$
$$\underbrace{0, a, b-a, b+a}_{4}, \underbrace{0, a}_{2}\Big). \quad (36)$$

The length of I is $2N(2N+1)$, because we have added $2N$ zeros according to the addition of one extra coordinate to $z^s_{a,k}$, as described above. The 4 elements $a/2, (b-c)/2, (b+c)/2, a/2$ of I^b are placed as a sequence in I, as shown in eq. (36).

As one can see we have in the partition function (35) simple Gaussian integrals over $z^s_{a,k}$. These can be evaluated and we are left with integrals over $m^{(d)}_{a,kk}$ and $\lambda_{a,k}$ only. It is convenient to rescale the Lagrange multipliers and introduce $\tilde{\lambda}_{a,k} = |m^{(d)}_{b,ii}|^{-2}\lambda_{a,k}$. Then, after performing the Gaussian integrals, we obtain

$$Z = \int \prod_{a=0,3;k=1}^{N} \mathrm{d}m_{a,kk}^{(d)} W(m_{a,kk}^{(d)})$$

$$\prod_{k=1}^{N} \frac{1}{|m_{0,kk}^{(d)}|^{4k-2} |m_{3,kk}^{(d)}|^{4k-4}} \int \prod_{a=0,3;k=1}^{N} \frac{\mathrm{d}\tilde{\lambda}_{a,k}}{(-i\tilde{\lambda}_{a,k})^{1/2}} \cdot$$

$$\prod_{a=0,3;k=1}^{N} e^{-\mathrm{Re}V(m_{a,kk}^{(d)})+i|m_{a,kk}^{(d)}|^2 \tilde{\lambda}_{a,k}} \prod_{a=0,3;k=1}^{N} Z_{a,k}(\tilde{\lambda}_{a,k}), \quad (37)$$

where

$$Z_{0,k}(\tilde{\lambda}_{0,k}) = \frac{1}{(a - i\tilde{\lambda}_{0,k})^{k-\frac{1}{2}}} \frac{1}{(b - c - i\tilde{\lambda}_{0,k})^{\frac{k}{2}}} \frac{1}{(b + c - i\tilde{\lambda}_{0,k})^{\frac{k}{2}}}$$
$$Z_{3,k}(\tilde{\lambda}_{3,k}) = \frac{1}{(a - i\tilde{\lambda}_{3,k})^{k-\frac{1}{2}}} \frac{1}{(b - c - i\tilde{\lambda}_{3,k})^{\frac{k-1}{2}}} \frac{1}{(b + c - i\tilde{\lambda}_{3,k})^{\frac{k-1}{2}}}. \quad (38)$$

4. Conclusions

We have defined a matrix model which reproduces the partition function of an integrable model (the XXZ model) on random surfaces. The random surfaces under consideration appears as random Manhattan lattices, which are dual to random surfaces embedded in a d dimensional regular Euclidean lattice. This formulations allows us to consider a new type of non-critical strings with $c > 1$.

We show how the matrix integral can be reduced to integrals over the eigenvalues of matrices. The important ingredient of the integration over angular parameters of the matrices (which usually defined by the Itzykson-Zuber integral) is the reduction of the problem to the adjoined representation.

In this paper we consider Heisenberg XXZ chain on random surfaces, but the approach can be applied to other integrable models. A standard matrix model approach to XXZ chain was considered earlier by P. Zinn-Justin in.[18]

References

1. D. Gross, F. Wilczek, Phys. Rev. Lett. **30**, 1343 (1973).
2. H. D. Politzer, Phys. Rev. Lett. **30**, 1346 (1973).
3. G. Savvidy, Phys.lett. B **71**,133 (1977),
   S. Matinyan, G. Savvidy, Nucl. Phys. B **134**, 539 (1978).

4. A. Polyakov, Phys. Lett. B. **103**, 207 (1981),
A. Polyakov, Phys. Lett. B. **103**, 211 (1981).
5. J. Ambjørn, B. Durhuus and J. Frölich, Nucl. Phys. **B257**[FS14], 433 (1985).
6. V.A. Kazakov, Phys. Lett. B **150**,282 (1985).
7. V.A. Kazakov, I.K. Kostov and A.A. Migdal, Phys. Lett. B **157**, 295 (1985).
8. F. David, Nucl. Phys. **B257** [FS14],45 (1985).
9. D. Gross and A.A. Migdal, Phys. Rev. Lett. **64**,127 (1990);
M. Douglas and S. Shenker, Nucl. Phys. B **335**, 635 (1990).
10. E. Brezin and V.A. Kazakov, Phys. Lett.B **236**, 144 (1990).
11. D. Knizhnik, A. Polyakov, A. Zamolodchikov, Mod. Phys. Lett. A **3**, 819 (1988).
12. A. Kavalov, A. Sedrakyan, Nucl. Phys. B **285**, 264 (1987).
13. A. Sedrakyan - Nucl.Phys. **B 554 [FS]**,514 (1999).
14. A. Sedrakyan, Phys. Rev. B **68**, 235329 (2003),
A. Sedrakyan, In Proceedings of Advanced NATO Workshop on Statistical Field Theories, edited by A. Capelli and G. Mussardo, Kluwer Academic, Amsterdam, 2002.
15. J. Chalker and P. Coddington, J. Phys. C 21, 2665 (1988).
16. P. Wiegmann, A. Zabrodin, J. Phys. A **36** 3411 (2003).
17. C. Itzikson, J. B. Zuber, J. Math. Phys. **21**, 411 (1980).
18. P. Zinn-Justin, Europhys.Lett. **64**, 734 (2003).
19. M. Brion,figure Lectures on the geometry of flag varieties, In "Topics in Cohomological Studies of Algebraic Varieties".
20. M. Brion, Lectures on the geometry of flag varieties, In "Topics in Cohomological Studies of Algebraic Varieties", Trends in Math., Birkhuser, 2005.

MAGNETIZATION AND CONCURRENCE PROPERTIES OF DIAMOND CHAIN: TWO APPROACHES

N. S. Ananikian*, H. A. Lazaryan and M. A. Nalbandyan
A.I. Alikhanyan National Science Laboratory,
Yerevan, 0036, Armenia
**E-mail: ananik@yerphi.am*

O. Rozas and S. M. de Souza
Departamento de Ciencias Exatas, Universidade Federal de Lavras,
Lavras, MG, 37200-000, Brazil

We present the results of magnetic properties and entanglement of the distorted diamond chain model for azurite using pure quantum exchange interactions. The magnetic properties and concurrence as a measure of pairwise thermal entanglement have been studied by means of variational mean-field like treatment based on Gibbs-Bogoliubov inequality. Such a system can be considered as an approximation of the natural material azurite, $Cu_3(CO_3)_2(OH)_2$. For values of exchange parameters, which are taken from experimental results, we study the thermodynamic properties, such as specific heat and magnetic susceptibility. We also have studied the thermal entanglement properties and magnetization plateau of the distorted diamond chain model for azurite.

We also consider entangled diamond chain with exactly solvable Ising and anisotropic Heisenberg (Ising-XXZ) model. The thermal entanglement (concurrence) is constructed for different values of anisotropic Heisenberg parameter, magnetic field and temperature.

Keywords: diamond chain; XXZ Heisenberg model; azurite; quantum entanglement; concurrence; magnetization plateau; mean-filed formalism

1. Introduction

In the last decade, several diamond chain structures have been discussed. Honecker et al.[1] studied the dynamic and thermodynamic properties for quantum antiferromagnetic Heisenberg model on a generalized diamond chain model. Besides, Pereira *et al.*[2,3] investigate the magnetization plateau of delocalized interstitial spins, as well as magnetocaloric effect in kinetically frustrated diamond chains. More recently, Lisnii[4] studied a distorted

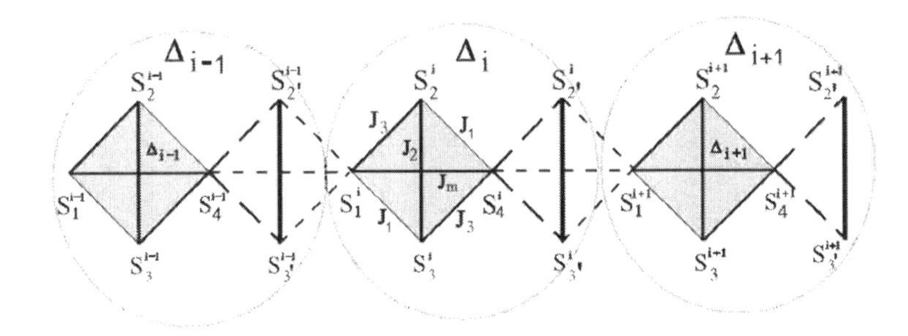

Fig. 1. Each Δ_i cluster consists of one rectangle of $\vec{S}_1^i$, $\vec{S}_2^i$, $\vec{S}_3^i$, $\vec{S}_4^i$ sites (grey rectangle) and dimer of $\vec{S}_{2'}^i$, $\vec{S}_{3'}^i$ sites (bold line).

diamond Ising-Hubbard chain and that the model also reveals the geometrical frustration. Thermodynamics of the Ising-Heisenberg model on a diamond-like chain is widely discussed in the Refs. 5–8. In Refs. 9,10 have discussed the thermal entanglement of the Ising-Heisenberg chain in the isotropic limit in the cluster approach.

Comprehensive investigation of quantum entanglement[11,12] in a variety of antiferromagnetic structures interesting and important, both theoretically and practically for the development of solid-state quantum computers. Entangled states constitute a valuable resource in quantum information processing,[13,14] for example the predicted capabilities of quantum computers rely on entanglement.

Numerous different methods of entanglement measuring have been proposed for its quantification.[15,16] In this paper we use concurrence[17,18] as entanglement measure of the spin-1/2 system.

In this paper, using Heisenberg model, we study the distorted diamond chain as approximation for natural material azurite[1,19–25] $Cu_3(CO_3)_2(OH)_2$. Azurite can be considered as one of the first experimental realizations of the 1D distorted ($J_1 \neq J_3$) diamond chain model (for the detailed structure of azurite see for example Ref. 19). It shows antiferromagnetic behavior at temperatures below Neel temperature 1.9 K.[26]

By means of variational mean-field like treatment based on Gibbs-Bogoliubov inequality[27–33] (Section 2), we study the magnetic (Section 3) and concurrence (Section 4) properties of distorted diamond chain and draw parallel between them and the specific heat ones.

We also study concurrence dependence on magnetic field and temperature of dimers in symmetric Ising-XXZ diamond chain (Section 5).

2. Gibbs-Bogoliubov approach

The Hamiltonian for the distorted diamond chain (Fig. 1) can be written as

$$H = \sum_i \left[J_1 \boldsymbol{\alpha}_i - \frac{h}{2} \left(S_1^{i\,z} + S_4^{i\,z} \right) - h \left(S_2^{i\,z} + S_3^{i\,z} \right) \right], \tag{1}$$

where $h = g\mu_B B_z$ and the g is set to 2,06 (see Ref. 25) and

$$\boldsymbol{\alpha}_i = \vec{S}_2^i \vec{S}_3^i + \delta_m \vec{S}_1^i \vec{S}_4^i + \delta_2 (\vec{S}_1^i \vec{S}_3^i + \vec{S}_2^i \vec{S}_4^i) + \delta_3 (\vec{S}_1^i \vec{S}_2^i + \vec{S}_3^i \vec{S}_4^i). \tag{2}$$

where we denote $\delta_2 = J_1/J_2$, $\delta_3 = J_3/J_2$ and $\delta_m = J_m/J_2$. Here and further exchange parameters (J_1, J_2, J_3) and the magnetic field h are taken in Boltzmann constant scaling *i.e.* Boltzmann constant is set to be $k_B = 1$.

Gibbs-Bogoliubov inequality[27–33] states that for free energy (Helmholtz potential) of the system we have

$$F \leq F_0 + \langle H - H_0 \rangle_0 \,, \tag{3}$$

where H is the real Hamiltonian which describes the system and H_0 is the trial one. Functions F and F_0 are free energies corresponding to H and H_0 respectively and $\langle \ldots \rangle_0$ denotes the thermal average over the ensemble defined by H_0.

We introduce a trial Hamiltonian H_0 as a set of noninteracting clusters of two types in the external self-consistent field: the rectangle and the line (see Fig. 1)

$$H_0 = \sum_{\Delta_i} ({H_0}^{(i)} + {H'_0}^{(i)}), \tag{4}$$

where the indexes of summation Δ_i label different noninteracting clusters and

$$\begin{aligned} H_0^{(i)} &= \lambda\,(\boldsymbol{\alpha}_i) - \gamma_1 S_1^{i\,z} - \gamma_2 S_2^{i\,z} - \gamma_3 S_3^{i\,z} - \gamma_4 S_4^{i\,z}, \\ {H'_0}^{(i)} &= \lambda' \left(\vec{S}_{2'}^i \vec{S}_{3'}^i \right) - \gamma'_2 S_{2'}^{i\,z} - \gamma'_3 S_{3'}^{i\,z}, \end{aligned} \tag{5}$$

where λ, λ' and γ_j, γ'_j are the variational parameters. Such a decomposition (by rectangles and lines) of diamond chain better reflect its structure and therefore gives better results than decompositions of more simple types, for example, by triangles.

Introducing trial Hamiltonian for our model containing unknown variational parameters we can minimize right hand side of Gibbs-Bogoliubov inequality (3) by variational parameters and get the values of those parameters.

We obtain the following values for the variational parameters: $\lambda = \lambda' = J_2$, $\gamma_2 = \gamma_3 = h$, $\gamma_1 = h - J_1 m_3' - J_3 m_2' - J_m m_4$, $\gamma_4 = h - J_3 m_3' - J_1 m_2' - J_m m_1$, $\gamma_2' = h - J_1 m_4 - J_3 m_1$, $\gamma_3' = h - J_3 m_4 - J_1 m_1$.

Using this values and the trial Hamiltonian one can calculate the value of any thermodynamical function of our system for the fixed h, J_i.

3. Magnetisation, specific heat and susceptibility

The results of the previous section can be used for investigation of the magnetic properties of the model. The magnetization of arbitrary site m_v of cluster Δ_i is defined as $m_v = Tr(S_v^{i\,z} e^{-H_0^{(i)}/T})/Z$, where $S_v^{i\,z}$ is the corresponding spin operator, $H_0^{(i)}$ is the Hamiltonian (5) and Z is the corresponding partition function of the cluster. $H_0^{(i)}$ also contains all six magnetizations $(m_2', m_3', m_1, \ldots m_4)$ and we can find them by solving these recurrent equation. The calculations of our paper is based on the effective diamond chain model for azurite with values of coupling constants taken from Ref. 19 and equal $J_2 = 33K$, $\delta_2 = 15.51/33$, $\delta_3 = 6.93/33$, $\delta_m = 4.62/33$.

The dependence of the average magnetization (for cluster) from external magnetic field is shown in Fig. 2(a). As it can be seen from the Fig. 2(a) the 1/3 magnetisation plateau at $T = 1.3K$ extends from 11 T to 29 T interval in $B_z = h/g\mu_B$ axes, while the experimental data[19,21] gave 11 T – 30 T interval.

The magnetic susceptibility is defined as follows: $\chi_0 = \left(\frac{\partial m}{\partial h}\right)_{h=0}$. The magnetic susceptibility measurements of azurite performed in Ref. 21 and it was found double-peak-like structure in the magnetic susceptibility curve, namely, in the temperature dependence of the magnetic susceptibilities the sharp peak appears at $5K$ and the rounded peak is observed at $23K$. The Fig. 2(b) shows the temperature dependence of magnetic susceptibility. It has first sharp peak at $4.4K$.

Analogous to the magnetic susceptibility, we also have calculated specific heat: $C = -T\left(\frac{\partial^2 f_0}{\partial T^2}\right)_{h=0}$. The specific heat measurements for azurite are performed in Ref. 21 and a sharp peak is observed at $T_N = 1.8K$ signaling an ordering transition and two anomalies have been observed in the specific heat at $T = 4K$ and $T = 18K$. The first peak is out of reach of a one-dimensional model. Our calculations gave a low-temperature peak for $H = 0$ at $T = 3$K and the second peak at $12K$ (see Fig. 2(c)). The obtained double-peak-like structure of specific heat reproduce the important features of the experimental results.[19,21] Density functional theory (DFT)[19] also

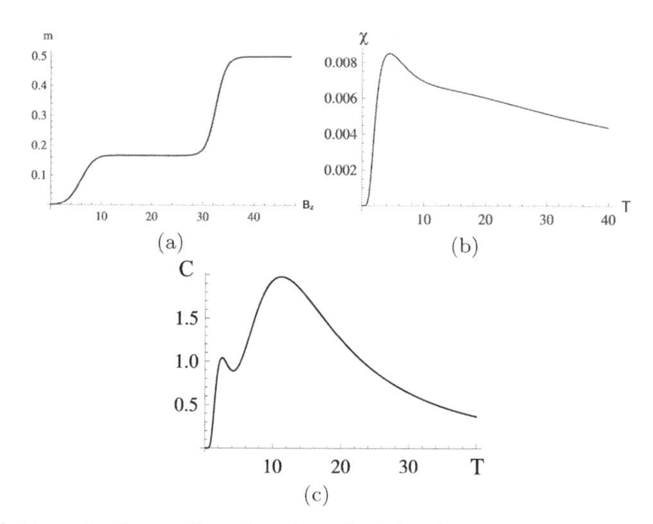

Fig. 2. (a) Magnetization m (in units of $g\mu_B$) of the cluster versus external magnetic field B_z (Tesla) for $J_2 = 33K$, $J_1 = 15.51\ K$, $J_3 = 6.93\ K$, $J_m = 4.62\ K$ at and T=1.3 K, (b) Zero field magnetic susceptibility versus temperature, (c) Zero field specific heat versus temperature.

gives the double-peak-like structure for the specific heat. Here we have reproduced the important features of the experimental results in the specific heat properties of azurite theoretically.

4. Concurrence and thermal entanglement

The mean-field like treatment transforms many-body system to the set of clusters in the effective self-consistent magnetic field. This allows to study, in particular, thermal entanglement properties of the system. The entanglement of formation[11,12] was proposed to quantify the entanglement of a bipartite system. Unfortunately the entanglement of formation extremely difficult to calculate, in general. However, in the special case of two spin 1/2 particles an analytical expression[17,18] can be obtained for the entanglement of formation of any density matrix of two spin 1/2 particles: $EF = H((1+\sqrt{1-C^2})/2)$, where $H(x) = -x\log_2(x) - (1-x)\log_2(1-x)$, and C is the quantity called *concurrence.*[17,18] In the case of diamond chain the concurrence $C_{i,j}$ corresponding to the pair (i,j) has the following form:

$$C_{i,j} = max\{\lambda_1 - \lambda_2 - \lambda_3 - \lambda_4, 0\},$$

where λ_k are the square roots of the eigenvalues (λ_1 is the maximal one) of the matrix $\tilde{\rho} = \rho_{ij} \cdot (\sigma_i^y \otimes \sigma_j^y) \cdot \rho_{ij}^* \cdot (\sigma_i^y \otimes \sigma_j^y)$, where ρ_{ij} is the reduced density matrix for (i, j) pair.

In Figure 3(a) it is shown the dependence of concurrences $C_{2,3} \equiv C_{J_2}$, $C_{1,2} = C_{3,4} \equiv C_{J_3}$, $C_{1,3} = C_{2,4} \equiv C_{J_1}$ and $C_{1,4} \equiv C_{J_m}$ on the magnetic field. The behavior of the concurrence can be used to analyze spin phases of azurite. As it can be seen from the Fig. 3(a) there is three regions in magnetic field axes with different ground states. For lower value of external magnetic field the opposite spins (S_2 and S_3) in diamond cluster is highly entangled. The neighboring spins with lower coupling constant are not entangled. For higher values of B_z when the magnetization has a plateau the entanglement of (S_1, S_4) pair is almost zero, *i. e.* practically unentangled, while the (S_2, S_3) pair is almost fully entangled. The concurrence of the neighboring spins on the plateau is small, comparing to (S_2, S_3) pair, moreover the neighboring spins with lower coupling constant (J_3) are unentangled.

Now, we consider the dependence of the concurrence on temperature. The Fig 3(b) shows the temperature dependence of the concurrence for different pairs of diamond chain at small value of the magnetic field ($B_z = 1T$). The neighboring spins with lower coupling constant (J_3) stay unentangled with increasing temperature while the concurrence of the bigger one (J_1) decreases with temperature to $4.5K$, where the entanglement vanishes. The concurrences of J_m pair J_1 behave similar and vanish almost at the same temperature, while the dominant J_2 pair stays entangled at higher temperatures (until $28K$). Almost the same behaviour shows the temperature dependence of the concurrence for different pairs of diamond chain at plateau phase $B_z = 18T$ (see Fig. 3(c)). The concurrences for J_m, J_3 and J_1 decrease with temperature and vanish sequential between $4K$ and $7K$ while the concurrence for J_2 pair stays entangled for the higher temperatures and vanishes at the same temperature as for small values of magnetic field ($28K$).

As it can be seen from the Figs. 3(a) and 3(c) in the plateau state dimers are almost fully entangled (lines labeled by C_{J_2} in the figures) whereas the monomer spins are weakly entangled (lines labeled by C_{J_1} and C_{J_3} in the figures).

Now we revert to the Fig. 3 (b) to notice that comparison of the Fig. 3(b) with the Fig. 2(c) shows that the $C(J_2)$ has a peak at nearly $T = 5K$ and it is located between two peaks of the specific heat. Roughly such a behavior can be understood as follows. As a result of interaction between the horizontal (J_m) and vertical (J_2) dimers and also as a result of an asymmetry ($J_1 > J_3$), decreasing of the concurrences (in comparison with

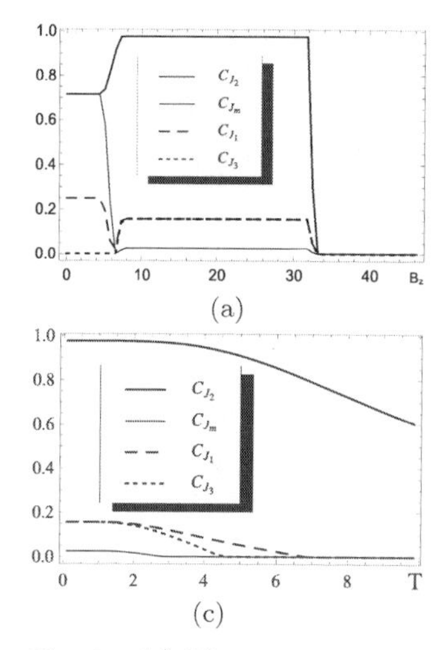

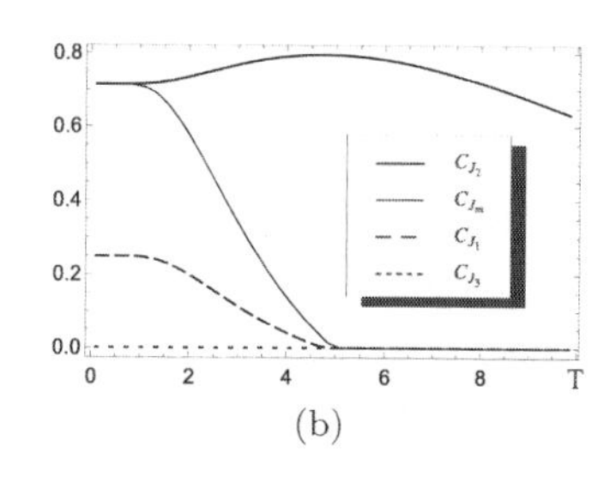

Fig. 3. (a) The concurrences $C_{J_2}, C_{J_1}, C_{J_3}$ and C_{J_m} versus external magnetic field B_z (Tesla) at $T = 0.1K$ for $J_2 = 33K$, $J_1 = 15.51\ K, J_3 = 6.93\ K, J_m = 4.62\ K$. The concurrences $C_{J_2}, C_{J_1}, C_{J_3}$ and C_{J_m} versus temperature for (b) $B_z = 0T$ and (c) $B_z = 18T$.

non-interacting case) of these dimers at zero temperature is observed. As temperature increases, energy "accumulates" in the horizontal dimer at first and in the vertical dimer later, which causes the double peak structure in the specific heat picture (see Fig. 2(c)). During the process of "energy accumulation" in the sites of the horizontal dimer (the first peak region) destruction of a quantum correlations between them takes place as a result of thermal fluctuations (so the concurrence $C(J_m)$ is decreasing, see Fig. 3(b)). And also destruction of a quantum correlations between the sites of horizontal dimer and vertical one occurs. As a consequence one can see an increasing of $C(J_2)$ from $T = 0K$ to $T = 5K$ (without above mentioned couplings (non-interacting dimers case) $C(J_2)$ gets its maximal value equals to 1 in this region). Further temperature increasing brings destruction of the quantum correlations between the sites of the vertical dimer, which is the result of "energy accumulation" on these sites (and as a consequence - decreasing of $C(J_2)$ and increasing the specific heat to the second peak).

Varying values J_1, J_2, J_3, J_m of the coupling constants, brings to analogical picture, except the symmetric case where $J_1 = J_3$. In the symmetric case the concurrences $C(J_2) = C(J_m) = 1$ (at zero temperature) and are decreasing with the temperature and as fast as are higher the values of the $J_1 = J_3$ coupling constants. Closer is the diamond to the symmetric case, worse is the appearance of the peak structure in the $C(J_2)$ picture.

5. Ising-XXZ diamond chain and entanglement

Now we are going to investigate concurrence dependence on magnetic field and temperature of dimers in Ising-XXZ diamond chain. Hamiltonian of the chain is written as

$$\begin{aligned} H = & \sum_{i=1}^{N} J\left(\boldsymbol{S}_{a,i}, \boldsymbol{S}_{b,i}\right)_{\Delta} + J_1\left(S_{a,i}^z + S_{b,i}^z\right)\left(\mu_i + \mu_{i+1}\right) \\ & - h\left(S_{a,i}^z + S_{b,i}^z\right) - \frac{h}{2}\left(\mu_i + \mu_{i+1}\right), \end{aligned} \tag{6}$$

where $(\boldsymbol{S}_{a,i}, \boldsymbol{S}_{b,i})_{\Delta} = S_{a,i}^x S_{b,i}^x + S_{a,i}^y S_{b,i}^y + \Delta S_{a,i}^z S_{b,i}^z$ corresponding to the dimers with anisotropic Heisenberg coupling, while nodal spins (μ_i, μ_{i+1}) are interacting with dimer spins by Ising type exchanges (J_1) and h is the external magnetic field. The schematic representation is shown in Fig. 4.

For particles i and j we write the reduced density matrix in the following form:

$$\begin{aligned} \rho_{i,j} = & \frac{1}{Z_N} \sum_{\{\mu\}} w(\mu_1, \mu_2) \ldots w(\mu_{r-1}, \mu_r) \\ & \varrho_{i,j}(\mu_r, \mu_{r+1}) w(\mu_{r+1}, \mu_{r+2}) \ldots w(\mu_N, \mu_1), \end{aligned} \tag{7}$$

where

$$\varrho(\mu, \mu') = \sum_{i=1}^{4} \mathrm{e}^{-\beta \varepsilon_i(\mu, \mu')} |\varphi_i\rangle\langle\varphi_i|, \tag{8}$$

which elements in standard up-down basis are:

$$\begin{aligned} \varrho_{1,1}(\mu, \mu') = & \mathrm{e}^{-\beta \varepsilon_1(\mu, \mu')}, \\ \varrho_{2,2}(\mu, \mu') = & \frac{1}{2}\left(\mathrm{e}^{-\beta \varepsilon_2(\mu, \mu')} + \mathrm{e}^{-\beta \varepsilon_3(\mu, \mu')}\right), \\ \varrho_{2,3}(\mu, \mu') = & \frac{1}{2}\left(\mathrm{e}^{-\beta \varepsilon_2(\mu, \mu')} - \mathrm{e}^{-\beta \varepsilon_3(\mu, \mu')}\right), \\ \varrho_{4,4}(\mu, \mu') = & \mathrm{e}^{-\beta \varepsilon_4(\mu, \mu')}, \end{aligned} \tag{9}$$

and the other elements are equal to zero. We also have

$$w(\mu, \mu') = \mathrm{tr}_{ab}\left(\varrho(\mu, \mu')\right) = \sum_{i=1}^{4} \mathrm{e}^{-\beta \varepsilon_i(\mu, \mu')}. \tag{10}$$

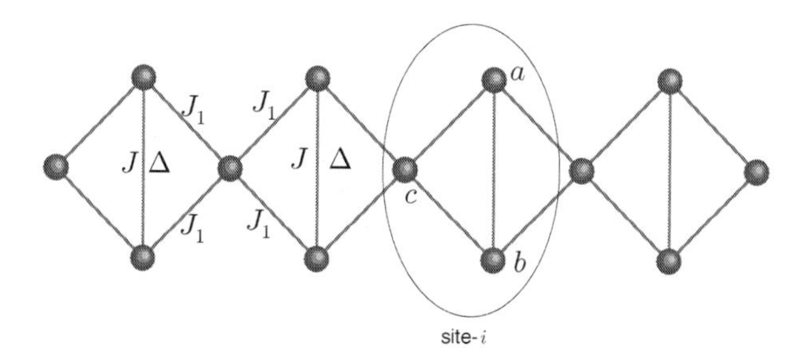

Fig. 4. The schematic representation of XXZ-Ising diamond chain, by red line we represent the quantum bipartite coupling.

Using the transfer matrix notation, we can write the reduced density operator as

$$\rho_{i,j} = \frac{1}{Z_N}\mathrm{tr}\left(W^{r-1}P_{i,j}W^{N-r}\right) = \frac{1}{Z_N}\mathrm{tr}\left(P_{i,j}W^{N-1}\right), \tag{11}$$

where

$$P_{i,j} = \begin{bmatrix} \varrho_{i,j}(\frac{1}{2},\frac{1}{2}) & \varrho_{i,j}(\frac{1}{2},-\frac{1}{2}) \\ \varrho_{i,j}(-\frac{1}{2},\frac{1}{2}) & \varrho_{i,j}(-\frac{1}{2},-\frac{1}{2}) \end{bmatrix}. \tag{12}$$

and the W transfer matrix elements were denoted by $w_{++} \equiv w(\frac{1}{2},\frac{1}{2})$, $w_{+-} \equiv w(\frac{1}{2},-\frac{1}{2})$ and $w_{--} \equiv w(-\frac{1}{2},-\frac{1}{2})$, which eigenvalues are

$$\Lambda_{\pm} = \frac{w_{++} + w_{--} \pm Q}{2}, \tag{13}$$

where

$$Q = \sqrt{(w_{++} - w_{--})^2 + 4w_{+-}^2}. \tag{14}$$

Partition function for finite chain under periodic boundary condition is given by

$$Z_N = \Lambda_+^N + \Lambda_-^N. \tag{15}$$

Real system is well represented in thermodynamic limit ($N \to \infty$), hence, the reduced density operator elements after some algebraic manipulation becomes

$$\begin{aligned}\rho_{i,j} =& \frac{1}{\Lambda_+}\left\{ \frac{\varrho_{i,j}(\frac{1}{2},\frac{1}{2})+\varrho_{i,j}(-\frac{1}{2},-\frac{1}{2})}{2} + \frac{2\varrho_{i,j}(\frac{1}{2},-\frac{1}{2})w_{+-}}{Q} \right.\\ &\left. + \frac{\left(\varrho_{i,j}(\frac{1}{2},\frac{1}{2})-\varrho_{i,j}(-\frac{1}{2},-\frac{1}{2})\right)(w_{++}-w_{--})}{2Q} \right\}.\end{aligned} \tag{16}$$

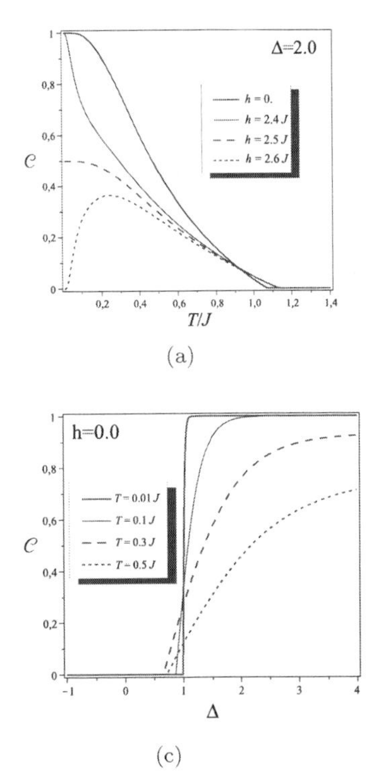

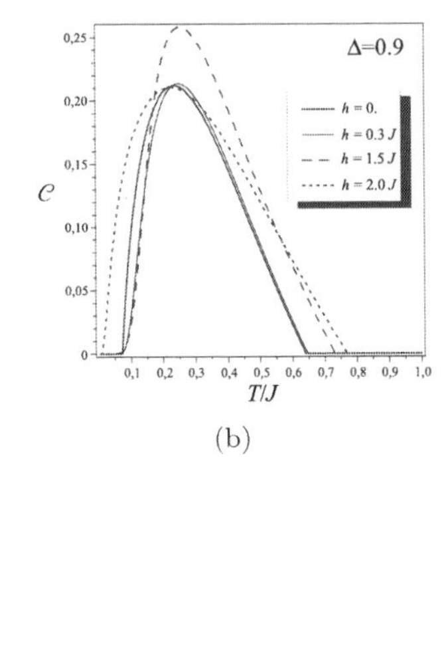

Fig. 5. Concurrence as a function of temperature for fixed value of J_1/J=1. In (a) is displayed for $\Delta = 2$, and in (b) for $\Delta = 0.9$, (c) concurrence as a function of anisotropic parameter Δ for zero magnetic field and several fixed values of temperature.

In Fig. 5 it is displayed explicitly the concurrence as a function of the temperature T/J for fixed value of $J_1/J = 1$, and for several values of magnetic field. In Fig. 5(a) we considered $\Delta = 2$, for null magnetic field and at low temperature the concurrence shows that the system is maximally entangled, however as temperature increases the entanglement is destroyed and exhibits a threshold temperature at around 1.1. For stronger magnetic field like $h/J = 2.4$ and when temperature increases, the entanglement is destroyed even faster than to that null magnetic field, although the systems becomes unentangled at around same temperature compared to the one with null magnetic field. When magnetic field takes the value $h/J = 2.5$, the largest value of entanglement is $C = 0.5$, while for $h/J = 2.6$ the system becomes unentangled at zero temperature, however for low temperature around $T/J \approx 0.3$ there is a maximum value of concurrence $C \approx 0.38$.

Additional interesting plot is illustrated in Fig. 5(b) for a fixed value of $\Delta = 0.9$, when anisotropy becomes weaker, the concurrence decreases ($C \lesssim 0.25$) and even for null magnetic field the system is unentangled at zero temperature, although for low temperature around $T/J \approx 0.3$ the concurrence have its maximum almost independent of temperature, as well as the threshold temperature is around $T/J \approx 0.7$.

In Fig. 5(c) we plot the concurrence as a function of anisotropy parameter Δ for null magnetic field, and for several fixed temperature. For $\Delta < 1$ the system becomes unentangled, whereas for low temperature and $\Delta \gtrsim 1$ the system becomes quickly entangled, while as temperature increases the maximally entangled region occurs only for large anisotropic parameter.

6. Conclusions

In this paper using Heisenberg model the distorted diamond chain was studied as approximation for natural material, azurite $Cu_3(CO_3)_2(OH)_2$. The magnetic properties and concurrence as a measure of pairwise thermal entanglement of the system was studied by means of variational mean-field like treatment based on Gibbs-Bogoliubov inequality. In our approach for the values of exchange parameters taken from theoretical and experimental results we have obtained the 1/3 magnetization plateau with correct critical values of magnetic field. We also studied the thermal entanglement properties of the distorted diamond chain and drew a parallel between them and the specific heat ones. We also studied the concurrence of Ising-XXZ diamond chain as a function of Hamiltonian parameters, temperature and external magnetic field.

References

1. A. Honecker, S. Hu, R. Peters J. Ritcher, *J. Phys.*: Condens. Matter **23,** 164211 (2011).
2. M. S. S. Pereira, F. A. B. F. de Moura, M. L. Lyra, *Phys. Rev. B* **77,** 024402 (2008).
3. M. S. S. Pereira, F. A. B. F. de Moura, M. L. Lyra, *Phys. Rev. B* **79,** 054427 (2009).
4. B. M. Lisnii, *Low Temp. Phys.* **37,** 296 (2011).
5. L. Canova, J. Strečka, M. Jascur, *J. Phys.*: Condens. Matter **18,** 4967 (2006).
6. B. M. Lisnii, Ukrainian Journal of Physics **56,** 1237 (2011).
7. O. Rojas, S.M. de Souza, V. Ohanyan, M. Khurshudyan, *Phys. Rev. B* **83** , 094430 (2011).
8. J.S. Valverde, O. Rojas, S.M. de Souza, *J. Phys.*: Condens. Matter **20,** 345208 (2008); O. Rojas, S.M. de Souza, *Phys. Lett. A* **375,** 1295 (2011).
9. N. S. Ananikian, L. N. Ananikyan, L. A. Chakhmakhchyan and O. Rojas, *J. Phys.*: Condens. Matter **24,** 256001 (2012).

10. L. Chakhmakhchyan, N. Ananikian, L. Ananikyan and C. Burdik, *J. Phys: Conf. Series* **343**, 012022 (2012).
11. C. H. Bennett, D. P. DiVincenzo, J. Smolin, W. K. Wootters, *Phys. Rev. A* **54,** 3824 (1996).
12. C. H. Bennett, G. Brassard, S. Popescu, B. Schumacher, J. Smolin, W. K. Wootters, *Phys. Rev. Lett.* **76,** 722 (1996).
13. C. H. Bennett, D. P. DiVincenzo, Nature **404**, 247 (2000).
14. D. Loss, D. P. DiVincenzo *Phys. Rev. A* **57,** 120 (1998).
15. L. Amico, R. Fazio, A. Osterloh, V. Vedral, *Rev. Mod. Phys.* **80,** 517 (2008).
16. R. Horodecki, P. Horodecki, M. Horodecki, K.Horodecki, *Rev. Mod. Phys.* **81,** 865 (2009).
17. S. Hill, W. K. Wootters, *Phys. Rev. Lett.* **78,** 5022 (1997).
18. W. K. Wootters, *Phys. Rev. Lett.*,**80,** 2245 (1998).
19. H. Jeschke, I. Opahle, H. Kandpal, R. Valenti, H. Das, T. Saha-Dasgupta, O. Janson, H. Rosner, A. Bruhl, B. Wolf, M. Lang, J. Richter, S. Hu, X. Wang, R. Peters, T. Pruschke, A. Honecker, *Phys. Rev. Lett.* **106**, 217201 (2011).
20. A. Honecker, A. Lauchli, *Phys. Rev. B* **63**, 174407 (2001).
21. H. Kikuchi, Y. Fujii, M. Chiba, S. Mitsudo, T. Idehara, T. Tonegawa, K. Okamoto, T. Sakai, T. Kuwai, H. Ohta, *Phys. Rev. Lett.* **94**, 227201 (2005).
22. B. Gu, G. Su, *Phys. Rev. Lett.* **97**, 089701 (2006).
23. K. C. Rule, A. U. B. Wolter, S. Süllow, D. A. Tennant, A. Brühl, S. Köhler, B. Wolf, M. Lang, J. Schreuer, *Phys. Rev. Lett.* **100**, 117202 (2008).
24. H.-J. Mikeska, C. Luckmann, Phys. Rev. B **77**, 054405 (2008).
25. H. Ohta, et.al, *J. Phys. Soc. Jpn.* **72**, 2464 (2003).
26. R. D. Spence, R.D. Ewing, *Phys. Rev.* **112**, 1544 (1958).
27. N. N. Bogoliubov *J. Phys.* (USSR) **11,** 23 (1947).
28. G. D. Mahan *Many-Particle Physics* (New York: Kluwer/Plenum 2000).
29. S-S Gong, G. Su, *Phys. Rev. A* **80,** 012323 (2009).
30. M. Asoudeh, V. Karimipour, *Phys. Rev. A.* **73,** 062109 (2006).
31. N. Canosa, J. M. Matera, R. Rossignoli, *Phys. Rev. A.* **76,** 022310 (2007).
32. N. S. Ananikian, L. N. Ananikyan, L. A. Chakhmakhchyan, A. N. Kocharian, *J. Phys. A* **44,** 025001 (2011).
33. L. Ananikyan, H. Lazaryan, Journal of Contemporary Physics (Armenian Academy of Sciences) **46,** 184 (2011).

COMPLETE (O_7, O_8) CONTRIBUTION TO $\bar{B} \to X_s\gamma$ AT $O(\alpha_s^2)$5 THEORY OF QHE

H.M.Asatrian[a], T.Ewerth[b,c], A.Ferroglia[d,f], C.Greub[e] and G.Ossola[f]

[a] *Yerevan Physics Institute, 0036 Yerevan, Armenia,* [b] *Institut für Theoretische Teilchenphysik, Karlsruhe Institute of Technology (KIT), D-76128 Karlsruhe, Germany,*
[c] *Dip. Fisica Teorica, Univ. di Torino & INFN Torino, I-10125 Torino, Italy,*
[d] *Institut für Physik (THEP), Johannes Gutenberg-Universität D-55099 Mainz, Germany,*
[e] *Albert Einstein Center for Fundamental Physics, Institute for Theoretical Physics, Univ. of Bern, CH-3012 Bern, Switzerland,*
[f] *Physics Department, New York City College of Technology, 300 Jay Street, Brooklyn NY 11201, USA*

We calculate the set of $O(\alpha_s^2)$ corrections to the branching ratio and to the photon energy spectrum of the decay process $\bar{B} \to X_s\gamma$ originating from the interference of diagrams involving the electromagnetic dipole operator O_7 with diagrams involving the chromomagnetic dipole operator O_8. The corrections evaluated here are one of the elements needed to complete the calculations of the $\bar{B} \to X_s\gamma$ branching ratio at next-to-next-to-leading order in QCD. We conclude that this set of corrections does not change the central value of the Standard Model prediction for $\text{Br}(\bar{B} \to X_s\gamma)$ by more than 1%.

1. Introduction

The first estimate of the $\bar{B} \to X_s\gamma$ branching ratio within the Standard Model at the next-to-next-to-leading order (NNLO) level was published some years ago:[1]

$$\text{Br}(\bar{B} \to X_s\gamma)_{\text{SM},\, E_\gamma > 1.6\,\text{GeV}} = (3.15 \pm 0.23) \times 10^{-4}\,. \tag{1}$$

This estimate combines a number of different corrections which were calculated by several groups.[2-12] The prediction given in Eq. (1) must be compared with the current world averages,

$$\text{Br}(\bar{B} \to X_s\gamma)_{\text{exp},\, E_\gamma > 1.6\,\text{GeV}} = \begin{cases} (3.55 \pm 0.24 \pm 0.09) \times 10^{-4}\,, & (\text{HFAG})^{13} \\ (3.50 \pm 0.14 \pm 0.10) \times 10^{-4}\,,^{14} & \end{cases} \tag{2}$$

which include measurements from CLEO, BaBar and Belle.[15–17] The central values of the theoretical prediction and of the HFAG average are compatible at the 1.2σ level, while both the theoretical and experimental uncertainties are very similar in size (about 7%). Since the experimental uncertainty is expected to decrease to 5% by the end of the B-factory era (which is already indicated by the average given in the second line of Eq. (2)), it is also desirable to reduce the theoretical uncertainty accordingly.

Unfortunately, at this level of accuracy, the theoretical uncertainty is dominated by non-perturbative contributions. As long as one restricts the analysis to processes mediated by the electromagnetic dipole operator $O_7 = \alpha_{\rm em}/(4\pi) m_b\, (\bar{s}\sigma^{\mu\nu} P_R b)\, F_{\mu\nu}$ alone, non-perturbative effects are well under control.[18–22] However, as soon as operators other than O_7 (such as the chromomagnetic dipole operator $O_8 = g_s/(16\pi^2) m_b\, (\bar{s}\sigma^{\mu\nu} P_R T^a b)\, G^a_{\mu\nu}$) are involved, one encounters non-perturbative effects of $O(\alpha_s \Lambda_{\rm QCD}/m_b)$. At present, the latter can only be estimated.[23] Hence a 5% uncertainty related to all of the unknown non-perturbative effects has been included in Eq. (1). A further reduction of the theoretical uncertainty below the 5% level seems to be rather difficult.[24] Still, given the importance of $\text{Br}(\bar{B} \to X_s\gamma)$ in constraining physics scenarios beyond the Standard Model,[25] it is worth to reduce the perturbative uncertainties as much as possible.

In particular, it would be desirable to reduce the uncertainty associated to the interpolation in m_c which was employed to obtain Eq. (1).[11] To get rid of the interpolation in m_c in the calculation of the branching ratio is a highly challenging task and it would represent a clear improvement of the theoretical prediction. Indeed, considering the work that has been done since the publication of,[1] and the work that is still in progress, an update of the estimate given in Eq. (1) will soon be warranted. Here we would like to mention that the effects of charm and bottom quark masses on gluon lines are now completely known (provided that one neglects on-shell amplitudes that are proportional to the small Wilson coefficients of the four-quark operators O_3-O_6).[26–29] Therefore this part could be removed from the interpolation. Also the $O(\alpha_s^2 \beta_0)$-effects in the (O_2, O_2), (O_2, O_7) and (O_7, O_8)-interference, which are known,[30] were not considered in.[1,11] Finally, the complete calculation of the (O_2, O_7)-interference for $m_c = 0$ is well underway.[31] The latter calculation in particular will help to fix the boundary for the m_c interpolation for vanishing m_c; this in turn would allow one to reduce the 3% uncertainty in Eq. (1) due to the interpolation. For complete up-to-date lists of needed perturbative and non-perturbative corrections to the branching ratio we refer the reader to the reviews.[32–35]

In this paper we calculate the complete (O_7, O_8)-interference corrections at $O(\alpha_s^2)$ to the photon energy spectrum $d\Gamma(b \to X_s^{\rm partonic}\gamma)/dE_\gamma$ and to the total decay width $\Gamma(b \to X_s^{\rm partonic}\gamma)|_{E_\gamma > E_0}$, where E_0 denotes the lower cut in the photon energy. The contributions containing massless and massive quark loops were already presented in[10,30] and,[28] respectively; the contributions which are not yet available in the literature are the ones proportional to the color factors C_F^2 and $C_F C_A$. From the technical point of view, the latter are the most complicated to evaluate and are the main subject of the present work.

The paper is organized as follows: In Sec. 2 we present our results for the (integrated) photon energy spectrum. The numerical impact of the (O_7, O_8) interference on the theoretical prediction for $\mathrm{Br}(\bar{B} \to X_s\gamma)$ at NNLO is estimated in Sec. 3. Finally, we present our conclusions in Sec. 4.

2. Results for the (integrated) photon energy spectrum

Within the low-energy effective theory, the partonic $b \to X_s\gamma$ decay rate can be written as

$$\Gamma(b \to X_s^{\rm parton}\gamma)_{E_\gamma > E_0} = \frac{G_F^2 \alpha_{\rm em} \overline{m}_b^2(\mu) m_b^3}{32\pi^4} |V_{tb}V_{ts}^*|^2 \sum_{i \le j} C_i^{\rm eff}(\mu)\, C_j^{\rm eff}(\mu) \int_{z_0}^1 dz\, \frac{dG_{ij}(z,\mu)}{dz}\,, \tag{3}$$

where m_b and $\overline{m}_b(\mu)$ denote the pole and the running $\overline{\rm MS}$ mass of the b quark, respectively, $C_i^{\rm eff}(\mu)$ indicates the effective Wilson coefficients at the low-energy scale, $z = 2E_\gamma/m_b$ is the rescaled photon energy, and $z_0 = 2E_0/m_b$ is the rescaled energy cut in the photon energy spectrum.*

As already anticipated in the introduction, we will focus on the function $dG_{78}(z,\mu)/dz$ corresponding to the interference of the electro- and the chromomagnetic dipole operators

$$O_7 = \frac{e}{16\pi^2}\, \overline{m}_b(\mu)\, (\bar{s}\sigma^{\mu\nu} P_R b)\, F_{\mu\nu}\,, \tag{4}$$

$$O_8 = \frac{g}{16\pi^2}\, \overline{m}_b(\mu)\, (\bar{s}\sigma^{\mu\nu} P_R T^a b)\, G_{\mu\nu}^a\,. \tag{5}$$

---

*In this paper we assume that the products $C_i^{\rm eff}(\mu)\, C_j^{\rm eff}(\mu)$ are real quantities. Therefore our formulas are not applicable to physics scenarios beyond the Standard Model which produce complex short distance couplings.

In NNLO approximation G_{78} can be rewritten as follows,

$$\frac{dG_{78}(z,\mu)}{dz} = \frac{\alpha_s(\mu)}{4\pi} C_F \widetilde{Y}^{(1)}(z,\mu) + \left(\frac{\alpha_s(\mu)}{4\pi}\right)^2 C_F \widetilde{Y}^{(2)}(z,\mu) + O(\alpha_s^3), \quad (6)$$

where $\alpha_s(\mu)$ indicates the running coupling constant in the $\overline{\text{MS}}$ scheme and

$$\widetilde{Y}^{(1)}(z,\mu) = \left[\frac{2}{9}\left(33 - 2\pi^2\right) + \frac{16}{3}\ln\left(\frac{\mu}{m_b}\right)\right]\delta(1-z) + \frac{2}{3}\left(z^2+4\right) - \frac{8}{3}\left(1-\frac{1}{z}\right)\ln(1-z). \quad (7)$$

The function $\widetilde{Y}^{(2)}(z,\mu)$ can be split further into a sum of contributions proportional to different color factors:

$$\widetilde{Y}^{(2)}(z,\mu) = C_F \widetilde{Y}^{(2,\mathrm{CF})}(z,\mu) + C_A \widetilde{Y}^{(2,\mathrm{CA})}(z,\mu) \\ T_R N_L \widetilde{Y}^{(2,\mathrm{NL})}(z,\mu) + T_R N_H \widetilde{Y}^{(2,\mathrm{NH})}(z,\mu) + T_R N_V \widetilde{Y}^{(2,\mathrm{NV})}(z,\mu). \quad (8)$$

Here, N_L, N_H and N_V denote the number of light ($m_q = 0$), heavy ($m_q = m_b$), and purely virtual ($m_q = m_c$) quark flavors, respectively; C_F, C_A and T_R are the SU(3) color factors with numerical values given by 4/3, 3 and 1/2, respectively. The expressions for the functions $\widetilde{Y}^{(2,i)}(z,\mu)$ with $i =$ NL, NH, NV can be found in.[28] The main result of the present work are the so far unknown functions $\widetilde{Y}^{(2,i)}(z,\mu)$ with $i =$ CF, CA, which are given by

$$\widetilde{Y}^{(2,\mathrm{CF})}(z,\mu) = \left(-37.1831 - \frac{64}{3}L_\mu - \frac{128}{3}L_\mu^2\right)\delta(1-z) \\ -11.7874\left[\frac{\ln(1-z)}{1-z}\right]_+ - 20.6279\left[\frac{1}{1-z}\right]_+ - 41.7874\ln(1-z) \\ -6.6667\ln^2(1-z) + f_1(z) - 12\,\widetilde{Y}^{(1)}(z,m_b)L_\mu + \frac{64}{3}H^{(1)}(z,m_b)L_\mu, \quad (9)$$

$$\widetilde{Y}^{(2,\mathrm{CA})}(z,\mu) = \left(4.7666 + \frac{808}{27}L_\mu + \frac{272}{9}L_\mu^2\right)\delta(1-z) \\ - 6.5024\ln(1-z) + f_2(z) + \frac{34}{3}\widetilde{Y}^{(1)}(z,m_b)L_\mu, \quad (10)$$

where

$$H^{(1)}(z,m_b) = -\left(\frac{5}{4} + \frac{\pi^2}{3}\right)\delta(1-z) - \left[\frac{\ln(1-z)}{1-z}\right]_+ \\ - \frac{7}{4}\left[\frac{1}{1-z}\right]_+ - \frac{z+1}{2}\ln(1-z) + \frac{7+z-2z^2}{4}, \quad (11)$$

$$f_1(z) = 20.6279 - 108.484\,z + 13.264\,z^2 + 16.1268\,z^3 - 33.2188\,z^4 + 69.8819\,z^5 - 111.088\,z^6 + 118.405\,z^7 - 79.6963\,z^8 + 29.929\,z^9 - 4.76579\,z^{10} - 56.8265\,(1-z)\ln(1-z) - 8.11265\,(1-z)\ln^2(1-z) - 5.77146\,(1-z)\ln^3(1-z)\,, \quad (12)$$

$$f_2(z) = 17.0559\,z + 20.9072\,z^2 - 0.471626\,z^3 + 10.1494\,z^4 - 17.4241\,z^5 + 24.7733\,z^6 - 20.4582\,z^7 + 8.47394\,z^8 - 0.173599\,z^9 - 0.657813\,z^{10} + 5.66536\,(1-z)\ln(1-z) - 11.1319\,(1-z)\ln^2(1-z) + 1.3999\,(1-z)\ln^3(1-z)\,. \quad (13)$$

Note that function $H^{(1)}(z,\mu)$ also appeared in Eq. (2.11) of Ref.,[9] and that we introduced the short-hand notation $L_\mu = \ln(\mu/m_b)$.

In Eqs. (9) and (10) the z-dependence of the μ-dependent terms and of those terms which become singular in the limit $z \to 1$ is exact. The functions $f_1(z)$ and $f_2(z)$ were instead obtained by making an ansatz for our numerical results of the non-singular parts, using the functional form

$$f_i(z) = \sum_{j=0}^{10} c_{i,j}\, z^j + c_{i,11}(1-z)\ln(1-z) + c_{i,12}(1-z)\ln^2(1-z) + c_{i,13}(1-z)\ln^3(1-z)\,. \quad (14)$$

The coefficients $c_0, \ldots, c_{13}$ were then determined by performing a least-square fit, using 100 specific ‘data’-points. We checked that the fit-functions remain essentially the same when changing the set of data-points. In particular, the integrals of the fit-functions, taken in an interval $[z_0, 1]$ ($0 \le z_0 < 1$), remain basically unchanged. The same holds true when changing the functional ansatz given in Eq. (14), e.g., to contain additional terms proportional to $(1-z)^2 \ln^n(1-z)$, with $n = 1, 2, 3$.

The plus distributions appearing in Eq. (9) are defined as

$$\int_0^1 dz \left[\frac{\ln^n(1-z)}{1-z}\right]_+ g(z) = \int_0^1 dz\, \frac{\ln^n(1-z)}{1-z}\,[g(z) - g(1)]\,, \quad (15)$$

where $g(z)$ is an arbitrary test function which is regular at $z = 1$, and $n = 0, 1$. In case the integration does not include the endpoint $z = 1$, we have ($c < 1$)

$$\int_0^c dz \left[\frac{\ln^n(1-z)}{1-z}\right]_+ g(z) = \int_0^c dz\, \frac{\ln^n(1-z)}{1-z} g(z)\,. \quad (16)$$

We observe that the plus distributions are present only in the part of the spectrum proportional to C_F^2, see Eq. (9). This is in agreement with the results reported in;[36] following the procedure presented in that work, it is possible to conclude that the plus distributions appearing in the (O_7, O_8) component of the photon energy spectrum at $\mathcal{O}(\alpha_s^2)$ must be the same ones as in the (O_7, O_7) component of the spectrum at $\mathcal{O}(\alpha_s)$ (up to an overall factor). In particular

$$\widetilde{Y}^{(2,\mathrm{CF})}(z, m_b)\Big|_{\text{plus distrib.}}$$
$$= -\frac{8}{9}\left(33 - 2\pi^2\right)\left\{\left[\frac{\ln(1-z)}{1-z}\right]_+ + \frac{7}{4}\left[\frac{1}{1-z}\right]_+\right\}$$
$$= -11.7874\left[\frac{\ln(1-z)}{1-z}\right]_+ - 20.6279\left[\frac{1}{1-z}\right]_+ . \qquad (17)$$

The structure in Eq. (17) emerges in our diagrammatic calculation from delicate cancellations among several contributions, and it provides a valuable test for our result.

3. Estimating the numerical impact on $\mathrm{Br}(\bar{B} \to X_s\gamma)$

In this section we investigate the numerical size of the (O_7, O_8)-interference at $O(\alpha_s^2)$ at the level of the branching ratio of the decay process $\bar{B} \to X_s\gamma$. In order to do so, we adopt the notation and conventions introduced in.[11] The $O(\alpha_s^2)$ correction to the function $K_{78}(E_0, \mu)$ in Eqs. (2.6) and (3.1) of[11] is given by

$$K_{78}^{(2)}(E_0, \mu) = \frac{C_F}{2}\int_{z_0}^{1} dz \left\{\widetilde{Y}^{(2)}(z, \mu) - C_F^2 \frac{8\pi}{3}\alpha_\Upsilon \delta(1-z)\right.$$
$$\left. - C_F\left[\frac{41}{2} - 2\pi^2 + 12\,L_\mu\right]\widetilde{Y}^{(1)}(z, \mu)\right\}, \qquad (18)$$

with $\alpha_\Upsilon = 0.22$. Note that Eq. (18) refers to the $1S$-scheme for the b-quark mass. The value used for this parameter in the numerics is $m_b^{1S} = 4.68\,\mathrm{GeV}$.

Combining the results of our present paper with those of[28] we are now able to write down the complete expression for $K_{78}^{(2)}$ containing all abelian and non-abelian contributions as well as the effects of the masses of the u, d, s, c and b quarks running in the bubbles inserted in gluon lines. This complete term affects the branching ratio by an amount

$$\Delta\mathrm{Br}(\bar{B} \to X_s\gamma)_{E_\gamma > E_0} = \mathrm{Br}(\bar{B} \to X_c e\bar{\nu})_{\mathrm{exp}} \left|\frac{V_{tb}V_{ts}^*}{V_{cb}}\right|^2 \frac{6\,\alpha_{\mathrm{em}}}{\pi\, C}\,\Delta P(E_0)\,, \qquad (19)$$

where

$$\Delta P(E_0) = 2C_7^{(0)\text{eff}}(\mu)C_8^{(0)\text{eff}}(\mu)\left(\frac{\alpha_s(\mu)}{4\pi}\right)^2 K_{78}^{(2)}(E_0,\mu)\,, \tag{20}$$

and C is the so-called semileptonic phase-space factor. In order to compare with Ref.,[11] we employ the numerical value for C which was obtained from a fit of the measured spectrum of the $\bar{B} \to X_c l\bar{\nu}$ decay in the 1S scheme†.[50,51]

One might think that $\Delta\text{Br}(\bar{B} \to X_s\gamma)_{E_\gamma > E_0}$ in Eq. (19) simply represents the shift due to $K_{78}^{(2)}$ of the theoretical prediction given in Eq. (1). This is, however, not the case because an approximated version of $K_{78}^{(2)}$ was already included in:[1] While the β_0-part of K_{78} (i.e. $K_{78}^{(2)\beta_0}$ when following the notation of[11]) was fully taken into account, the remaining piece, $K_{78}^{(2)\text{rem}}$, was calculated for large values of $\rho = m_c^2/m_b^2$ and then interpolated (combined with contributions not related to the (O_7, O_8)-interference) to the physical value of ρ. To remove $K_{78}^{(2)\text{rem}}$ from the interpolation procedure and to replace it by the exact result obtained by us, is beyond the scope of the present paper; this issue will be correctly treated in a systematic update of Eq. (1) in the near future. To get nevertheless an idea of the numerical size of the $O(\alpha_s^2)$ contribution of the (O_7, O_8)-interference at the level of branching ratio for $\bar{B} \to X_s\gamma$, we can ignore this issue and simply discuss a few numerical aspects of the quantity $\Delta\text{Br}(\bar{B} \to X_s\gamma)_{E_\gamma > E_0}$, based on $K_{78}^{(2)}$ which we calculated in this paper.

Fig. 1 shows $\Delta\text{Br}(\bar{B} \to X_s\gamma)_{E_\gamma > E_0}$ as a function of z_0 for $\mu = 2.5\,\text{GeV}$, $\alpha_s(2.5\,\text{GeV}) = 0.271$, $C_7^{(0)\text{eff}}(2.5\,\text{GeV}) = -0.369$, $C_8^{(0)\text{eff}}(2.5\,\text{GeV}) = -0.171$, $N_L = 3$ and $N_H = N_V = 1$. (Note that the scale $\mu = 2.5\,\text{GeV}$ defines the central value of the branching ratio given in Eq. (1).) The remaining numerical input parameters are taken from.[11] The thick red solid line shows the complete result, the thin red solid line corresponds to the contribution proportional to C_F^2 (including the numerical value of C_F^2), the green dashed line indicates the contribution proportional to $C_F C_A$ (again including $C_F C_A$), and the blue dotted line is the contribution stemming from massless and massive quark bubbles. The dash-dotted blue line indicates the corrections obtained by applying the large-β_0 approximation.[53,54] The vertical line indicates the value of z_0 corresponding to the choice $E_0 = 1.6$ GeV. It can be seen from the figure that the contributions proportional to C_F^2 and $C_F C_A$ cancel each other over almost the whole range of z_0,

†Using the numerical value for C as obtained from a fit in the kinetic scheme raises the central value given in Eq. (1) by approximately 3%.[52]

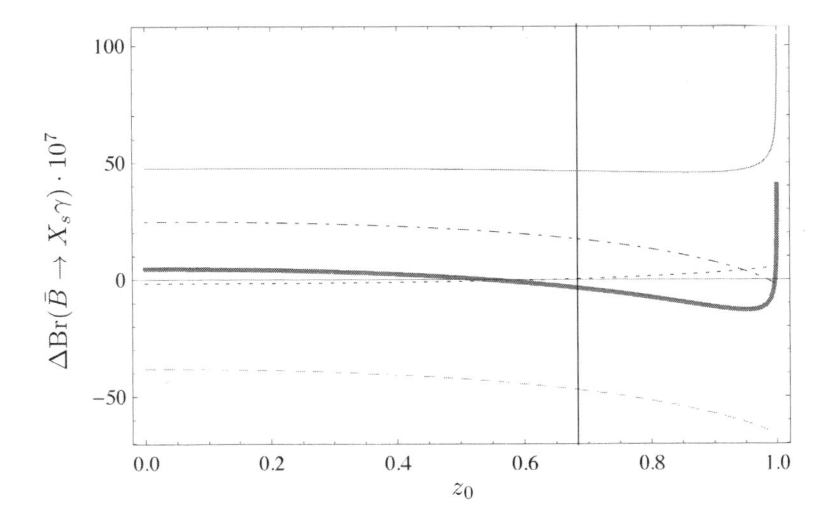

Fig. 1. $\Delta Br(\bar{B} \to X_s\gamma)_{E_\gamma > E_0}$ *as a function of* z_0. *See text for more details.*

resulting in a contribution of size similar to the one due to the fermionic corrections. Only in the region very close to the endpoint the contribution proportional to C_F^2 dominates, due to its singular behavior for $z_0 \to 1$. We stress, however, that the cancellations mentioned above refer to a value of $\mu = 2.5$ GeV; they do not occur anymore when going to smaller values of μ.

For a photon energy cut-off $E_0 = 1.6\,\mathrm{GeV}$, as the one employed in Eq. (1), we find the following numerical value for the quantity in Eq. (19) (using $\mu = 2.5$ GeV):

$$\begin{aligned}\Delta\mathrm{Br}(\bar{B} \to X_s\gamma)_{E_\gamma > 1.6\,\mathrm{GeV}} \\ = \frac{C_F}{2}\big(52.21\,C_F - 23.57\,C_A - 2.66\,C_F^2 - 2.28\,N_L \\ 8.73\,N_H - 1.42\,N_V\big) \times 10^{-7} = -3.57 \cdot 10^{-7}, \quad (21)\end{aligned}$$

where in the last step we inserted numerical values for the color factors and we set $N_L = 3$ and $N_H = N_V = 1$. By comparing this with the central value of the estimate given in Eq. (1), one sees that the $O(\alpha_s^2)$ corrections $K_{78}^{(2)}$ have an impact of -0.11% at the level of the branching ratio. For the analogous effect generated by the large-β_0 approximation of $K_{78}^{(2)}$ only, $K_{78}^{(2)\beta_0}$, we find $+0.56\%$. These results are, however, very strongly dependent on the scale μ. For $\mu = 1.25,\ 2.34,\ 5\,\mathrm{GeV}$ the corresponding numbers read −5.15, −0.23, −0.07% (full) and +0.10, +0.49, +1.15% (large-β_0 approximation), respectively. From these results we conclude that the large-β_0 approximation does not provide a good estimate of the full $O(\alpha_s^2)$ correction of the (O_7, O_8)-contribution for $\mu \in [1.25, 5]$ GeV. As already mentioned, a more detailed analysis of the effect of the complete calculation of $K_{78}^{(2)}$ on the central value of Eq. (1) would require to repeat the interpolation procedure of.[11] While this is beyond the scope of the present work, we can conclude that the correction originating from the (O_7, O_8)-interference at $O(\alpha_s^2)$ will not alter the central value of Eq. (1) by more than 1%.

4. Summary and conclusions

In the present work we calculated the set of the $O(\alpha_s^2)$ corrections to the partonic decay process $b \to X_s\gamma$ which originates from the interference of diagrams involving the electromagnetic dipole operator O_7 with diagrams involving the chromomagnetic dipole operator O_8. These corrections are one of the elements needed in order to complete the calculation of the branching ratio for the radiative decay $\bar{B} \to X_s\gamma$ up to NNLO in QCD.

To carry out the calculation, we mapped the interference of diagrams contributing to the processes $b \to s\gamma$, $b \to s\gamma g$, $b \to s\gamma gg$ and $b \to s\gamma s\bar{s}$ onto 2-, 3- , and 4-particle cuts for the three-loop b-quark self-energy diagrams which include insertions of the operators O_7 and O_8. Subsequently, we evaluated each single cut by employing the Cutkosky rules. From the technical point of view, the calculation was made possible by the use of the Laporta Algorithm[55] to identify the needed Master Integrals, and of the differential equation method and sector decomposition method to solve the Master Integrals.

From the phenomenological point of view, it is interesting to estimate the effect of these corrections on the theoretical prediction for the $\bar{B} \to X_s\gamma$ branching ratio. Our conclusion is that they will not change its central value given in Eq. (1) by more than 1%.

At present, the largest theoretical uncertainty affecting the prediction in Eq. (1) is of non-perturbative origin. It is expected to set a lower limit of about 5% on the total theoretical uncertainty for the prediction of the $\bar{B} \to X_s\gamma$ branching ratio in the near future. The perturbative $O(\alpha_s^2)$ corrections of the (O_7, O_8)-interference presented in this paper are a further contribution to make the perturbative uncertainty negligible with respect to the non-perturbative one.

References

1. M. Misiak and et al., *Phys. Rev. Lett.* **98** (2007) 022002 [arXiv:hep-ph/0609232].
2. M. Misiak and M. Steinhauser, *Nucl. Phys. B* **683** (2004) 277 [arXiv:hep-ph/0401041].
3. M. Gorbahn and U. Haisch, *Nucl. Phys. B* **713** (2005) 291 [arXiv:hep-ph/0411071].
4. M. Gorbahn, U. Haisch and M. Misiak, *Phys. Rev. Lett.* **95** (2005) 102004 [arXiv:hep-ph/0504194].
5. M. Czakon, U. Haisch and M. Misiak, *JHEP* **0703** (2007) 008 [arXiv:hep-ph/0612329].
6. I. Blokland, A. Czarnecki, M. Misiak, M. Slusarczyk and F. Tkachov, *Phys. Rev. D* **72** (2005) 033014 [arXiv:hep-ph/0506055].
7. K. Melnikov and A. Mitov, *Phys. Lett. B* **620** (2005) 69 [arXiv:hep-ph/0505097].
8. H.M. Asatrian, A. Hovhannisyan, V. Poghosyan, T. Ewerth, C. Greub and T. Hurth, *Nucl. Phys. B* **749** (2006) 325 [arXiv:hep-ph/0605009].
9. H.M. Asatrian, T. Ewerth, A. Ferroglia, P. Gambino and C. Greub, *Nucl. Phys. B* **762** (2007) 212 [arXiv:hep-ph/0607316].
10. K. Bieri, C. Greub and M. Steinhauser, *Phys. Rev. D* **67** (2003) 114019 [arXiv:hep-ph/0302051].
11. M. Misiak and M. Steinhauser, *Nucl. Phys. B* **764** (2007) 62 [arXiv:hep-ph/0609241].
12. M. Misiak and M. Steinhauser, arXiv:1005.1173.
13. E. Barberio and et al., [Heavy Flavor Averaging Group], arXiv:0808.1297, and on-line update at http://www.slac.stanford.edu/xorg/hfag/rare/winter10/radll/btosg.pdf
14. M. Artuso, E. Barberio and S. Stone, *PMC Phys. A* **3** (2009) 3 [arXiv:0902.3743].
15. S. Chen and et al., [CLEO Collab.], *Phys. Rev. Lett.* **87** (2001) 251807 [arXiv:hep-ex/0108032].
16. B. Aubert and et al., [BaBar Collab.], *Phys. Rev. D* **77** (2008) 051103 [arXiv:0711.4889].
17. A. Limosani *et al.* [Belle Collab.], *Phys. Rev. Lett.* **103** (2009) 241801 [arXiv:0907.1384].
18. A.F. Falk, M.E. Luke and M.J. Savage, *Phys. Rev. D* **49** (1994) 3367 [arXiv:hep-ph/9308288].
19. I.I. Y. Bigi, B. Blok, M.A. Shifman, N.G. Uraltsev and A.I. Vainshtein, arXiv:hep-ph/9212227.
20. C.W. Bauer, *Phys. Rev. D* **57** (1998) 5611 [Erratum-ibid. D **60** (1999) 099907] [arXiv:hep-ph/9710513].
21. M. Gremm and A. Kapustin, *Phys. Rev. D* **55** (1997) 6924 [arXiv:hep-ph/9603448].
22. T. Ewerth, P. Gambino and S. Nandi, *Nucl. Phys. B* **830** (2010) 278 [arXiv:0911.2175].
23. S.J. Lee, M. Neubert and G. Paz, *Phys. Rev. D* **75** (2007) 114005 [arXiv:hep-

ph/0609224].
24. M. Benzke, S.J. Lee, M. Neubert and G. Paz, arXiv:1003.5012.
25. See, e.g., F. Domingo and U. Ellwanger, *JHEP* **0712** (2007) 090 [arXiv:0710.3714].
26. H.M. Asatrian, T. Ewerth, H. Gabrielyan and C. Greub, *Phys. Lett. B* **647** (2007) 173 [arXiv:hep-ph/0611123].
27. R. Boughezal, M. Czakon and T. Schutzmeier, *JHEP* **0709** (2007) 072 [arXiv:0707.3090].
28. T. Ewerth, *Phys. Lett. B* **669** (2008) 167 [arXiv:0805.3911].
29. A. Pak and A. Czarnecki, *Phys. Rev. Lett.* **100** (2008) 241807 [arXiv:0803.0960].
30. Z. Ligeti, M.E. Luke, A.V. Manohar and M.B. Wise, *Phys. Rev. D* **60** (1999) 034019 [arXiv:hep-ph/9903305].
31. R. Boughezal, M. Czakon and T. Schutzmeier, in preparation; M. Czakon, T. Huber and T. Schutzmeier, in preparation.
32. M. Misiak, arXiv:0808.3134.
33. A. Ferroglia, *Mod. Phys. Lett. A* **23** (2008) 3123 [arXiv:0812.0082].
34. T. Ewerth, arXiv:0909.5027.
35. M. Misiak, *Acta Phys. Polon. B* **40** (2009) 2987 [arXiv:0911.1651].
36. M. Neubert, *Eur. Phys. J. C* **40** (2005) 165 [arXiv:hep-ph/0408179].
37. R. E. Cutkosky, *J. Math. Phys.* **1** (1960) 429.
38. M.J.G. Veltman, *Physica* **29** (1963) 186.
39. E. Remiddi, *Helv. Phys. Acta* **54** (1982) 364.
40. C. Anastasiou, L.J. Dixon, K. Melnikov and F. Petriello, *Phys. Rev. D* **69** (2004) 094008 [arXiv:hep-ph/0312266].
41. C. Anastasiou and K. Melnikov, *Nucl. Phys. B* **646** (2002) 220 [arXiv:hep-ph/0207004].
42. C. Anastasiou, K. Melnikov and F. Petriello, *Phys. Rev. D* **69** (2004) 076010 [arXiv:hep-ph/0311311].
43. C. Anastasiou and A. Lazopoulos, *JHEP* **0407** (2004) 046 [arXiv:hep-ph/0404258].
44. E. Remiddi, *Nuovo Cim. A* **110**, (1997) 1435 [arXiv:hep-th/9711188].
45. M. Argeri and P. Mastrolia, *Int. J. Mod. Phys. A* **22** (2007) 4375 [arXiv:0707.4037].
46. T. Binoth and G. Heinrich, *Nucl. Phys. B* **585** (2000) 741 [arXiv:hep-ph/0004013].
47. Z. Nagy and D.E. Soper, *Phys. Rev. D* **74** (2006) 093006 [arXiv:hep-ph/0610028].
48. A. Lazopoulos, K. Melnikov and F. Petriello, *Phys. Rev. D* **76** (2007) 014001 [arXiv:hep-ph/0703273].
49. C. Anastasiou, S. Beerli and A. Daleo, *JHEP* **0705** (2007) 071 [arXiv:hep-ph/0703282].
50. C.W. Bauer, Z. Ligeti, M. Luke, A.V. Manohar and M. Trott, *Phys. Rev. D* **70**, (2004) 094017 [arXiv:hep-ph/0408002].
51. A.H. Hoang and A.V. Manohar, *Phys. Lett. B* **633**, (2006) 526 [arXiv:hep-ph/0509195].

52. P. Gambino and P. Giordano, *Phys. Lett. B* **669** (2008) 69 [arXiv:0805.0271].
53. S.J. Brodsky, G.P. Lepage and P.B. Mackenzie, *Phys. Rev. D* **28** (1983) 228.
54. M. Beneke and V.M. Braun, *Phys. Lett. B* **348** (1995) 513 [arXiv:hep-ph/9411229].
55. S. Laporta, *Int. J. Mod. Phys. A* **15** (2000) 5087 [arXiv:hep-ph/0102033].

NON-TRIVIAL HOLONOMY AND CALORONS

P. van Baal

Instituut-Lorentz for Theoretical Physics, University of Leiden,
P.O. Box 9506, Nl-2300 RA Leiden, The Netherlands

The progress on calorons (finite temperature instantons) is sketched. In particular there is some interest for confining temperatures, where the holonomy (the asymptotic value of the Polyakov loop) is non-trivial. In the last section I give more recent results by others.

Keywords: Holonomy; Finite Temperature; Instantons.

1. Introduction

First of all I would like to congratulate Sergei Matinyan with his 80th birthday. I met him actually in the States where I gave a seminar in Duke on June 1, 1993 on my finite volume work. He emailed me on June 10 with a suggestion, making the zero-momentum modes stochastic. The last sentence of my reply ended with: "But if some day this might be achieved, your idea with stochastic behavior in the low-energy modes only, might very well be the appropriate one." Unfortunately I became stuck in finite volumes and want to present now "non-trivial holonomy and calorons", which is at finite temperature, but infinite volumes.

There is an other reason why I love to be back in Armenia. I was in Nor Amberd for an INTAS meeting 15 years earlier (May 26-29, 1996), where I apparently gave two talks and Ara Sedrakyan showed us around (only me and Charlotte Kristjansen, who now gives two talks, were representing the West). I have good memories of Armenia.

2. The setting

There has been a revised interest in studying instantons at finite temperature T, so-called calorons,[1,2] because new explicit solutions could be obtained where the Polyakov loop at spatial infinity (the so-called holonomy) is non-trivial. They reveal more clearly the monopole constituent nature of these calorons.[3] Non-trivial holonomy is therefore expected to play a role in the confined phase (i.e. for $T < T_c$) where the trace of the Polyakov loop fluctuates around small values. The properties of instantons are therefore directly coupled to the order parameter for the deconfining phase transition.

At finite temperature A_0 plays in some sense the role of a Higgs field in the adjoint representation, which explains why magnetic monopoles occur as constituents of calorons. Since A_0 is not necessarily static it is better to consider the Polyakov loop as the analog of the Higgs field, $P(t,\vec{x}) = \text{Pexp}\left(\int_0^\beta A_0(t+s,\vec{x})ds\right)$, which transforms under a periodic gauge transformation $g(x)$ to $g(x)P(x)g^{-1}(x)$, like an adjoint Higgs field. Here $\beta = 1/kT$ is the period in the imaginary time direction, under which the gauge field is assumed to be periodic. Finite action requires the Polyakov loop at spatial infinity to be constant. For SU(n) gauge theory this gives $\mathcal{P}_\infty = \lim_{|\vec{x}|\to\infty} P(0,\vec{x}) = g^\dagger \exp(2\pi i \text{diag}(\mu_1,\mu_2,\ldots,\mu_n))g$, where g brings $\mathcal{P}_\infty$ to its diagonal form, with n eigenvalues being ordered according to $\sum_{i=1}^n \mu_i = 0$ and $\mu_1 \leq \mu_2 \leq \ldots \leq \mu_n \leq \mu_{n+1} \equiv 1 + \mu_1$. In the algebraic gauge, where $A_0(x)$ is transformed to zero at spatial infinity, the gauge fields satisfy the boundary condition $A_\mu(t+\beta,\vec{x}) = \mathcal{P}_\infty A_\mu(t,\vec{x})\mathcal{P}_\infty^{-1}$.

Caloron solutions are such that the total magnetic charge vanishes. A single caloron with topological charge one contains $n-1$ monopoles with a unit magnetic charge in the i-th U(1) subgroup, which are compensated by the n-th monopole of so-called type $(1,1,\ldots,1)$, having a magnetic charge in each of these subgroups.[4] At topological charge k there are kn constituents, k monopoles of each of the n types. Monopoles of type j have a mass $8\pi^2\nu_j/\beta$, with $\nu_j \equiv \mu_{j+1} - \mu_j$. The sum rule $\sum_{j=1}^n \nu_j = 1$ guarantees the correct action, $8\pi^2 k$.

Prior to their explicit construction, calorons with non-trivial holonomy were considered irrelevant,[2] because the one-loop correction gives rise to an infinite action barrier. However, the infinity simply arises due to the integration over the finite energy density induced by the perturbative fluctuations in the background of a non-trivial Polyakov loop.[5] The calculation of the non-perturbative contribution was performed in.[6] When added to this perturbative contribution, with minima at center elements, these minima turn unstable for decreasing temperature right around the expected value of T_c. This lends some support to monopole constituents being the relevant degrees of freedom which drive the transition from a phase in which the center symmetry is broken at high temperatures to one in which the center symmetry is restored at low temperatures. Lattice studies, both using cooling[7] and chiral fermion zero-modes[8] as filters, have also conclusively confirmed that monopole constituents do dynamically occur in the confined phase.

3. Some properties of caloron solutions

Using the classical scale invariance we can always arrange $\beta = 1$, as will be assumed throughout. A remarkably simple formula for the SU(n) action density exists,[4]

$$\mathrm{Tr}F_{\alpha\beta}^2(x) = \partial_\alpha^2\partial_\beta^2 \log\psi(x), \quad \psi(x) = \tfrac{1}{2}\mathrm{tr}(\mathcal{A}_n\cdots\mathcal{A}_1) - \cos(2\pi t),$$
$$\mathcal{A}_m \equiv \frac{1}{r_m}\begin{pmatrix} r_m & |\vec{\rho}_{m+1}| \\ 0 & r_{m+1}\end{pmatrix}\begin{pmatrix}\cosh(2\pi\nu_m r_m) & \sinh(2\pi\nu_m r_m) \\ \sinh(2\pi\nu_m r_m) & \cosh(2\pi\nu_m r_m)\end{pmatrix},$$

with $r_m \equiv |\vec{x} - \vec{y}_m|$ and $\vec{\rho}_m \equiv \vec{y}_m - \vec{y}_{m-1}$, where $\vec{y}_m$ is the location of the mth constituent monopole with a mass $8\pi^2\nu_m$. Note that the index m should be considered mod n, such that e.g. $r_{n+1} = r_1$ and $\vec{y}_{n+1} = \vec{y}_1$ (there is one exception, $\mu_{n+1} = 1 + \mu_1$). It is sufficient that only one constituent location is far separated from the others, to show that one can neglect the $\cos(2\pi t)$ term in $\psi(x)$, giving rise to a static action density in this limit.[4]

In Fig. 1 we show how for SU(2) there are two lumps, except that the second lump is absent for trivial holonomy. Fig. 2 demonstrates for SU(2) and SU(3) that there are indeed n lumps (for SU(n)) which can be put anywhere. These lumps are constituent monopoles, where one of them has a winding in the temporal direction (which cannot be seen from the action density).

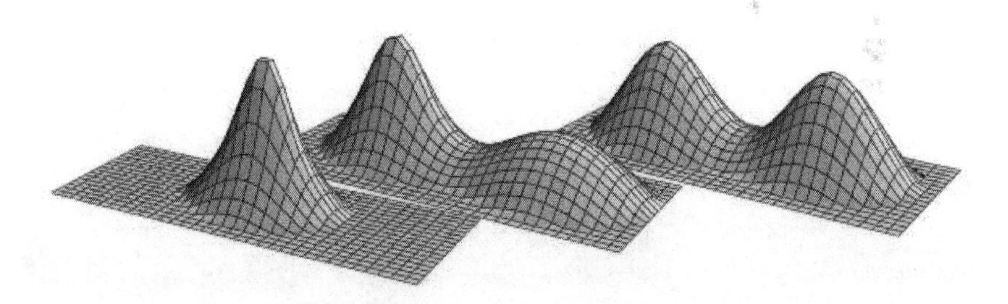

Fig. 1. Shown are three charge one SU(2) caloron profiles at $t = 0$ with $\beta = 1$ and $\rho = 1$. From left to right for $\mu_2 = -\mu_1 = 0$ ($\nu_1 = 0, \nu_2 = 1$), $\mu_2 = -\mu_1 = 0.125$ ($\nu_1 = 1/4, \nu_2 = 3/4$) and $\mu_2 = -\mu_1 = 0.25$ ($\nu_1 = \nu_2 = 1/2$) on equal logarithmic scales, cutoff below an action density of $1/(2e)$.

3.1. *Fermion zero-modes*

An essential property of calorons is that the chiral fermion zero-modes are localized to constituents of a certain charge only. The latter depends

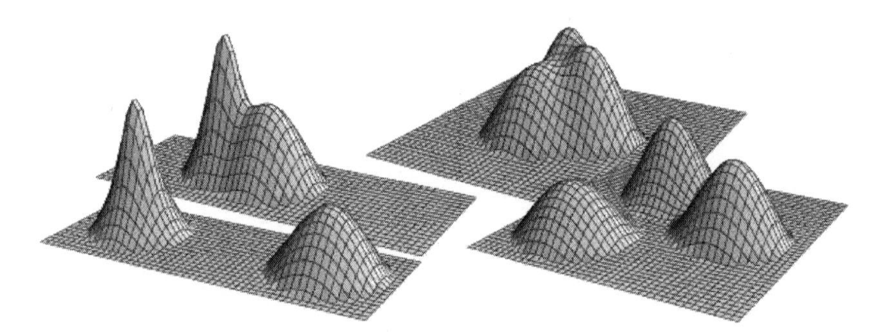

Fig. 2. On the left are shown two charge one SU(2) caloron profiles at $t = 0$ with $\beta = 1$ and $\mu_2 = -\mu_1 = 0.125$, for $\rho = 1.6$ (bottom) and 0.8 (top) on equal logarithmic scales, cutoff below an action density of $1/(2e^2)$. On the right are shown two charge one SU(3) caloron profiles at $t = 0$ and $(\nu_1, \nu_2, \nu_3) = (1/4, 7/20, 2/5)$, implemented by $(\mu_1, \mu_2, \mu_3) = (-17/60, -1/30, 19/60)$. The bottom configuration has the location of the lumps scaled by 8/3. They are cutoff at $1/(2e)$.

on the choice of boundary condition for the fermions in the imaginary time direction (allowing for an arbitrary U(1) phase $\exp(2\pi i z)$).[9] This provides an important signature for the dynamical lattice studies, using chiral fermion zero-modes as a filter.[8] To be precise, the zero-modes are localized to the monopoles of type m provided $\mu_m < z < \mu_{m+1}$. Denoting the zero-modes by $\hat{\Psi}_z(x)$, we can write $\hat{\Psi}_z^\dagger(x)\hat{\Psi}_z(x) = -(2\pi)^{-2}\partial_\mu^2 \hat{f}_x(z, z)$, where $\hat{f}_x(z, z')$ is a Green's function which for $z \in [\mu_m, \mu_{m+1}]$ satisfies $\hat{f}_z(z, z) = \pi < v_m(z)|\mathcal{A}_{m-1} \cdots \mathcal{A}_1 \mathcal{A}_n \cdots \mathcal{A}_m|w_m(z) > /(r_m \psi)$, where the spinors v_m and w_m are defined by $v_m^1(z) = -w_m^2(z) = \sinh(2\pi(z - \mu_m)r_m)$, and $v_m^2(z) = w_m^1(z) = \cosh(2\pi(z - \mu_m)r_m)$.

To obtain the finite temperature fermion zero-mode one puts $z = \frac{1}{2}$, whereas for the fermion zero-mode with periodic boundary conditions one takes $z = 0$. From this it is easily seen that in case of well separated constituents the zero-mode is localized only at $\vec{y}_m$ for which $z \in [\mu_m, \mu_{m+1}]$. To be specific, in this limit $\hat{f}_x(z, z) = \pi \tanh(\pi r_m \nu_m)/r_m$ for SU(2), and more generally $\hat{f}_x(z, z) = 2\pi \sinh[2\pi(z - \mu_m)r_m] \sinh[2\pi(\mu_{m+1} - z)r_m]/(r_m \sinh[2\pi\nu_m r_m])$. We illustrate in Fig. 3 the localization of the fermion zero-modes for the case of SU(3).

3.2. *Calorons of higher charge*

We have been able to use a "mix" of the ADHM and Nahm formalism,[10] both in making powerful approximations, like in the far field limit (based on our ability to identify the exponentially rising and falling terms), and

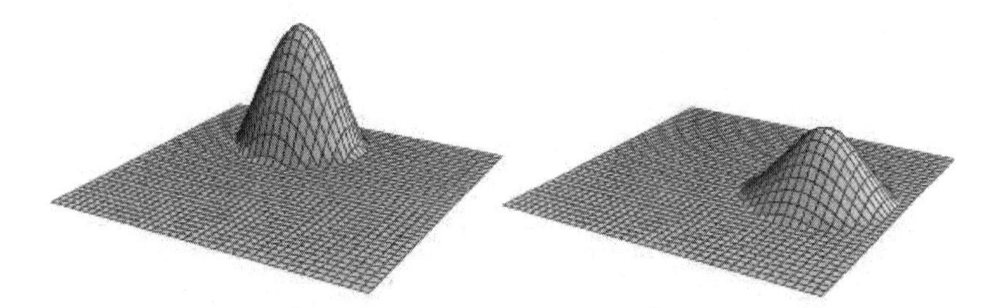

Fig. 3. For the lower right SU(3) configuration in Fig. 2 we have determined on the left the zero-mode density for fermions with anti-periodic boundary conditions in time and on the right for periodic boundary conditions. They are plotted at equal logarithmic scales, cut off below $1/e^5$.

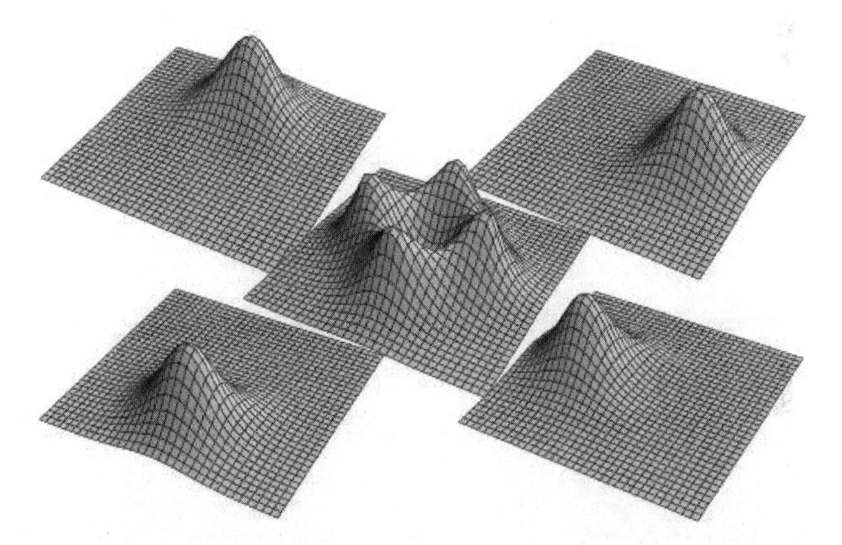

Fig. 4. In the middle is shown the action density in the plane of the constituents at $t = 0$ for an SU(2) charge 2 caloron with $\mathrm{tr}\,\mathcal{P}_\infty = 0$, where all constituents strongly overlap. On a scale enhanced by a factor $10\pi^2$ are shown the densities for the two zero-modes, using either periodic (left) or anti-periodic (right) boundary conditions in the time direction.

for finding exact solutions through solving the homogeneous Green's function.[11] We found axially symmetric solutions for arbitrary k, as well as for $k = 2$ two sets of non-trivial solutions for the matching conditions that interpolate between overlapping and well-separated constituents. For this task we could make use of an existing analytic result for charge-2 monopoles,[12] adapting it to the case of carolons. An example is shown in Fig. 4.

4. More recent results

There are more recent lectures by Bruckmann[13] and Diakonov.[14] Also, Diakonov and Petrov made some progress on constructing the hyperKähler metric which approximates the metric for an arbitrary number of calorons. They claim that this already gives confinement.[14,15] But some cautionary remarks can be made.[16] Also multi-calorons were revisited,[17] and the authors claim to have the full SU(2) moduli space for $k = 2$.

The calorons have also adjoint fermionic zero-modes, and they are now known in analytical form.[18] Finally, Ünsal has published a paper concerning the mechanism of confinement in QCD-like theories,[19] for example SU(2) with $1 \leq n_f \leq 4$ adjoint Majorana fermions. He argues that there are BPS and KK monopoles (precisely the constituents of the caloron), which have zero-modes under the adjoint fermions. They then make BPS-$\overline{\text{KK}}$ bound states (instead of BPS-KK).

Acknowledgments

Thanks to Ara Sedrakyan for inviting me again to Nor Amberd in Armenia and to Tbilisi in Georgia, and doing such a good job in hosting people in spite of difficult times. The content of this talk is old (see arXiv:0901.2853), so there is no arXiv number for it. I went also to Lattice 2011 at Squaw Valley, Lake Tahoe, July 11-16 and thank the organizers, Poul Damgaard, Kim Splittorff and Jac Verbaarschot for inviting me to the ECT* workshop on "Chiral dynamics with Wilson fermions" in Trento, October 24-28, 2011 and to Robijn Bruinsma and Terry Tomboulis of UCLA, where amongst other things, I gave a seminar on January 24, 2012. Finally, I would like to thank FOM for financial support.

References

1. B.J. Harrington and H.K. Shepard, Phys. Rev. D17 (1978) 2122; Phys. Rev. D18 (1978) 2990.
2. D.J. Gross, R.D. Pisarski and L.G. Yaffe, Rev. Mod. Phys. 53 (1981) 43.
3. T.C. Kraan and P. van Baal, Phys. Lett. B428 (1998) 268 [hep-th/9802049]; Nucl. Phys. B533 (1998) 627 [hep-th/9805168]; K. Lee, Phys. Lett B426 (1998) 323 [hep-th/9802012]; K. Lee and C. Lu, Phys. Rev. D58 (1998) 025011 [hep-th/9802108].
4. T.C. Kraan and P. van Baal, Phys. Lett. B435 (1998) 389 [hep-th/9806034].
5. N. Weiss, Phys. Rev. D24 (1981) 475.
6. D. Diakonov, N. Gromov, V. Petrov and S. Slizovskiy, Phys. Rev. D70 (2004) 036003 [hep-th/0404042]; D. Diakonov and N. Gromov, Phys. Rev. D72 (2005) 025003 [hep-th/0502132].

7. E.-M. Ilgenfritz, B.V. Martemyanov, M. Müller-Preussker, S. Shcheredin and A.I. Veselov, Phys. Rev. D66 (2002) 074503 [hep-lat/0206004]; F. Bruckmann, E.-M. Ilgenfritz, B.V. Martemyanov and P. van Baal, Phys. Rev. D70 (2004) 105013 [hep-lat/0408004]; P. Gerhold, E.-M. Ilgenfritz and M. Müller-Preussker, Nucl. Phys. B760 (2007) 1 [hep-ph/0607315].
8. C. Gattringer and S. Schaefer, Nucl. Phys. B654 (2003) 30 [hep-lat/0212029]; C. Gattringer and R. Pullirsch, Phys.Rev. D69 (2004) 094510 [hep-lat/0402008].
9. M. García Pérez, A. González-Arroyo, C. Pena and P. van Baal, Phys. Rev. D60 (1999) 031901 [hep-th/9905016]; M.N. Chernodub, T.C. Kraan and P. van Baal, Nucl. Phys. B(Proc.Suppl.)83-84 (2000) 556.
10. M.F. Atiyah, N.J. Hitchin, V. Drinfeld and Yu.I. Manin, Phys. Lett. 65A (1978) 185; W. Nahm, "Self-dual monopoles and calorons," in: Lecture Notes in Physics, 201 (1984) 189.
11. F. Bruckmann and P. van Baal, Nucl. Phys. B645 (2002) 105 [hep-th/0209010]; F. Bruckmann, D. Nógrádi and P. van Baal, Nucl. Phys. B666 (2003) 197 [hep-th/0305063]; Nucl. Phys. B698 (2004) 233 [hep-th/0404210].
12. H. Panagopoulos, Phys. Rev. D28 (1983) 380.
13. F. Bruckmann, Eur. Phys. J. Special Topics 152 (2007) 61 [arXiv:0706.2269 [hep-th]].
14. D. Diakonov, Acta Phys. Polon. B39 (2008) 3365 [arXiv:0807.0902 [hep-th]]; Nucl. Phys. B(Proc.Suppl.)195 (2009) 5 [arXiv:0906.2456 [hep-ph]].
15. D. Diakonov and V. Petrov, Phys. Rev. D76 (2007) 056001 [arXiv:0704.3181 [hep-th]]; AIP (Conf. Proc.)1134 (2009) 190 [arXiv:0809.2063 [hep-th]].
16. F. Bruckmann, S. Dinter, E.-M. Ilgenfritz, M. Müller-Preussker and M. Wagner, Phys. Rev. D79 (2009) 116007 [arXiv:0903.3075 [hep-ph]].
17. A. Nakamula and J. Sakaguchi, Multi-Calorons Revisited, J. Math. Phys. 51 (2010) 043503 [arXiv:0909.1601 [hep-th]].
18. M. García Perez and A. Gozález-Arroyo, JHEP 0611 (2006) 091 [hep-th/0609058]; M. García Perez, A. Gozález-Arroyo and A. Sastre, Phys. Lett. B668 (2008) 340 [arXiv:0807.2285 [hep-th]]; JHEP 0906 (2009) 065 [arXiv:0905.0645 [hep-th]].
19. M. Ünsal, Phys. Rev. D80 (2009) 065001 [arXiv:0709.3269 [hep-th]].

$SU(N)$ AND $O(N)$ OFF-SHELL NESTED BETHE ANSATZ AND FORM FACTORS

H. Babujian

Yerevan Physics Institute, Alikhanian Brothers 2,
Yerevan, 375036 Armenia
E-mail: babujian@yerphi.am

A. Foerster

Instituto de Física da UFRGS, Av. Bento Gonçalves 9500,
Porto Alegre, RS - Brazil
E-mail: angela@if.ufrgs.br

M. Karowski

Institut für Theoretische Physik, Freie Universität Berlin, Arnimallee 14,
14195 Berlin, Germany
E-mail: karowski@physik.fu-berlin.de

The purpose of the "bootstrap program" for integrable quantum field theories in 1+1 dimensions is to construct explicitly a model in terms of its Wightman functions. In this article, the "bootstrap program" is briefly reviewed and mainly illustrated in terms of the $SU(N)$ Gross-Neveu model and the $O(N)$ σ-model. Systems of $SU(N)$- and $O(N)$-matrix difference equations are solved by means of the off-shell version of the nested algebraic Bethe ansatz. In the nesting process for the $O(N)$ case a new object, the Π-matrix, is introduced to overcome the complexities of the $O(N)$ group structure. Some explicit examples are discussed.

Keywords: Integrable quantum field theory, Form factors.

1. Introduction

The *bootstrap program* to formulate particle physics in terms of the scattering data, i.e. in terms of the S-matrix goes back to Heisenberg[1] and Chew.[2] Remarkably, this approach works very well for integrable quantum field theories in 1+1 dimensions.[3–8] The program does *not* start with any classical Lagrangian. Rather it classifies integrable quantum field theoretic models and in addition provides their explicit exact solutions in terms of all Wightman functions. We achieve contact with the classical models only, when at the end we compare our exact results with Feynman graph (or other) expansions which are usually based on Lagrangians.

One of the authors (M.K.) et al.[4] formulated the on-shell program i.e. the exact determination of the scattering matrix using the Yang-Baxter equations. The concept of generalized form factors was introduced by one of the authors (M.K.) et al.[7] In this article consistency equations were formulated which are expected to be satisfied by these quantities. Thereafter this approach was developed further and studied in the context of several explicit models by Smirnov.[9] In the present article we apply the form factor program for $SU(N)$ and $O(N)$ invariant S-matrices.[10–14] We have to apply the nested "off-shell"[a] Bethe ansatz to get the vectorial part of the form factors. We compare the $1/N$ expansions for the chiral $SU(N)$ Gross-Neveu model[15] and the nonlinear $O(N)$ σ-model[16,17] with our exact results for the form factors finding full agreement. The $SU(N)$ Gross-Neveu model is in particular interesting because the particles are anyons. Both models exhibits asymptotic freedom. Finally the Wightman functions should be obtained by taking integrals and sums over intermediate states. The explicit evaluation of all these integrals and sums remains an open challenge for almost all models, except the scaling Ising model.

2. The "bootstrap program"

The 'bootstrap program' for integrable quantum field theories in 1+1-dimensions provides the solution of a model in term of all its Wightman functions. The result is obtained in three steps:

(1) The S-matrix is calculated by means of general properties such as unitarity and crossing, the Yang-Baxter equations (which are a consequence of integrability) and the additional assumption of 'maximal analyticity'. This means that the two-particle S-matrix is an analytic function in the physical plane (of the Mandelstam variable $(p_1 + p_2)^2$) and possesses only those poles there which are of physical origin. The only input which depends on the model is the assumption of a particle spectrum with an underlining symmetry. A *classification* of all S-matrices obeying the given properties is obtained.

[a] "Off-shell" in the context of the Bethe ansatz means that the spectral parameters in the algebraic Bethe ansatz state are not fixed by Bethe ansatz equations in order to get an eigenstate of a Hamiltonian, but they are integrated over.

(2) Generalized form factors which are matrix elements of local operators

$$^{out}\langle\, \theta'_m, \ldots, \theta'_1 \,|\mathcal{O}(x)|\, \theta_1, \ldots, \theta_n \,\rangle^{in}$$

are calculated by means of the S-matrix. More precisely, the "form factor equations" $(i) - (v)$ as listed in section 3 are solved.

(3) The Wightman functions are obtained by inserting a complete set of intermediate states. In particular the two point function for a hermitian operator $\mathcal{O}(x)$ reads

$$\langle\, 0 \,|\mathcal{O}(x)\, \mathcal{O}(0)|\, 0 \,\rangle = \sum_{n=0}^{\infty} \frac{1}{n!} \int \frac{d\theta_1}{4\pi} \ldots \left| \langle\, 0 \,|\mathcal{O}(0)|\, \theta_1, \ldots, \theta_n \,\rangle^{in} \right|^2 e^{-ix \sum p_i} .$$

Up to now a direct proof that these sums converge exists only for the scaling Ising model,[8,18–21] however, it was shown[22] that models with factorizing S-matrices exist within the framework of algebraic quantum field theory.

Integrability

Integrability in (quantum) field theories means that there exist infinitely many local (or non-local) conservation laws

$$\partial_\mu J^\mu_L(t,x) = 0 \quad (L = \pm 1, \pm 3, \ldots) .$$

A consequence of such conservation laws in 1+1 dimensions is that there is no particle production and the n-particle S-matrix is a product of 2-particle S-matrices

$$S^{(n)}(p_1, \ldots, p_n) = \prod_{i<j} S_{ij}(p_i, p_j) .$$

If backward scattering occurs the 2-particle S-matrices will not commute and one has to specify the order. In particular for the 3-particle S-matrix there are two possibilities

$$S^{(3)} = S_{12} S_{13} S_{23} = S_{23} S_{13} S_{12}$$

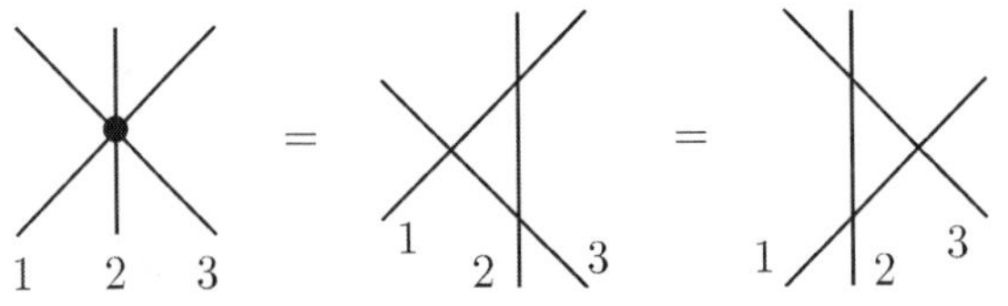

which yield the **"Yang-Baxter Equation"**.

The two particle S-matrix is of the form $S_{\alpha\,\beta}^{\beta'\alpha'}(\theta_{12})$ where α, β etc. denote the type of the particles and the rapidity difference $\theta_{12} = \theta_1 - \theta_2$ is defined by $p_i = m_i(\cosh\theta_i, \sinh\theta_i)$. We also use the short hand notation $S_{12}(\theta_{12})$. It satisfies unitarity and crossing

$$S_{21}(\theta_{21})S_{12}(\theta_{12}) = 1 \tag{1}$$

$$S_{12}(\theta_1 - \theta_2) = \mathbf{C}^{2\bar{2}}\, S_{\bar{2}1}(\theta_2 + i\pi - \theta_1)\, \mathbf{C}_{\bar{2}2} = \mathbf{C}^{1\bar{1}}\, S_{2\bar{1}}(\theta_2 - (\theta_1 - i\pi))\, \mathbf{C}^{\bar{1}1} \tag{2}$$

where $\mathbf{C}^{1\bar{1}}$ and $\mathbf{C}_{1\bar{1}}$ are charge conjugation matrices.

Examples of integrable models in 1+1-dimensions are

- the $SU(N)$ Gross-Neveu[15] model described by the Lagrangian

$$\mathcal{L} = \bar{\psi}\, i\gamma\partial\, \psi + \frac{g^2}{2}\left((\bar{\psi}\psi)^2 - (\bar{\psi}\gamma^5\psi)^2\right),$$

  where the Fermi fields form an $SU(N)$ multiplet.
- the nonlinear $O(N)$ σ-model defined by the Lagrangian and the constraint

$$\mathcal{L} = \frac{1}{2}\sum_{\alpha=1}^{N}(\partial_\mu\varphi_\alpha)^2 \qquad \text{with} \quad g\sum_{\alpha=1}^{N}\varphi_\alpha^2 = 1$$

  where $\varphi_\alpha(x)$ is an isovector N-plet set of bosonic fields.

Further integrable quantum field theories are: the sine-Gordon, the Toda, the scaling Z_N-Ising, the $O(N)$ Gross-Neveu models etc.

The S-matrix

- The two particle S-matrix of the $SU(N)$ Gross-Neveu model is[23–26]

$$S_{\alpha\beta}^{\delta\gamma}(\theta) = \quad \begin{array}{c} \delta \qquad \gamma \\ p_4\ \ p_3 \\ \times \\ p_1\ \ p_2 \\ \alpha \qquad \beta \end{array} \quad = \delta_{\alpha\gamma}\delta_{\beta\delta}\, b(\theta) + \delta_{\alpha\delta}\delta_{\beta\gamma}\, c(\theta) \tag{3}$$

  where due to Yang-Baxter $c(\theta) = -\frac{2\pi i}{N\theta}\, b(\theta)$ holds and the highest weight amplitude is given as

$$a^{SU(N)}(\theta) = b(\theta) + c(\theta) = -\frac{\Gamma\left(1 - \frac{\theta}{2\pi i}\right)\Gamma\left(1 - \frac{1}{N} + \frac{\theta}{2\pi i}\right)}{\Gamma\left(1 + \frac{\theta}{2\pi i}\right)\Gamma\left(1 - \frac{1}{N} - \frac{\theta}{2\pi i}\right)} \tag{4}$$

There is a bound state pole at $\theta = i\eta = 2\pi i/N$ in the antisymmetric tensor sector which agrees with Swieca's[27] picture that the bound state of $N-1$ particles is to be identified with the anti-particle.

- The two particle S-matrix of the nonlinear $O(N)$ σ-model is[6]

$$S^{\delta\gamma}_{\alpha\beta}(\theta) =$$

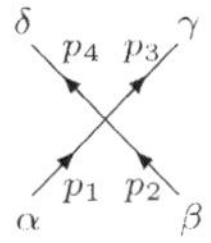

$$= \delta^\gamma_\alpha \delta^\delta_\beta \, b(\theta) + \delta^\delta_\alpha \delta^\gamma_\beta \, c(\theta) + \delta^{\delta\gamma}\delta_{\alpha\beta} \, d(\theta) \tag{5}$$

where due to Yang-Baxter $c(\theta) = -\frac{2\pi i}{(N-2)\theta} b(\theta)$ and crossing $b(\theta) = b(i\pi-\theta)$, $d(\theta) = c(i\pi-\theta)$ hold. The highest weight amplitude $b(\theta)+c(\theta)$ is given as

$$a^{O(N)}(\theta) = -\frac{\Gamma\left(\frac{1}{2}+\frac{\theta}{2\pi i}\right)\Gamma\left(\frac{1}{2}+\frac{1}{N-2}-\frac{\theta}{2\pi i}\right)\Gamma\left(1-\frac{\theta}{2\pi i}\right)\Gamma\left(\frac{1}{N-2}+\frac{\theta}{2\pi i}\right)}{\Gamma\left(\frac{1}{2}-\frac{\theta}{2\pi i}\right)\Gamma\left(\frac{1}{2}+\frac{1}{N-2}+\frac{\theta}{2\pi i}\right)\Gamma\left(1+\frac{\theta}{2\pi i}\right)\Gamma\left(\frac{1}{N-2}-\frac{\theta}{2\pi i}\right)}. \tag{6}$$

For the Bethe ansatz used below it is more convenient to use instead of the real basis $|\alpha\rangle_r$, $(\alpha = 1, 2, \dots, N)$ a complex basis

$$\begin{cases} |\alpha\rangle = \frac{1}{\sqrt{2}}\left(|2\alpha-1\rangle_r + i|2\alpha\rangle_r\right), \; \alpha = 1, \dots, [N/2] \\ |\bar\alpha\rangle = \frac{1}{\sqrt{2}}\left(|2\alpha-1\rangle_r - i|2\alpha\rangle_r\right), \; \bar\alpha = \bar 1, \dots, \overline{[N/2]}. \end{cases}$$

For N odd there is in addition $|0\rangle = |\bar 0\rangle = |N\rangle_r$. In (5) $\delta^{\delta\gamma}\delta_{\alpha\beta}$ is then replaced by $\mathbf{C}^{\delta\gamma}\mathbf{C}_{\alpha\beta}$ with the charge conjugation matrix $\mathbf{C}$.

3. Form factors

For a local operator $\mathcal{O}(x)$ the generalized form factors[7] are defined as

$$F^{\mathcal{O}}_{\alpha_1\dots\alpha_n}(\theta_1, \dots, \theta_n) = \langle 0 \,|\, \mathcal{O}(0) \,|\, p_1, \dots, p_n \rangle^{in}_{\alpha_1\dots\alpha_n} \tag{7}$$

for $\theta_1 > \dots > \theta_n$. For other orders of the rapidities they are defined by analytic continuation. The index α_i denotes the type of the particle with momentum p_i. We also use the short notations $F^{\mathcal{O}}_{\underline\alpha}(\underline\theta)$ or $F^{\mathcal{O}}_{1\dots n}(\underline\theta)$.

We assume 'maximal analyticity' for the form factors which means that they are meromorphic and all poles in the 'physical strips' $0 \le \operatorname{Im}\theta_{ij} \le \pi$ have a physical interpretation. Together with the usual LSZ-assumptions[28] of local quantum field theory the following form factor equations can be derived:

(i) The Watson's equations describe the symmetry property under the permutation of both, the variables θ_i, θ_j and the spaces $i, j = i+1$ at the same time

$$F^{\mathcal{O}}_{\dots ij \dots}(\dots,\theta_i,\theta_j,\dots) = F^{\mathcal{O}}_{\dots ji \dots}(\dots,\theta_j,\theta_i,\dots)\, S_{ij}(\theta_{ij}) \tag{8}$$

for all possible arrangements of the θ's.

(ii) The crossing relation which implies a periodicity property under the cyclic permutation of the rapidity variables and spaces

$$\begin{aligned} &{}^{\text{out},\bar{1}}\langle\, p_1 \,|\, \mathcal{O}(0) \,|\, p_2,\dots,p_n \,\rangle^{\text{in,conn.}}_{2\dots n} \\ &= \mathbf{C}^{\bar{1}1} \sigma_1^{\mathcal{O}} F^{\mathcal{O}}_{1\dots n}(\theta_1 + i\pi, \theta_2,\dots,\theta_n) = F^{\mathcal{O}}_{2\dots n1}(\theta_2,\dots,\theta_n,\theta_1 - i\pi)\mathbf{C}^{1\bar{1}} \end{aligned} \tag{9}$$

where $\sigma^{\mathcal{O}}_{\alpha}$ takes into account the statistics of the particle α with respect to $\mathcal{O}$ (e.g., $\sigma^{\mathcal{O}}_{\alpha} = -1$ if α and $\mathcal{O}$ are both fermionic, these numbers can be more general for anyonic or order and disorder fields[29]).

(iii) There are poles determined by one-particle states in each sub-channel given by a subset of particles of the state in (7).
In particular the function $F^{\mathcal{O}}_{\underline{\alpha}}(\underline{\theta})$ has a pole at $\theta_{12} = i\pi$ such that

$$\operatorname*{Res}_{\theta_{12}=i\pi} F^{\mathcal{O}}_{1\dots n}(\theta_1,\dots,\theta_n) = 2i\,\mathbf{C}_{12}\, F^{\mathcal{O}}_{3\dots n}(\theta_3,\dots,\theta_n)\left(\mathbf{1} - \sigma^{\mathcal{O}}_2 S_{2n}\dots S_{23}\right). \tag{10}$$

(iv) If there are also bound states in the model the function $F^{\mathcal{O}}_{\underline{\alpha}}(\underline{\theta})$ has additional poles. If for instance the particles 1 and 2 form a bound state (12), there is a pole at $\theta_{12} = i\eta$ $(0 < \eta < \pi)$ such that

$$\operatorname*{Res}_{\theta_{12}=i\eta} F^{\mathcal{O}}_{12\dots n}(\theta_1,\theta_2,\dots,\theta_n) = F^{\mathcal{O}}_{(12)\dots n}(\theta_{(12)},\dots,\theta_n)\,\sqrt{2}\Gamma^{(12)}_{12} \tag{11}$$

where $\Gamma^{(12)}_{12}$ is the bound state intertwiner.[30,31]

(v) Naturally, since we are dealing with relativistic quantum field theories we finally have

$$F^{\mathcal{O}}_{1\dots n}(\theta_1 + \mu,\dots,\theta_n + \mu) = e^{s\mu}\, F^{\mathcal{O}}_{1\dots n}(\theta_1,\dots,\theta_n) \tag{12}$$

if the local operator transforms under Lorentz transformations as $F^{\mathcal{O}} \to e^{s\mu} F^{\mathcal{O}}$ where s is the "spin" of $\mathcal{O}$.

These equations have been proposed by Smirnov[9] as generalizations of equations derived in the original articles.[7,8,32] They have been proven[33] by means of the LSZ-assumptions and 'maximal analyticity'.

We will now provide a constructive and systematic way of how to solve the equations (i) – (v) for the co-vector valued function $F^{\mathcal{O}}_{1\dots n}$ once the scattering matrix is given.

3.1. *Two-particle form factors*

For the two-particle form factors the form factor equations (i) and (ii) are

$$\begin{aligned} F(\theta) &= F(-\theta)\, S(\theta) \\ F(i\pi-\theta) &= F(i\pi+\theta) \end{aligned} \tag{13}$$

for all eigenvalues of the two-particle S-matrix. For general theories Watson's[34] equations only hold below the particle production thresholds. However, for integrable theories there is no particle production and therefore they hold for all complex values of θ. It has been shown[7] that these equations together with "maximal analyticity" have a unique solution.

As an example we write the (highest weight) $SU(N)$ and $O(N)$ form factor functions[10,14]

$$F^{SU(N)}(\theta) = \exp \int_0^\infty \frac{dt}{t\sinh^2 t} e^{\frac{t}{N}} \sinh t\,(1-1/N)\,(1-\cosh t\,(1-\theta/(i\pi))) \tag{14}$$

$$F^{O(N)}(\theta) = \exp \int_0^\infty \frac{dt}{t\sinh t} \frac{1-e^{-2t/(N-2)}}{1+e^{-t}}\,(1-\cosh t\,(1-\theta/(i\pi))) \tag{15}$$

which are the minimal solution of (13) with $S(\theta) = a(\theta)$ as given by (4) and (6), respectively. In particular for $O(3)$ we have $F^{O(3)}(\theta) = (\theta - i\pi)\tanh\frac{1}{2}\theta$.

3.2. *The general form factor formula*

As usual[7] we split off the minimal part and write the form factor for n particles as

$$F^{\mathcal{O}}_{\alpha_1\ldots\alpha_n}(\theta_1,\ldots,\theta_n) = K^{\mathcal{O}}_{\alpha_1\ldots\alpha_n}(\underline{\theta}) \prod_{1\le i<j\le n} F(\theta_{ij})\,. \tag{16}$$

By means of the following "off-shell Bethe ansatz" for the (co-vector valued) K-function

$$\boxed{K^{\mathcal{O}}_{\alpha_1\ldots\alpha_n}(\underline{\theta}) = \int_{\mathcal{C}_{\underline{\theta}}} dz_1 \cdots \int_{\mathcal{C}_{\underline{\theta}}} dz_m\, h(\underline{\theta},\underline{z})\, p^{\mathcal{O}}(\underline{\theta},\underline{z})\, \Psi_{\alpha_1\ldots\alpha_n}(\underline{\theta},\underline{z})} \tag{17}$$

we transform the complicated form factor equations $(i)-(v)$ into simple ones for the p-functions which are scalar and simple functions of $e^{\pm z_i}$. The **"off-shell Bethe ansatz"** state $\Psi_{\alpha_1\ldots\alpha_n}(\underline{\theta},\underline{z})$ is obtained as a product of

S-matrix elements and the integration contour $\mathcal{C}_{\underline{\theta}}$ encircles poles of $h(\underline{\theta},\underline{z})$ (see below). The scalar function

$$h(\underline{\theta},\underline{z})=\prod_{i=1}^{n}\prod_{j=1}^{m}\phi_j(\theta_i-z_j)\prod_{1\le i<j\le m}\tau_{ij}(z_i-z_j) \tag{18}$$

depends only on the S-matrix (see below), whereas the p-function $p^{\mathcal{O}}(\underline{\theta},\underline{z})$ depends on the operator.

In case of higher rank r the **"nested Bethe ansatz state"** Ψ is of the form

$$\Psi_{\alpha_1\dots\alpha_n}(\underline{\theta},\underline{z})=L_{\beta_1\dots\beta_m}(\underline{z})\Phi^{\beta_1\dots\beta_m}_{\alpha_1\dots\alpha_n}(\underline{\theta},\underline{z})\,. \tag{19}$$

For the co-vector valued function L which belongs to rank $r-1$ one makes an ansatz analogous to (17). Nesting means that one repeats this up to $SU(2)$, respectively $O(4)$ or $O(3)$. The number of Bethe ansatz levels is equal to the rank, i.e. $\operatorname{rank}(SU(N))=N-1$ and $\operatorname{rank}(O(N))=[N/2]$.

- For $SU(N)$ the basic Bethe ansatz co-vectors Φ of equation (19) may be depicted as

$$\Phi^{\underline{\beta}}_{\underline{\alpha}}(\underline{\theta},\underline{z})= \quad \text{with } \begin{cases}2\le\beta_i\le N\\ 1\le\alpha_i\le N\,.\end{cases}$$

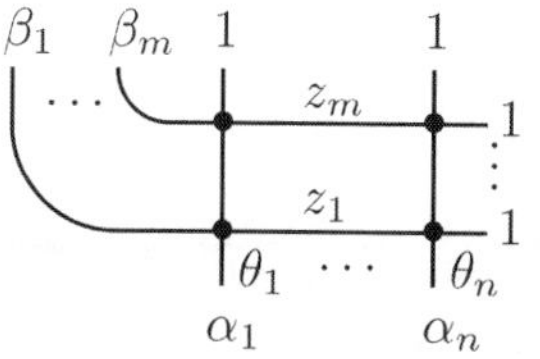

It means that $\Phi^{\underline{\beta}}_{\underline{\alpha}}(\underline{\theta},\underline{z})$ is a product of S-matrix elements as given by the picture where at all crossing points of lines there is an S-matrix (3) and the sum over all indices of internal lines is to be taken.

- For $O(N)$ the basic Bethe ansatz co-vectors Φ is more complicated

$$\Phi^{\underline{\beta}}_{\underline{\alpha}}(\underline{\theta},\underline{z})= \quad \text{with } \begin{cases}\beta_i=2,\dots,(0),\dots,\bar{2}\\ \alpha_i=1,2,\dots,(0),\dots,\bar{2},\bar{1}\end{cases}$$

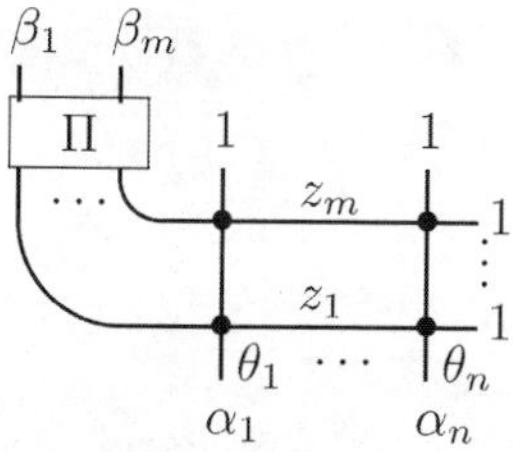

The matrix Π maps the $O(N)$ S-matrix to the $O(N-2)$ one where the rank decreases by 1

$$S^{O(N-2)}_{ij}\Pi_{\dots ij\dots}=\Pi_{\dots ji\dots}S^{O(N)}_{ij}\,.$$

For $SU(N)$ the matrix Π is trivial because the S-matrix elements do not depend on N (for a suitable normalization and parameterization).

We concentrate here on the results for $O(N)$, the results for $SU(N)$ have been published[10–12] previously. For general N the functions ϕ_j and τ_{ij} in (18) depend on whether $i,j = e,o$ are even or odd. The form factor equations (ii) and (iii) imply equations for $\tilde{\phi}_j(z) = a(z)\phi_j(z)$ and $\tau_{ij}(z)$

$$(\mathrm{i}) : \begin{cases} \tilde{\phi}_j(z) = \tilde{b}(z+2\pi i)\tilde{\phi}_j(z+2\pi i) \\ \tau_{ij}(z-2\pi i)/\tilde{b}(2\pi i - z) = \tau_{ij}(z)/\tilde{b}(z) \end{cases}$$

$$(\mathrm{ii}) : \tilde{\phi}_e(-z)\tilde{\phi}_o(-z-i\pi+i\eta)F(z)F(i\pi+z) = 1 \ , \qquad \eta = \frac{2\pi}{N-2}$$

with $\tilde{b}(z) = b(z)/a(z)$. The equation for τ_{ij} from (iii) is more complicated[35] and skipped here. For $O(3)$ the functions ϕ_j do not depend on j and solutions for $\tilde{\phi}$ and τ are simple

$$\tilde{\phi}(z) = \frac{1}{z} \ , \quad \tau(z) = z^2 .$$

In order that the form factors $F^{\mathcal{O}}(\theta)$ satisfy the form factor equations (i)–(ii) the p-function $p^{\mathcal{O}}(\underline{\theta}, \underline{\underline{z}})$ have to satisfy some simple equations, e.g.

$$\begin{aligned} p^{\mathcal{O}}(\underline{\theta}, \underline{\underline{z}}) &= p^{\mathcal{O}}(\theta_1 + 2\pi i, \theta_2, \ldots, \underline{\underline{z}}) \\ &= p^{\mathcal{O}}(\underline{\theta}, \ldots, z_i^{(l)} + 2\pi i, \ldots) \end{aligned}$$

For $SU(N)$ there are some additional phase factors because of the anyonic statistics of the fields and particles.

4. Examples:

4.1. *The chiral SU(N)-Gross-Neveu model*

The classical Lagrangian density is

$$\mathcal{L} = \sum_{\alpha=1}^{N} \bar{\psi}_\alpha i\gamma\partial\psi_\alpha + \frac{1}{2}g^2 \left(\left(\sum_{\alpha=1}^{N} \bar{\psi}_\alpha \psi_\alpha \right)^2 - \left(\sum_{\alpha=1}^{N} \bar{\psi}_\alpha \gamma^5 \psi_\alpha \right)^2 \right) \tag{20}$$

where ψ_α is an $SU(N)$ isovector N-plet of Fermi fields. The quantum version of this model exhibit anyonic statistics.[24]

The p-function which gives the exact $SU(N)$ form factors for the field component $\psi^{(\pm)}(x) = \psi_1^{(\pm)}(x)$ is[10]

$$p^{\psi_1^{(\pm)}}(\underline{\theta}, \underline{z}) = \exp \pm \frac{1}{2} \left(\sum_{i=1}^{m} z_i - \left(1 - \frac{1}{N}\right) \sum_{i=1}^{n} \theta_i \right) \tag{21}$$

and the 1-particle form factor is

$$\langle\, 0\,|\, \psi^{(\pm)}(0)\,|\, \theta\,\rangle_\alpha = \delta_{\alpha 1}\, e^{\mp\frac{1}{2}(1-\frac{1}{N})\theta}\,.$$

The 3-particle form factor given by (16), (17) and (21) can be expressed in terms of Meijer's G-functions. The $1/N$ expansion of the exact result for the operator $\mathcal{O}(x) = -i\,(i\gamma\partial - m)\,\psi(x)$ is[12]

$$\begin{aligned} {}^{\gamma}_{out}\langle\, \theta_3\,|\, \mathcal{O}(0)\,|\, \theta_1, \theta_2\,\rangle^{in}_{\alpha\beta} &= \frac{2i\pi}{N} m \\ \times \left(\delta^1_\alpha \delta^\gamma_\beta \frac{\sinh\theta_{23}}{\theta_{23}} \left(\frac{1}{\cosh\frac{1}{2}\theta_{23}} - \gamma^5 \frac{1}{\sinh\frac{1}{2}\theta_{23}} \right) \right.&\left. u(\theta_1) - (1, \alpha \leftrightarrow 2, \beta) \right) \end{aligned} \tag{22}$$

which agrees with the $1/N$ expansion in terms of Feynman graphs starting from the Lagrangian (20).

4.2. *The nonlinear $O(N)$ σ-model*

The model is defined by the Lagrangian and the constraint

$$\mathcal{L} = \frac{1}{2}\sum_{\alpha=1}^{N} (\partial_\mu \varphi_\alpha)^2\,, \qquad g\sum_{\alpha=1}^{N} \varphi_\alpha^2 = 1 \tag{23}$$

where $\varphi_\alpha(x)$ is an isovector N-plet set of bosonic fields.

The p-function for the field $\varphi(x) = \varphi_1(x)$ is

$$p^\varphi(\underline{\theta}, \underline{z}) = 1$$

and the 1-particle form factor is

$$\langle 0|\varphi(0)|\theta\rangle_\alpha = \delta_{\alpha 1}$$

The exact 3-particle form factor of $\varphi(x)$ for $O(3)$ can be calculated from

$$K^\varphi_{\underline{\alpha}}(\underline{\theta}) = \int_{C_{\underline{\theta}}} dz_1 \int_{C_{\underline{\theta}}} dz_2\, \tilde{h}(\underline{\theta}, \underline{z}) p^\varphi(\underline{\theta}, \underline{z})\, L(z_{12}) \tilde{\Phi}_{\underline{\alpha}}(\underline{\theta}, \underline{z})$$

with $L(z) = \frac{(z-i\pi)}{z(z-2\pi i)} \tanh\frac{1}{2}z$. as[35]

$$F^\varphi_{\alpha\beta\gamma}(\underline{\theta}) = \left(\theta_{23}\delta^1_\alpha \mathbf{C}_{\beta\gamma} - (\theta_{13} - 2\pi i)\, \delta^1_\beta \mathbf{C}_{\alpha\gamma} + \theta_{12}\delta^1_\gamma \mathbf{C}_{\alpha\beta}\right) G(\theta_{12}) G(\theta_{13}) G(\theta_{23})$$

where $G(\theta) = \frac{(\theta-i\pi)}{\theta(\theta-2\pi i)} \tanh^2\frac{1}{2}\theta$. This agrees with results of Balog et al.[36] obtained using different techniques. The $1/N$ expansion of 3-particle form factor of the operator $\mathcal{O}(x) = i(\Box + m^2)\varphi(x)$ is[35]

$$\begin{aligned} &F^{\mathcal{O}}{}_{\alpha\beta\gamma}(\theta_1, \theta_2, \theta_3) \\ &= -\frac{8\pi i}{N} m^2 \left(\delta^1_\alpha \mathbf{C}_{\beta\gamma} \frac{\sinh\theta_{23}}{i\pi - \theta_{23}} + \delta^1_\beta \mathbf{C}_{\alpha\gamma} \frac{\sinh\theta_{13}}{i\pi - \theta_{13}} + \delta^1_\gamma \mathbf{C}_{\alpha\beta} \frac{\sinh\theta_{12}}{i\pi - \theta_{12}} \right) \end{aligned}$$

which agrees with the $1/N$ expansion in terms of Feynman graphs starting from the Lagrangian (23).

Acknowledgments

The authors have profited from discussions with A. Fring, R. Schrader, F. Smirnov and A. Belavin. H.B. thanks R. Flume, R. Poghossian and P. Wiegmann for valuable discussions. H.B. and M.K. were supported by Humboldt Foundation and H.B. is also supported by the Armenian grant 11-1_c028. A.F. acknowledges support from DAAD (Deutscher Akademischer Austausch Dienst) and CNPq (Conselho Nacional de Desenvolvimento Científico e Tecnológico).

References

1. W. Heisenberg, *Zeitschrift für Naturforschung* **1**, 608 (1946).
2. G. Chew, *W. A. Benjamin, Inc. New York* (1961).
3. B. Schroer, T. T. Truong and P. Weisz, *Phys. Lett.* **B63**, 422 (1976).
4. M. Karowski, H. J. Thun, T. T. Truong and P. Weisz, *Phys. Lett.* **B67**, 321 (1977).
5. M. Karowski and H. J. Thun, *Nucl. Phys.* **B130**, 295 (1977).
6. A. B. Zamolodchikov and A. B. Zamolodchikov, *Annals Phys.* **120**, 253 (1979).
7. M. Karowski and P. Weisz, *Nucl. Phys.* **B139**, 455 (1978).
8. B. Berg, M. Karowski and P. Weisz, *Phys. Rev.* **D19**, 2477 (1979).
9. F. Smirnov, *Adv. Series in Math. Phys. 14, World Scientific* (1992).
10. H. M. Babujian, A. Foerster and M. Karowski, *J. Phys.* **A41**, p. 275202 (2008).
11. H. M. Babujian, A. Foerster and M. Karowski, *SIGMA* **2**, paper 082, 16 pages (2007).
12. H. M. Babujian, A. Foerster and M. Karowski, *Nucl. Phys.* **B825**, 396 (2010).
13. H. M. Babujian, A. Foerster and M. Karowski, *Theor. Math. Phys.* **155**, 512 (2008).
14. H. M. Babujian, A. Foerster and M. Karowski, *J. Phys.* **A45**, p. 055207 (2012).
15. D. J. Gross and A. Neveu, *Phys. Rev.* **D10**, p. 3235 (1974).
16. A. M. Polyakov, *Phys. Lett.* **B59**, 79 (1975).
17. K. Pohlmeyer, *Commun. Math. Phys.* **46**, 207 (1976).
18. R. Z. Bariev, *Phys. Lett.* **A55**, 456 (1976).
19. B. M. McCoy, C. A. Tracy and T. T. Wu, *Phys. Rev. Lett.* **38**, 793 (1977).
20. M. Sato, T. Miwa and M. Jimbo, *Proc. Japan Acad.* **53**, 6 (1977).
21. V. P. Yurov and A. B. Zamolodchikov, *Int. J. Mod. Phys.* **A6**, 4557 (1991).
22. G. Lechner, *Commun.Math.Phys.* **277**, 821 (2008).
23. B. Berg, M. Karowski, V. Kurak and P. Weisz, *Nucl. Phys.* **B134**, 125 (1978).
24. R. Koberle, V. Kurak and J. A. Swieca, *Phys. Rev.* **D20**, 897 (1979).

25. B. Berg and P. Weisz, *Nucl. Phys.* **B146**, 205 (1978).
26. E. Abdalla, B. Berg and P. Weisz, *Nucl. Phys.* **B157**, 387 (1979).
27. V. Kurak and J. A. Swieca, *Phys. Lett.* **B82**, 289 (1979).
28. H. Lehmann, K. Symanzik and W. Zimmermann, *Nuovo Cim.* **1**, 205 (1955).
29. H. Babujian, A. Foerster and M. Karowski, *Nucl. Phys.* **B736**, 169 (2006).
30. M. Karowski, *Nucl. Phys.* **B153**, 244 (1979).
31. H. Babujian and M. Karowski, *Nucl. Phys.* **B620**, 407 (2002).
32. M. Karowski, *in: W. Rühl (Ed.), Field theoretic methods in particle physics, Plenum, New York, (1980)* , 307 Presented at Kaiserslautern NATO Inst. 1979.
33. H. M. Babujian, A. Fring, M. Karowski and A. Zapletal, *Nucl. Phys.* **B538**, 535 (1999).
34. K. M. Watson, *Phys. Rev.* **95**, 228 (1954).
35. H. M. Babujian, A. Foerster and M. Karowski, *to be published* .
36. J. Balog and M. Niedermaier, *Nucl. Phys.* **B500**, 421 (1997).

SPONTANEOUS BREAKING OF LORENTZ-INVARIANCE AND GRAVITONS AS GOLDSTONE PARTICLES

Z. Berezhiani*

Dipartimento di Fisica, Università di L'Aquila, 67010 Coppito, AQ, and INFN, Laboratori Nazionali del Gran Sasso, 67010 Assergi, AQ, Italy
**E-mail: zurab.berezhiani@aquila.infn.it*

O.V. Kancheli

Institute of Theoretical and Experimental Physics, 117259 Moscow, Russia
E-mail: oleg.kancheli@gmail.com

We consider some aspects of spontaneous breaking of Lorentz Invariance in field theories, discussing the possibility that the certain tensor operators may condensate in the ground state in which case the tensor Goldstone particles would appear. We analyze their dynamics and discuss to which extent such a theory could imitate the gravity. We are also interested if the universality of coupling of such 'gravitons' with other particles can be achieved in the infrared limit.

We are pleased to publish this article in a Volume dedicated to the famous physicist and wonderful personality Sergey Matinyan, with whom we have so many links

1. Introduction

The bad ultraviolet behavior of the General Relativity asks for different descriptions of gravity at small distances – as in String Theory or Loop Gravity approaches. A possibility to represent the gravitons as the Goldstone particles connected with fluctuations of a tensor condensate which spontaneously breaks the Lorentz-Invariance (LI) can be quite promising, especially if this condensate effectively emerges in the context of some renormalizable gauge theory in the strong coupling regime. Such a theory might be free of the short distance problems.

The idea that the Spontaneous Lorentz-violation by a vector condensate giving rise to the non-scalar Goldstone bosons in the particle spectrum was first discussed by Bjorken in his seminal paper[1] where these goldstones were associated with the photon. This idea was further extended also to tensor condensates and corresponding goldstones were interpreted as graviton-like objects.[2,3]

In particular, one can try to incorporate the goldstone gravitons in the following picture. Imagine that there is no fundamental "geometrical" gravity we live in a flat 3+1-dimensional Minkowskian space-time. Our fundamental physics is described by some renormalizable field theory based on a gauge symmetry G_S which can be, for example, the Standard Model or some its generalization like $SU(5)$ or $SO(10)$. This "stardard" gauge sector is supplemented by an *additional* non-Abelian gauge sector G_C and there are some fields (perhaps in mixed representations) mediating between G_S and G_C sectors. The latter becomes strongly coupled at some very large scale Λ at which the Lorentz-breaking tensor condensates are formed and the corresponding Goldstone bosons emerge. It is natural that then one is tempted to consider them as the the mediators of gravity. Respectively, the confinement scale Λ should be related to the Planck scale $m_p \simeq 10^{19}$ GeV. Such a picture singles out the three-dimensional space since the introduction of additional dimensions would spoil the renormalizability. In addition, gauge theories in higher dimensions, instead of being asymptotically free at small distances and getting stronger at the infrared limit, become weakly coupled in the infrared region which makes difficult the formation of the composite states and condensates. An important feature of such Goldstone gravity may be that it does not interact with the Lorentz-invariant part of the energy-momentum tensor which opens up a way for understanding the cosmological term problem.[3]

It is rather unclear how far can one advance towards replacing the geometrical gravity by the composite goldstone tensor field. One must answer crucial questions:

- is it possible to make the interaction of such goldstone gravitons universal, at least in the infra-red (IR) limit? The universality of gravitation, automatic in the General Relativity, is experimentally tested with very high precision
- is it possible in the presence of vector and/or tensor condensates to reproduce the Lorentz invariance of the relevant measurable quantities at the experimentally acceptable level?

These questions are not independent. If Lorentz invariance is restored at large distances, then perhaps the universality of the goldstone interactions will be also restored. There are arguments that the only consistent Lorentz-invariant theory of interacting massless spin two particles is the general relativity,[4] but these arguments are cannot be directly applied to the Lorentz-violation case. Perhaps the inverse statement can be true - if at large distances the gravitational interactions become universal, then Lorentz non-invariant terms can die out in the IR limit, or turn into gauge fixing terms giving no contributions to the measurable quantities.

Unfortunately, there is no good theoretical understanding of these issues and there are no more or less reliable model calculations. But if these hypotheses are correct and in the IR limit the non-universal and LI-violating contributions to the physical amplitudes are indeed suppressed, then such an approach to the gravity could open up a number of interesting possibilities.

In this paper, which is mainly based on ref.,[5] we discuss some of these questions and also outline the main peculiarities of corresponding cosmological models.

2. Tensor condensates and spontaneous Lorentz-violation

In the quantum field theory composite condensates can arise from the enough strong attraction between virtual particles. The non-Abelian gauge theories with adequately restricted multiplet content usually become strongly coupled below certain 'confinement' scale Λ, and various composite condensates can appear, like the quark $\langle \bar{q}q \rangle$ and gluonic $\langle GG \rangle$ condensates in the QCD. Unfortunately, about the formation of condensates we know mainly from the experiment (and lattice calculations) and not directly from theory. Only some special cases, e.g. in gauge theories in the large N_c limit or in theories with large supersymmetry this phenomenon can be analyzed in more details. We also know (only from the experiment) that in the QCD case the condensates are Lorentz scalars, and yet we do not understand the dynamical reasons for this. But may be for other gauge groups with specific multiplet contents, different from QCD, the nonscalar condensates can also appear along with "usual" scalar condensates.

In a more phenomenological approach, applicable for energy scales below the confinement scale, one can consider the composite states as effective fields and analyze their effective Lagrangian taking into the account that some of the effective composite fields get negative m^2. As a result the ground state is coherently filled with such tachyons which in addition can have non-trivial quantum numbers. The last has no advantage for precise calculations but gives a clue for the possible symmetry breaking channels and represents a way usually followed when considering various applications of spontaneous symmetry breaking.

Let us consider a theory based on some non-abelian gauge symmetry G_C in a minskowskian space-time and discuss a general composite symmetric tensor operator with the "vacuum" quantum numbers:[a]

$$\tau_{\mu\nu} = \mathrm{Tr}_C\big[A \cdot \bar{\varphi}\partial_\mu\partial_\nu\varphi + B \cdot \bar{\chi}(\partial_\mu\gamma_\nu + \partial_\nu\gamma_\mu)\chi + B' \cdot (\bar{\chi}\gamma_\mu\chi)(\bar{\chi}\gamma_\nu\chi) \\ + C \cdot G_{\mu\rho}G_{\rho\nu} + C' \cdot G_{\mu\rho}G_{\rho\lambda}G_{\lambda\nu} + \ldots\big], \quad (1)$$

where φ and χ respectively are the scalar and fermion fields in some representations of G_C and $B_{\mu\nu}$ are the field-strength tensors of the gauge fields. We take all these fields to be dimensionless and for reproducing their canonical dimensions one can use confinement scale Λ at which G_C sector becomes strongly coupled. In the following, all fields belonging to G_C sector of the theory will be generically denoted as Ψ, and their 'color' indices will be omitted. The dimensionless coefficients $A, B, C, \ldots$ generally can be some scalar functions of Ψ-fields. The trace in (1) is taken over the indices of the internal gauge group so that all terms are gauge singlets of G_C. Let us suppose that in certain conditions the vacuum expectation value[b]

$$\langle \tau_{\mu\nu}(x) \rangle \ = \ \int D\Psi \ \tau_{\mu\nu} \ e^{i\int d^4x L(\Psi)} \quad (2)$$

can be nonzero, and let us study the behavior of the system in a neighborhood of $\langle \tau_{\mu\nu} \rangle$. Let us introduce, as usual, the external current $J_{\mu\nu}(x)$ linearly coupled with $\tau_{\mu\nu}(x)$. Then the response of the system to the excitation of the degrees of freedom related to $\tau_{\mu\nu}(x)$ can be described by the generating functional

$$Z[J_{\mu\nu}(x)] = \int D\Psi \, e^{i\int d^4x[L(\Psi)+\tau_{\mu\nu}(\Psi)J^{\mu\nu}(x)]}$$

$$= \int D\Psi \, Dt_{\mu\nu} \, \delta_x[t_{\mu\nu}(x) - \tau_{\mu\nu}(\Psi(x))] \, e^{i\int d^4x[L(\psi)+t_{\mu\nu}(x)J^{\mu\nu}(x)]}$$

$$= \int Dt_{\mu\nu} \, e^{i\int d^4x[\mathcal{L}(t_{\mu\nu})+t_{\mu\nu}J^{\mu\nu}]}, \quad (3)$$

---

[a] As far as we are in a Minkowski space, the indices can be moved up and down by means of $\eta_{\mu\nu} = \eta^{\mu\nu} = \mathrm{diag}(1,1,1,-1)$, e.g. $\tau_\mu^\nu = \eta^{\nu\rho}\tau_{\rho\mu}$. In the following, the mute indices, if they are not properly (up and down) spaced, will imply a 'correct' way of summation with $\eta_{\mu\nu}$, e.g. $G_{\mu\rho}G_{\rho\nu}$ means $G_{\mu\rho}\eta^{\rho\lambda}G_{\lambda\nu}$.

[b] We do not discuss here a more general case when $\tau_{\mu\nu}$ could have also some nontrivial internal quantum numbers, so that its condensation would break also internal symmetries. In addition, we assume for simplicity that no vector condensate or the condensate of other tensor structures emerge in this system.

where we introduced the integration with the δ_x functional over the auxiliary field $t_{\mu\nu}(x)$ connected with the operator $\tau_{\mu\nu}$. In this way, by integrating out the fundamental degrees of freedom with the constraint, one can define the effective lagrangian for the effective tensor field $t_{\mu\nu}$:

$$i\int d^4x\mathcal{L}(t_{\mu\nu}(x)) \;=\; \ln\Big[\int D\psi\, e^{i\int d^4x L(\Psi)}\,\delta_x[t_{\mu\nu}(x)-\tau_{\mu\nu}(\Psi)]\Big], \qquad (4)$$

As far as we start with initially Lorentz-invariant theory, the effective Lagrangian $\mathcal{L}(t_{\mu\nu})$ must be Lorentz-invariant. Next, we suppose that for a slowly changing $t_{\mu\nu}(x)$ it is approximately local and can be expanded in powers of $t_{\mu\nu}$ and its derivatives $t_{\mu\nu,\alpha}(x) = \partial_\alpha t_{\mu\nu}$; $t_{\mu\nu,\alpha\beta}(x) = \partial_\alpha\partial_\beta t_{\mu\nu}$ etc., as:

$$\mathcal{L}(t_{\mu\nu}) \;=\; -V(t_{\mu\nu}) + \Gamma^{\alpha\beta\gamma\delta\rho\sigma}t_{\alpha\beta,\gamma}t_{\delta\rho,\sigma} + W^{\alpha\beta\gamma\delta\rho\sigma\lambda\varsigma\tau}t_{\alpha\beta,\gamma}t_{\delta\rho,\sigma}t_{\lambda\varsigma,\tau} + \ldots \qquad (5)$$

where $V(t_{\mu\nu})$ is effective potential and generically also the coefficients $\Gamma, W, \ldots$ are functions of $t_{\mu\nu}(x)$. We are interested in a situation when near to the confinement scale Λ_C the nontrivial vacuum expectation value (VEV) can be generated:

$$\langle\tau_{\mu\nu}\rangle \;=\; \langle t_{\mu\nu}\rangle \;=\; n_{\mu\nu}, \qquad (6)$$

where $n_{\mu\nu}$ is a constant symmetric tensor ($n_{\mu\nu} \neq \eta_{\mu\nu}$) with the elements of the order of 1. Notice, however, that $t_{\mu\nu}(x) = s(x)\eta_{\mu\nu} + \tilde{t}_{\mu\nu}(x)$ is not an irreducible representation of Lorentz group and it consists of a traceless tensor field $\tilde{t}_{\mu\nu}(x)$ (with condition $\mathrm{Tr}\,\tilde{t} = \eta^{\mu\nu}\tilde{t}_{\mu\nu} = 0$) and a scalar field $s(x)$. We are not interested in the case when LI remains unbroken, i.e. $n_{\mu\nu} \propto \eta_{\mu\nu}$, when only the scalar part of $t_{\mu\nu}$ gets a VEV $\langle s\rangle = \sigma \neq 0$. The Lorentz breaking needs that the VEV of the traceless $\tilde{t}_{\mu\nu}$ part, $\langle\tilde{t}_{\mu\nu}\rangle = \tilde{n}_{\mu\nu}$, is non-zero.

Possible configurations of $n_{\mu\nu}$ can be classified in terms of four eigenvalues λ^a:

$$n_{\mu\nu}S^a_\nu \;=\; \lambda^a S^a_\mu\,, \qquad (7)$$

where S^a_μ are the orthonormal eigenvectors:

$$S^a_\mu S^b_\mu \;=\; \delta^{ab}, \qquad \sum_a S^a_\mu S^a_\nu \;=\; \eta_{\mu\nu}, \qquad \sum_a S^a_\mu S^a_\nu \lambda^a \;=\; n_{\mu\nu}\;.$$

This means that by means of rotations and boosts one can chose a coordinate system in wich $n_{\mu\nu}$ is diagonal:[c]

$$n_{\mu\nu} = \begin{pmatrix} \lambda_1 & 0 & 0 & 0 \\ 0 & \lambda_2 & 0 & 0 \\ 0 & 0 & \lambda_3 & 0 \\ 0 & 0 & 0 & -\lambda_0 \end{pmatrix}, \quad \text{or} \quad \tilde{n}_{\mu\nu} = \begin{pmatrix} \tilde{\lambda}_1 & 0 & 0 & 0 \\ 0 & \tilde{\lambda}_2 & 0 & 0 \\ 0 & 0 & \tilde{\lambda}_3 & 0 \\ 0 & 0 & 0 & \tilde{\lambda}_4 \end{pmatrix} \tag{8}$$

where $\lambda_{1,2,3} = \sigma + \tilde{\lambda}_{1,2,3}$ and $\sigma - \lambda_0 = \tilde{\lambda}_4 = \tilde{\lambda}_1 + \tilde{\lambda}_2 + \tilde{\lambda}_3$. The eigenvectors in this frame are also diagonal: $S^a_\mu \sim \delta^a_\mu \lambda^a$.

The general form of potential $V(t_{\mu\nu}) = V(s, \tilde{t}_{\mu\nu})$, supposing that it is analytic at the origin, can be taken as a polynomial function of scalar $s(x) = I_0$ and three independent Lorentz-invariant combinations of a tensor $\tilde{t}_{\mu\nu}$: $I_1 = \mathrm{Tr}(\tilde{t}^2) = \tilde{t}^\nu_\mu \tilde{t}^\mu_\nu$, $I_2 = \mathrm{Tr}(\tilde{t}^3) = \tilde{t}^\nu_\mu \tilde{t}^\rho_\nu \tilde{t}^\mu_\rho$ and $I_3 = \mathrm{Tr}(\tilde{t}^4) = \tilde{t}^\nu_\mu \tilde{t}^\rho_\nu \tilde{t}^\sigma_\rho \tilde{t}^\mu_\sigma$. Namely, we take $V(t_{\mu\nu}) = \Lambda^4_C \mathcal{P}(I_0, I_1, I_2, I_3)$ where Λ is the confinement scale, and the dimensionless coefficients of the polynomial $\mathcal{P}(I_i)$ are generally expected to be of the order of 1. Since the invariants I_i are not positively defined such potentials can lead to nontrivial minima for very broad area of these coefficients. Conditions for extremum $\partial V/\partial t_{\mu\nu} = 0$ are reduced to $\partial V/\partial I_i = 0$ and the stability conditions for condensate is $\partial^2 V/\partial I_i \partial I_k > 0$. The roots I_i of equation $\partial V/\partial I_i = 0$ are related to the eigenvalues of matrix $n_{\mu\nu}$:

$$I_0 = \sigma, \quad I_1 = \sum_{i=1}^3 \tilde{\lambda}_i^2, \quad I_2 = \sum_{i=1}^3 \tilde{\lambda}_i^3, \quad I_3 = \sum_{i=1}^3 \tilde{\lambda}_i^4 \,. \tag{9}$$

The simplest model example of potential, such as $\mathcal{P}(I_i) = \sum_{i=0}^3(-2a_i I_i + I_i^2)$ already leads to general form of spontaneous LI breaking. In this case the roots, coming from conditions $\partial V/\partial I_i = 0$, are $I_i = a_i$, $V(I_i) = -\Lambda^4 \sum_{i=0}^3 a_i$, and, depending from values of a_i, one can have configurations (8) with $\langle s \rangle \neq 0$ and $\langle \tilde{n}_{\mu\nu} \rangle \neq 0$ with all $\tilde{\lambda}_i$ being different. The degenerate case, when some of eigenvalues $\tilde{\lambda}_i$ may coincide, can lead to number of simplifications. For example the case of maximal degeneracy $\tilde{\lambda}_1 = \tilde{\lambda}_2 = \tilde{\lambda}_3 \neq 0$ leaves unbroken rotational invariance. In this case on can go to a system where the space is isotropic and this will lead to additional restrictions on possible goldstone-gravity configurations. But there is no reason to expect such a degeneracy for the general potentials $\mathcal{P}(I_i)$.

Let us suppose now that the particles $\Psi = (\chi, \varphi, G)$ of the strongly coupled sector G_C are coupled also to a sector of 'ordinary' particles S (quarks and leptons, standard gauge fields, etc.), described by the Gauge

[c]Here we do not consider possibility of non-diagonal light-like configurations of $n_{\mu\nu}$.

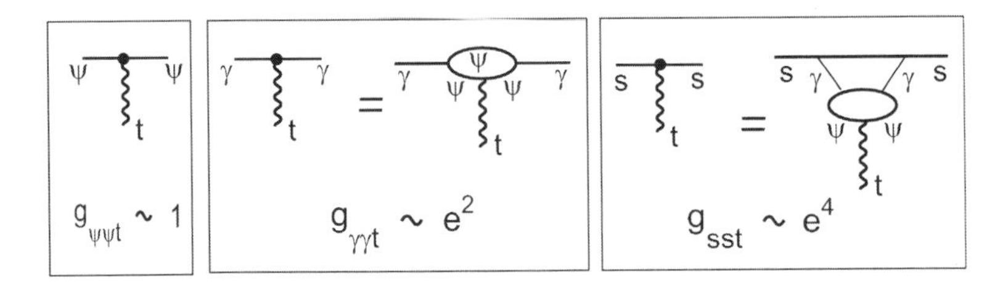

Fig. 1. A schematic (Ψ) and weekly coupled sectors (S) and mediator particles (γ). Vertexes, the main diagrams, and their orders in e^2 are shown

group $G_{\rm S}$, which can be the Standard Model $SU(3)\times SU(2)\times U(1)$ or some of its grand unified generalization as e.g. $SU(5)$ or $SO(10)$, and the latter remains weakly coupled at the energies $\sim\Lambda$. The contact between the sectors $G_{\rm C}$ and $G_{\rm S}$ can be mediated via some mediator fields γ which have small couplings $\sim e$ with both Ψ and S fields. These γ fields could be e.g. fields of some gauge symmetry. Imagine indeed a gauge theory based on a product $G_{\rm C}\times G_{\rm S}\times G_{\rm med}$ where Ψ and S fields are in nontrivial representations of $G_{\rm med}$ which is spontaneously broken at the scale $\sim\Lambda$, perhaps by the condensates of the same Ψ fields that spontaneously breaks also LI. In other words, $G_{\rm C}$ sector acts as a technicolor for the breaking of $G_{\rm med}$ gauge bosons of which become massive and induce couplings of the composite tensor field $t_{\mu\nu}(\Psi)$ to the effective tensor combinations $T_{\mu\nu}({\rm S})$ containing the standard fields of $G_{\rm S}$ sector, via the effective operators of the type $t_{\mu\nu}T_{\mu\nu}$, $\partial_\mu\partial_\rho t_{\rho\nu}T_{\mu\nu}$, etc. (See scheme given in Fig. 1.) The combinations of S fields entering in $T_{\mu\nu}$ belong to the same universality class as the energy momentum tensor $\mathcal{T}_{\mu\nu}$ of S fields. It is natural to assume that the presence of $G_{\rm med}$ and $G_{\rm S}$ sectors cannot affect the dynamics of the strongly coupled secttor $G_{\rm C}$ and in particular the formation of the tensor condensate (6).

We have tree classes of vertices of the Goldstone graviton, $t\Psi\Psi$, $t\gamma\gamma$, tSS, corresponding to diagrams in Fig. 1 respectively with the couplings ~ 1, $\sim e^2$, $\sim e^4$. Clearly, e cannot be taken very small on the scale Λ, otherwise one can have very big nonuniversality of $t_{\mu\nu}$ with different types of fields since. If $e\sim 1$, then it would be more natural that some general universalization mechanism could equalize couplings at very large distances (see discussion in last section).

When LI is spontaneously broken and $\langle t_{\mu\nu}\rangle \neq t_{\mu\nu}$, the mean value of energy-momentum tensor for the ψ_i fields in vacuum can be also nontrivial and has the general form :

$$\langle\, T_{\mu\nu}\rangle \;=\; a_0\eta_{\mu\nu} \;+\; a_1\tilde{n}_{\mu\nu} \qquad (10)$$

In the Lorentz frames, where the $\tilde{n}_{\mu\nu}$ is diagonal, $T_{\mu\nu}$ is also diagonal. In this case the $\tilde{n}_{\mu\nu}$ condensate can be interpreted as an anisotropic "solid media" filling the space with the energy density T_{00} and anisotropic pressures T_{kk}.

The a_0 term in (10) is usually perturbatively divergent in renormalizable theories, and this leads to problems in General relativity, In our case, the Lorentz-invariant part $\sim\ \eta_{\mu\nu}$ is irrelevant it does not interact with the goldstone gravitons (it can be freely subtracted it is usually done in field theories without gravity).

On the other hand, the non-LI contributions in $T_{\mu\nu}$ are probably non-universal and the coefficients a_1 are scale-dependent. Near the confinement scale $a_1 \sim \Lambda_c^4$, but perhaps they can become small on large distances as e.g. in $\mathcal{G}^{(i)}_{\mu\nu}$ (31). The $\tau_{\mu\nu}$ has the same symmetries as energy-momentum tensor $T_{\mu\nu}$, and so $\tau_{\mu\nu}$ and $T_{\mu\nu}$ can partially enter in the same universality class with respect to a scale change - therefore one can hope that in the IR limit their main parts will become universal.

3. Tensor condensate oscillations and Goldstone gravitons

So let us suppose that the Ψ field system induces a non-trivial tensor condensate $\langle t_{\mu\nu}\rangle$. As discussed in previous section this means that the minimum of potential $V(t_{\mu\nu})$, defined by the condition $\partial V/\partial t_{\mu\nu} = 0$ is at the non-trivial point $\langle t_{\mu\nu}\rangle = n_{\mu\nu} = \sigma\eta_{\mu\nu} + \tilde{n}_{\mu\nu}$. Since $V(t_{\mu\nu})$ does not depend on the derivatives of $t_{\mu\nu}$, the field configurations $t_{\mu\nu}(x)$ which can be obtained from $n_{\mu\nu}$ by the arbitrary local Lorentz transformation

$$n_{\mu\nu}(x) \;=\; \Omega^{\lambda}_{\mu}(x)\; n_{\lambda\sigma}\; \Omega^{\sigma}_{\nu}(x) \;=\; \sigma\eta_{\mu\nu} \;+\; \Omega^{\lambda}_{\mu}(x)\; \tilde{n}_{\lambda\sigma}\; \Omega^{\sigma}_{\nu}(x), \qquad (11)$$

correspond to the same value of the potential as $n_{\mu\nu}$: $V(n_{\mu\nu}(x)) = V(n_{\mu\nu})$. The local $O(3,1)$ rotation matrixes

$$\Omega^{\nu}_{\mu}(x) \;=\; \left(\exp[\frac{1}{2}\omega_{ab}(x)\Sigma^{ab}]\right)^{\nu}_{\mu} \qquad (12)$$

depend on antisymmetric "angular" fields $\omega_{ab}(x)$ which represent six independent flat directions. Here Σ^{ab} is the spin part of generators of Lorentz rotation in the vector representation: $\left(\Sigma^{ab}\right)^{\nu}_{\mu} \;=\; \eta^{a\nu}\delta^{b}_{\mu} - \eta^{b\nu}\delta^{a}_{\mu}$, where the local (a,b) and tensor (μ,ν) indices are 'mixed'. Thus, the variables $\Omega^{\nu}_{\mu}(x)$ are connected with the massless Goldstone-like degrees of freedom, whose Lagrangian are given only by the derivative terms with $\partial_\lambda t_{\mu\nu}(x)$ in (5).

The full infinitesimal action of these operators on a tensor, $\left(\Sigma_{\varrho\sigma}\right)_{\mu\nu}^{\alpha\beta} t_{\alpha\beta}$, is given by relation $\left(\Sigma_{\varrho\sigma}\right)_{\mu\nu}^{\alpha\beta} = \eta_{\sigma\mu}\, \delta_\rho^\alpha\, \delta_\nu^\beta - \eta_{\rho\mu}\, \delta_\sigma^\alpha\, \delta_\nu^\beta + \eta_{\sigma\nu}\, \delta_\mu^\alpha\, \delta_\rho^\beta - \eta_{\rho\nu}\, \delta_\mu^\alpha\, \delta_\sigma^\beta$. Therefore, taking all this into account, one can divide dynamical variables $t_{\mu\nu}(x)$ in the massless Goldstone-like "rotational" modes connected with $\Omega_\mu^\lambda(x)$ and the massive modes $\varrho^a(x)$ connected with the excitations of the eigenvalues λ^a (7):

$$t_{\mu\nu}(x) \;=\; \sum_{a=1}^{4} w_\mu^a(x)\; \left(\lambda^a + \varrho^a(x)\right)\; w_\nu^a(x) \;\; , \tag{13}$$

where $w_\mu^a(x) \;=\; \Omega_\mu^\lambda(x) S_\lambda^a$ are the analogues of local tetrad vectors, often used for representing the gravitational field. Here these tetrads $w_\mu^a(x)$ are restricted by a "gauge" condition that in every point of the space-time they can be reduced by the local Lorentz-transformation to the predefined constant vectors S_λ^a. The masses of the fields $\varrho^a(x)$ are defined by the second derivatives $\partial^2 V/\partial t_{\mu\nu}^2$ and are expected to be order Λ. Note that in general one should not expect that they correspond directly to some new massive stable particles formed on the scale Λ. They contain the operator combinations of the fundamental fields $\Psi = \chi, \varphi, \ldots$ in approximately the same proportion as in the decomposition (1) for $\tau_{\mu\nu}$, and can be highly unstable.

One can also represent (2) in a slightly different form

$$\begin{aligned} t_{\mu\nu}(x) &= \Omega_\mu^\lambda(x)\; \left[n_{\lambda\sigma} + t_{\lambda\sigma}^m(x)\right]\; \Omega_\nu^\sigma(x) \\ &= s(x)\eta_{\mu\nu} + \Omega_\mu^\lambda(x)\; \left[\tilde{n}_{\lambda\sigma} + \tilde{t}_{\lambda\sigma}^m(x)\right]\; \Omega_\nu^\sigma(x) \;\; , \end{aligned} \tag{14}$$

where $\tilde{n}_{\mu\nu}$ is a traceless part of $n_{\mu\nu}$ ($\eta^{\mu\nu}\tilde{n}_{\mu\nu} = 0$) and $t_{\mu\nu}^m(x) = S_\mu^a S_\nu^a \varrho^a(x)$ is a part of $t_{\mu\nu}(x)$ that contains only the massive modes (which splits in a scalar $s(x)$ and traceless part $\tilde{t}_{\mu\nu}^m(x)$), while the massless modes are encoded in $\Omega_\mu^\lambda(x)$. (Here and below the summation over the "eigenvalue" index a is assumed as in (2).) For small $\omega_{ab}(x)$ the massless fields ($\omega_{\mu\nu} = -\omega_{\nu\mu}$) enter in linear form $\Omega_{\mu\nu}(x) = \eta_{\mu\nu} + \omega_{\mu\nu}$, and hence the week field decomposition of $t_{\mu\nu}(x)$ in massless fields is given by

$$t_{\mu\nu}(x) \;\simeq\; n_{\mu\nu} \;+\; h_{\mu\nu}(x), \qquad |h_{\mu\nu}| \ll |n_{\mu\nu}| \;, \tag{15}$$

where the symmetric tensor $h_{\mu\nu}(x)$ entering in usual linearized description of weak gravitational field can be represented by antisymmetric goldstone fields as $\omega_{\alpha\beta}$

$$h_{\mu\nu} \;\simeq\; \omega_{\mu\beta} n_\nu^\beta \;-\; \omega_{\nu\beta} n_\mu^\beta \;, \tag{16}$$

In fact, this representation of $h_{\mu\nu}$ in terns of $\omega_{\mu\nu}$ corresponds to a definite gauge choice in linear approximation, with fixed values of the invariants I_i. Note, that this gauge in which the goldstone gravitons appear is similar to the axial gauge $n_\mu A^\mu = 0$ for vector goldstone particles, where n_μ is the vector condensate.

Between the degrees of freedom contained in $t_{\mu\nu}(x)$ in the neighborhood of $n_{\mu\nu}$ should be no tachyons, as far we are already in the minimum of $V(t)$ with respect to all independent variations of components of $t_{\mu\nu}$. One should not expect any ghosts contributions in terms with derivatives $t_{\alpha\beta,\gamma}$, so that all signs of corresponding kinetic terms in (5) should be correct – otherwise there will be a noncausal propagation of $t_{\mu\nu}$ fields, but it is impossible in terms of primary fields $\Psi = \chi, \varphi, ...$ These conditions restricts the form of coefficients $\Gamma^{\alpha\beta\gamma\delta\rho\sigma}(t)$, ... , entering effective lagrangian (5) near the $t_{\mu\nu} = n_{\mu\nu}$. And just this second term

$$\mathcal{L}_2 = \Gamma^{\alpha\beta\gamma\delta\rho\sigma} t_{\alpha\beta,\gamma} t_{\delta\rho,\sigma} , \tag{17}$$

in the Lagrangian (5) defines the minimal dynamics for the goldstone fields $\omega_{\mu\nu}$. The stability conditions for $n_{\mu\nu}$ can be also be formulated in terms of corresponding effective Hamiltonian

$$\mathcal{H} = t_{\alpha\beta,0}\, \partial\mathcal{L}_2/\partial(t_{\alpha\beta,0}) - \mathcal{L}_2 .$$

For this one must require that for all small variations $|\omega_{\alpha\beta}| \ll 1$, $|\varrho^a| \ll \Lambda$ of $t_{\mu\nu}$ fields

$$\delta t_{\mu\nu}(x) \simeq \varrho^a(x)\, S^a_\alpha S^a_\nu + \omega_{\alpha\beta}(x)\, (g_{\mu\alpha} n_{\nu\beta} - n_{\mu\alpha} g_{\nu\beta} + g_{\nu\alpha} n_{\mu\beta} - n_{\nu\alpha} g_{\mu\beta}) \tag{18}$$

the Hamiltonian remains positively defined, and the corresponding variation $\delta\mathcal{H} \geq 0$ in the neighborhood of $t_{\mu\nu}(x) = n_{\mu\nu}$.

Note that the equation of motion for the Goldstone field $\delta\mathcal{L}/\delta\omega_{ab}(x) = 0$ follow from the terms with derivatives in (5). On the other hand these equations must be also contained in the conservation laws $\partial_\mu J_{[\alpha\beta]\mu} = 0$ for the Noetter currents, connected with the symmetry which is spontaneously broken. In our case these currents are the full angular momentum densities $J_{[\alpha\beta]\mu}$ for field system $\Psi = \chi, \varphi, ...$, or for Lagrangian (5) for the effective fields $t_{\mu\nu}$. Since the Lagrangians are translation invariant these equations reduce to

$$T_{\alpha\beta} - T_{\beta\alpha} - \partial_\mu S_{[\alpha\beta]\mu} = 0 ,$$

where $T_{\alpha\beta}$ is canonical energy-momentum tensor, and $S_{\alpha\beta\mu}$ - is the spin part of $J_{[\alpha\beta]\mu}$.

To find explicitly the form of the Lagrangian $\mathcal{L}_2$ for weak fields, including also their minimal interactions, we need the expressions for the derivatives of $t_{\alpha\beta}$ up to the second order in physical fields

$$\begin{aligned}\partial_\lambda t_{\mu\nu} &\simeq \partial_\lambda \varrho^a \Big(S^a_\mu S^a_\nu \;+\; \omega_{\mu\alpha} S^a_\alpha S^a_\nu \;+\; \omega_{\nu\alpha} S^a_\alpha S^a_\mu \Big) \\ +\partial_\lambda \omega_{\alpha\beta} & \left(\eta_{\mu\alpha} t^m_{\nu\beta} \;+\; \eta_{\nu\alpha} t^m_{\mu\beta} \;+\; n_{\mu\alpha}\omega_{\beta\nu} + n_{\nu\alpha}\omega_{\beta\mu} \right) \;+\cdots \qquad (19)\end{aligned}$$

In the week field approximation tensor $\Gamma^{\cdots\cdots}$ can be taken as a constant, depending only from η_{ik} and n_{ik}. Then the general form of the Lagrangian $\mathcal{L}_2$ (17) reads as

$$\begin{aligned}\mathcal{L}_2 \;\simeq\; & \left(P^{\gamma\alpha\beta\sigma\delta\rho} + \omega_{ab} Q^{ab\;\gamma\alpha\beta\sigma\delta\rho} \right) \partial_\gamma t^m_{\alpha\beta} \partial_\sigma t^m_{\delta\rho} \\ & + \left(\partial_\gamma \omega_{ab} \partial_\sigma \omega_{mn} \right) \left(t^m_{\alpha\beta} t^m_{\delta\rho} \right) \left(H_1^{\gamma ab\;\cdots\;\rho} + \omega_{..} H_2^{\cdots\cdots\cdots} \right) \\ & \quad + \partial_\gamma \omega_{ab} \cdot \left(t^m_{\alpha\beta} \partial_\gamma t^m_{\delta\rho} \right) U^{\gamma ab\;\alpha\;\cdots\;\rho} \;+\cdots\;, \qquad (20)\end{aligned}$$

where $P^{\cdots}, Q^{\cdots}, H^{\cdots}$ are constant tensors, constructed from η^{ik} and n^{ik}, so that to fulfill the $\mathcal{H}$ stability conditions. In (20) the first and second lines correspond to kinetic terms for the massive and goldstone modes and the third line describes their interaction in which, as is usual for goldstone particles, fields $\omega_{\alpha\beta}$ enter only through the derivative terms. Notice that the states described by ω_{ab} and $t^m_{\alpha\beta}$ are orthogonal and so there should be no direct mixing between them, i.e. no terms of the type $\partial_\mu \omega_{ab}\; \partial_\mu t^m_{ab}$. It is very essential and specific for the non-scalar goldstone particles that the massless fields $\omega_{\alpha\beta}$ can enter also in the coefficient of the kinetic term for massive mode without derivatives (first line in (20)).

For the scalar goldstones such terms are cancelled in all orders and their interactions with other fields enter only through the derivative terms - like in the third line in (20)). Namely, for a scalar Lagrangian $L = \partial_\mu \phi^\dagger \partial_\mu \phi - V(\phi)$ where ϕ - arbitrary column vector we can localize Ω - the global symmetry of $V(\phi)$ separating the massive ρ and the "massless" modes $\partial_\mu(\Omega\rho)^\dagger \partial_\mu(\Omega\rho)$ where $\Omega(x) = \exp(i\vec{T}\vec{\omega}(x))$. For $\langle\rho\rangle \neq 0$ the goldstone fields $\vec{\omega}(x)$ fully decouple from massive kinetic term and enter in L only trough derivative $L = \partial_\mu \rho^\dagger \partial_\mu \rho + (\rho^\dagger \rho) \partial_\mu \vec{\omega} \partial_\mu \vec{\omega} + \partial_\mu \vec{\omega} \vec{J}_\mu - V(\rho)$ where the current $\vec{J}_\mu = i(\partial_\mu \rho^\dagger \vec{T} \rho - \rho^\dagger \vec{T} \partial_\mu \rho)$. So the interaction of such a $\vec{\omega}$-goldstones is switched off when their momenta is going to zero.

The 'diagonal' terms in $\mathcal{L}_2$ of type $\partial_\lambda t_{\alpha\beta}\; \partial_\lambda t_{\alpha\beta}$ analogously do not give the contribution to nonderivative goldstone coupling, but the asymmetrical terms like $\partial_\lambda t_{\alpha\beta}\; \partial_\alpha t_{\lambda\beta}$ - already can give. In fact these are the same terms in the kinetic part of $\mathcal{L}_2$ that contribute to the spin operators. From all possible field combinations, entering (20) and linear in $\omega_{\alpha\beta}$, only the term

$$\omega^{\alpha\beta} J_{\alpha\beta}, \text{ where } J_{\alpha\beta} = \big(S^a_\alpha \partial_\beta \varrho^a - S^a_\beta \partial_\alpha \varrho^a\big)\big(S^a_\mu \partial_\mu \varrho^a\big) \\ + \partial_{[\alpha} \omega_{..} n_{..} \partial_{.} \omega_{\beta]} \tag{21}$$

gives the finite contribution in the soft limit. Note that terms (21) in $\mathcal{L}_2$, although linear in $\omega_{\alpha\beta}$, does not contribute to the potential $V(t_{\mu\nu})$ even after taking into account loop corrections. Expression (21), representing the soft interaction of goldstone-gravitons with the rest of the fields (matter), can be written in a more "standard" way using (16):

$$h_{\mu\nu} T^{\mu\nu} \ , \quad \text{where} \quad T^{\mu\nu} \ = \ 2 n^\mu_\lambda J^{\lambda\nu}, \quad h_{\mu\nu} \ \simeq \ \omega_{\mu\beta} n_{\beta\nu} \ - \ \omega_{\beta\nu} n_{\beta\mu}$$

The kinetic terms for the goldstone fields $\omega_{\alpha\beta}$ (the second line in (20)) can be expressed also by means of $h_{\alpha\beta}$:

$$\big(\partial_\gamma \omega_{ab} \partial_\sigma \omega_{mn}\big)\big(n_{\alpha\beta} n_{\delta\rho}\big) H^{\gamma ab \, \cdots \, \rho} \ \Rightarrow \ \big(\partial_\gamma h_{ab} \partial_\sigma h_{mn}\big) \ \tilde{H}^{\gamma ab\sigma mn}, \tag{22}$$

where the symmetric "metric" fluctuations $h_{\alpha\beta}$ are given by (16) and $\tilde{H}^{\gamma ab\sigma mn}$ in low energy limit is a constant tensor constructed from $n_{\alpha\beta}$ and $\eta_{\alpha\beta}$.

If LI is restored at long distances then expression (22) should reduce to sum of the Pauli-Fierz Lagrangian

$$L_{pf} \ = \ \partial_\alpha h_{\mu\nu} \partial^\mu h^{\nu\alpha} - \frac{1}{2} \partial_\alpha h_{\mu\nu} \partial^\alpha h^{\mu\nu} - f^\alpha \partial_\alpha h + \frac{1}{2} \partial^\mu h \partial_\mu h \ , \tag{23}$$

where $h = \eta^{\mu\nu} h_{\mu\nu}$, $f_\mu = \partial^\nu h_{\mu\nu}$ and terms $\tilde{L}$, depending from $n_{\mu\nu}$ in such a form that they can be considered as a gauge fixing terms (depending only on the valus of the invariants I_i in (9). This non LI gauge can be defined by a condition which implies that $h_{\mu\nu}$ can be represented in the form $h_{\mu\nu} \ = \ \omega_{\mu\beta} n_{\beta\nu} \ - \ \omega_{\beta\nu} n_{\beta\mu}$. Now $\omega_{\nu\beta}$ are new variables in (23), $h = 0$, and only first two terms in right hand side of (23) remain.

The antisymmetric fields ω_{mn} contain 6 local quantities, connected with the massless spin two fields. Two of them represent the transverse spin-two massless particles (gravitons). The remaining four degrees of freedom are not propagating. In the GR the corresponding quantities correspond to a Coulomb (Newton) field coupled to the energy of sources, and to the 3-vector $\vec{g}$ with components $\sim g_{0i}/g_{00}$ connected with the angular velocity ($\sim \vec{\nabla} \times \vec{g}$) of the local system. For comparison, in the case of vector condensate $n_\mu \ = \ \langle A_\mu \rangle$ we have 3 goldstone objects, formed from n_μ by local transformations $\omega_{mn}(x)$ which rotate n_μ. Two of them represent the transverse vector particles (photons), and third corresponds to the classical Coulomb field.

3.1. *Degenerate case*

After spontaneous breaking of LI we have in fact a theory with two metrics: one is a flat background metric $\eta_{\mu\nu}$ and another is a dynamical metric $t_{\mu\nu}(x)$. It is instructive to consider briefly the simplified case when the condensate $n_{\mu\nu}$ is maximally degenerate: $\lambda_1 = \lambda_2 = \lambda_3 \neq -\lambda_0$. Then in a coordinate system where $n_{\mu\nu}$ is diagonal one has $n_{11} = n_{11} = n_{11} = \lambda_1$, $n_{00} = \lambda_0$, and there emerge only tree Goldstone modes. In the weak field limit these are the components $h_{i0} = \omega_{i0}(\lambda_0 - \lambda_1)$.

This means that not all gravitational configurations can be reproduced in this case, but only the ones that correspond to the ansatz

$$ds^2 \ = \ dt^2 \ - \ 2v_i \ dx^0 dx^i \ - \ (\delta_{ik} + v_i v_k) \ dx^i dx^k \ , \tag{24}$$

where $v_k(x^i, x^0)$ are some functions of x^i and x^0. Let us consider briefly the case of weak field, with $v_i \simeq h_{i0} \ll 1$. Then the "electric" components of the curvature tensor read as $R^{0i0k} \sim \partial^0(\partial^i h^{0k} + \partial^k h^{0i})$ while the other components are small in a weak field limit.

The gravitational field of static mass m follows in this case from equations $R^{00} \sim m_p^{-2} T^{00} \quad \Rightarrow \quad \partial^0 \partial^i h^{0i} \sim m_p^{-2} m \delta^3(x)$, with a simple solution

$$\omega^{i0} \ \simeq \ h^{i0} \ \simeq \ \frac{m}{m_p^2}\frac{x^0 x^i}{|x|^3} \ , \qquad R^{0i0k} \sim (m/m_p^2)\frac{1}{|x|^3}\Big(\delta^{ik} - 3\frac{x^i x^k}{|x|^2}\Big)$$

which leads to the Newton law.[d] The force acting on a test particle is $f^k \sim \Gamma^k_{00} \sim \partial^0 h^{k0} \sim x^k/|x|^3$.

For a uniformly distributed energy sources $T^{00}(x) \ = \ \varrho_0$ the equation $R^{00} \ \sim \ \partial^0 \partial^i h^{0i} \ \sim \ \varrho_0$ gives regular solution $h^{i0} \ \ \sim \ \ \varrho_0 x^i(x^0 \pm c)$, with nonzero components of curvature tensor $R^{0i0k} \sim \varrho_0/m_p^2$. This corresponds to a spatial metric at small x^i

$$g_{ik} = \eta_{ik} + h_{0i}h_{0k} = \eta_{ik} + \frac{x^i x^k}{a^2(x^0)}, \quad a(x^0) \sim \frac{m_p^2}{(c \pm x^0)\varrho_0} \sim a_0(1 \pm \frac{\varrho_0}{m_p^2} a_0 x^0 + ...)$$

with the time varying "cosmological" scale factor $a(x^0)$.[e] The other way to "measure" this scale variation can be found from the equation for

[d]Generalization to strong fields can also be found in this gauge. It corresponds to Schwarzschild solution in some special synchronous system connected with form (24), and i.e. applicable up to distances $\sim |x^0| \sim |x^i| \sim \Lambda^{-1}$ where the goldstones can disappear due to the melting of the tensor condensate $n_{\mu\nu}$.

[e]Comparing this behavior of $a(x^0)$ with the case of the cosmological scale factor in the GR $(\partial^0 a/a)^2 \sim \varrho_0 m_p^{-2}$, one can try to fix the Hubble parameter as $H \sim \sqrt{\varrho_0}/m_p$. But this is the result of linearized approximation. Evidently, all these solutions are applicable only in limited intervals of x^0 when h_{0i} are small.

geodesic deviation, which in the case of two standing test particles, separated by the distance δx^i, reduces to $\partial^0\partial^0\delta x^i \sim R^{0i0k}\delta x^k$. Because $(\partial^0\partial^0\delta x^i)/\delta x^i = (\partial^0\partial^0 a)/a$, this corresponds to the standard evolution equation $(\partial^0\partial^0 a)/a \sim \varrho_0/m_p^2$ for homogenous cosmology.

4. Einstein equations

Going to the general case of strong goldstone fields it is interesting to find under what conditions the first derivative term (17), without massive fields ϱ^a, can be reduced to the Einstein-Hilbert Lagrangian, with $t_{\mu\nu}(x)$ playing the role of metric. The answer to this question, without entering into the details, can be formulated in a rather simple and natural form.

The **first condition** is that tensor $\Gamma^{\alpha\beta\gamma\delta\rho\sigma}$ must be expressed only through the "contravariant" $t^{\mu\nu}$, which should be defined as inverse to $t_{\mu\nu}$, i.e. through $t_{\gamma\nu}t^{\gamma\mu} = \delta^\mu_\nu$. Solving this equation, we have

$$t^{\alpha\beta} = \frac{1}{4!\,t}\epsilon^{\alpha\lambda\rho\gamma}\epsilon^{\beta\sigma\mu\nu}t_{\lambda\sigma}t_{\rho\mu}t_{\gamma\nu}\ , \tag{25}$$

where $t = \det[t_{\mu\nu}(x)]$.

Using such $t^{\alpha\beta}$ the general expression (17) for $\mathcal{L}_2$ can be presented in the form:

$$\mathcal{L}_2 = \mathcal{L}_2^{eh} + \tilde{\mathcal{L}}_2,$$

$$\mathcal{L}_2^{(eh)} = t_{\alpha\beta,\gamma}t_{\delta\rho,\sigma}\left(c_1 t^{\alpha\delta}\,t^{\beta\sigma}\,t^{\gamma\rho} + c_2 t^{\alpha\delta}\,t^{\beta\rho}\,t^{\gamma\sigma} + c_3 t^{\alpha\gamma}\,t^{\beta\rho}\,t^{\delta\sigma}\right.$$
$$\left. + c_4 t^{\alpha\gamma}\,t^{\delta\rho}\,t^{\beta\sigma} + c_5 t^{\alpha\beta}\,t^{\delta\sigma}\,t^{\gamma\rho}\right)\ , \tag{26}$$

where c_i are some scalar functions from invariants $I_{0,1,2,3}$ formed out of $t_{\mu\nu}$. All such invariants should be taken at extremal values as far as we have excluded heavy fields ρ_i. In this way, all c_i in (26) are constants. The part $\tilde{\mathcal{L}}_2$ contains terms with the same structure as in (26) but in which some "contravariant" $t^{\mu\nu}$ used for index contraction are changed by $t_{\mu\nu}$ or $\eta_{\mu\nu}$.

Then the first condition is that $\tilde{\mathcal{L}}_2 = 0$, or that it can be represented as a function of the invariants I_i and of their derivatives in x^μ. Such a $\tilde{\mathcal{L}}_2$ can be considered as gauge fixing term which does not change the dynamics of $t_{\mu\nu}$ fields. Note that this condition, although simply formulated, is dynamically very restrictive and it hardly can be fulfilled without fine tuning for some parameters in the Ψ field system.

The **second condition** is that in the "weak field limit"

$$t_{\mu\nu}(x) = n_{\mu\nu} + h_{\mu\nu}(x)\ , \qquad h_{\mu\nu} \to 0$$

the expression (26) transforms into the Pauli-Fierz Lagrangian (23) for massless spin-two particles in a nonorthogonal coordinates with metric $n_{\mu\nu}$.

In fact such a condition removes the spin zero massless modes, which are present in general in (26). As it can be simply shown it leads to relations for coefficients in (26):

$$c_2 = -2c_1, \qquad c_3 = c_4 = c_5 = 0 .$$

These two conditions fix the form of $\mathcal{L}_2$ up to overall constant ($c_1 \sim \Lambda^2 \to m_p^2$) and after that we obtain finally that

$$\mathcal{L}_2^{(eh)} = c_1 \, t_{\alpha\beta,\gamma} \, t_{\delta\sigma,\rho} \left(2t^{\alpha\delta}t^{\beta\rho}t^{\gamma\sigma} - t^{\alpha\delta}t^{\beta\sigma}t^{\gamma\rho}\right) , \qquad (27)$$

where $t^{\alpha\beta}$ is given by (25). Possibly this condition can be relaxed, because in fact what we need here is to truncate the contributions from spin 0 and spin 1 components under specific gauge choice.

This expression (27) precisely coincides with the Hilbert-Einstein Lagrangian in gauges which fixed value of $t(x) = \det[t_{\mu\nu}(x)] = const.$ On the other hand, here we have just this case - the "gauge" of goldstones in $t_{\mu\nu}(x)$ is fixed by fixing the invariants I_i to constant values derived by minimization of the potetial $V(t_{\mu\nu})$. and which coincide with roots of equations $\partial V/\partial I_k = 0$ (See (9)) and are connected to eigenvalues of Λ_a. As a result, $\det[t_{\mu\nu}(x)] = \lambda^{(1)}\lambda^{(2)}\lambda^{(3)}\lambda^{(4)} = const.$ Because of this the Goldstone modes from $t_{\mu\nu}$ correspond to the 4-volume preserving fluctuations of metric. This (in the other way) explains why the goldstone gravitons does not interact with the terms corresponding to cosmological constant.

The Hilbert-Einstein Lagrangian $\mathcal{L}_2$ in (27) is polynomial in $t_{\mu\nu}(x)$, and in fact such a theory looks like some special σ-model.[f] It takes even more simple form, when we substitute in (27) and (25) $t_{\mu\nu}$ in form (11) and use Ω^μ_ν as a main variables, describing the gravitational field.

The terms with higher then two powers of derivatives in (5) can lead to higher order in "curvature" terms in $\mathcal{L}$, and near to the "Planck" scale where $\partial_\mu \sim \Lambda$ all such terms can be of same order as $\mathcal{L}_2$. Here again as for $\mathcal{L}_2$ we have two type of terms. Terms that we construct by using only the covariant $t^{\mu\nu}$ for contraction of indices and other terms - in which $\eta^{\mu\nu}$ and $t_{\mu\nu}$ are also used. Then from the first we come to expressions containing invariants constructed only from higher powers of the curvature tensor and taken in gauge with fixed values of I_i.[g] The other terms are

[f]The accidental scale-invariance of $\tilde{L}$ in (27) of type $x \to a\,x$, $t_{\mu\nu} \to a^{2/11} t_{\mu\nu}$ is probably fictitious because gauge fixing conditions with fixed values of I_i are not invariant under such transformation.

[g]In fact the same situation takes place for the usual geometrical gravity, where the "natural" general form of the Lagrangian is $\sum_n c_n R_n$ where $c_n \sim m_p^{2n}$ and the invariants $R_n \sim (R_{abcd})^n$ are constructed from n-th powers of the curvature tensor.

different, and perhaps they can be interpreted as the gauge fixing terms for the "invariant" part of the Lagrangian. Such a possibility is attractive but unfortunately we do not see clear reasons for its realization. Less restrictive is the possibility that the coefficients entering in such an expansion can be scale dependent, and it is not excluded that at distances $x \gg \Lambda^{-1}$ only the invariant parts survives and the additional terms becomes comparatively small, for example like $\sim (1/x\Lambda)^2$ or turn into "gauge fixing" expressions.

5. Lorentz-violation in observables

Let us discuss couplings of the effective tensor field $t_{\mu\nu}$ to different fields of the observable sector S. Let us take, for example, a scalar field ϕ (it can be e.g. the ordinary Higgs doublet of the Standard Model). Its effective low-energy Lagrangian, along with the kinetic term $\partial_\mu\phi\partial_\nu\phi$ or possible derivative couplings $\partial_\mu\phi\partial_\nu\phi$, must include the following couplings

$$\partial_\mu\phi\partial_\nu\phi\; t_{\mu\nu}\;, \quad \partial_\mu\phi\partial_\nu\phi\; t_{\mu\lambda}t_{\lambda\nu}\;, \quad \phi^k\partial_\mu\phi\partial_\nu\phi\; t_{\mu\nu}\;, \quad \cdots \tag{28}$$

which after the spontaneous LI breaking give rise to the 'gravitational' couplings with the Goldstone graviton. On the other hand, condensation of $\tilde{t}_{\mu\nu}$ (or better to say of its traceless part $\tilde{t}_{\mu\nu}$) can bring to Lorentz-violation in various observable quantities. At the distances $x \gg \Lambda^{-1}$ the condensate $\langle t_{\mu\nu}\rangle = n_{\mu\nu}$ can be considered as a constant external tensor field. Therefore, Therefore, in general, all Green functions and scattering amplitudes may depend on $n_{\mu\nu}$. In any physical amplitude $A_m(k_1, k_2, ..., k_m, \; n_{\mu\nu}$ with the external momenta $k_i \ll \Lambda$, the tensor $n_{\mu\nu}$ can be considered as external spurion field with the imprint of the Lorentz-violation.

In particular, propagators will get tree level corrections depending on $n_{\mu\nu}$

$$G^{-1}(k) = k^2 - m^2 + \; k_\mu k_\nu N_{\mu\nu} + O(k^4) = \mathcal{G}_{\mu\nu}k_\mu k_\nu - m^2 \; + \; O(k^4)$$
$$\mathcal{G}_{\mu\nu} \;=\; \eta_{\mu\nu} + N_{\mu\nu}\;, \qquad N_{\mu\nu} = c_1 n_{\mu\nu} + c_2 n_{\mu\lambda}n_{\lambda\nu} \; + \cdots \;, \tag{29}$$

where coefficients c_i are proportional to the respective coupling constants of the operators (28) in the effective Lagrangian. The term of type (28) come also from expansion of the vertices with loops in external momenta. There are also induced vertices of same general type (28), where some $n_{\mu\nu}$ are replaced by Minkowski metric $\eta_{\mu\nu}$. So these contributions can be represented in a combined form as[h]

$$\partial_\mu\phi\partial_\nu\phi\; \mathcal{G}^{(1)}_{\mu\nu}\;, \quad \phi^k\partial_\mu\phi\partial_\nu\phi\; \mathcal{G}^{(2)}_{\mu\nu}\;, \quad \cdots \tag{30}$$

[h]Even for slightly nonuniversal $\mathcal{G}^{(i)}_{\mu\nu}$ we can have the nonstandard $(\varepsilon, \vec{k})$ dispersion re-

where the coefficients $\mathcal{G}^{(i)}_{\mu\nu} = \eta_{\mu\nu} + c^{(i)}_1 n_{\mu\nu} + c^{(i)}_2 n_{\mu\lambda} n_{\lambda\nu} + \cdots$ can be recasted as $\mathcal{G}^{(i)}_{\mu\nu} \propto \eta_{\mu\nu} + a^{(i)}_1 \tilde{n}_{\mu\nu} + a^{(i)}_2 \tilde{n}_{\mu\lambda}\tilde{n}_{\lambda\nu} + \cdots$. Now, if coefficients $\mathcal{G}^{(i)}_{\mu\nu}$ are expected to be universal, i.e. $\mathcal{G}_{\mu\nu} = \mathcal{G}^{(1)}_{\mu\nu} = \mathcal{G}^{(2)}_{\mu\nu}$ and so on, then $\mathcal{G}_{\mu\nu}$ can be interpreted as a new metric corresponding to a flat space with non-orthonormalized coordinates. This 'geometrical interpretation can be simply generalized to $n_{\mu\nu} \to t_{\mu\nu}(x)$ which slowly varies with x at the scales $\gg \Lambda^{-1}$.

Inn general the coefficients $\mathcal{G}^{(i)}_{\mu\nu}$ are not universal, at least near to the scale Λ, and they depend on the particle ϕ and vertex type. So it is probably impossible to interpret the additional terms in (29) and (28) as coming from motion of ϕ particles in nonorthogonal coordinates, represented by pure gauge gravitational field $\mathcal{G}_{\mu\nu}$, if this motion is measured at distances $\sim \Lambda^{-1}$. For such measurements vacuum will look like a highly anisotropic media. However, it is possible that at large distances $x \gg \Lambda^{-1}$ the difference between various $\mathcal{G}^{(i)}_{\mu\nu}$ can become very small. This can be represented as:

$$\mathcal{G}^{(i)}_{\mu\nu} = \mathcal{G}_{\mu\nu} + \frac{k^2}{\Lambda^2} Z^{(i)}_{\mu\nu} + \frac{k_\alpha k_\beta n_{\alpha\beta}}{\Lambda^2} Y^{(i)}_{\mu\nu} + \ldots \quad , \tag{31}$$

where the nonuniversal contributions to $\mathcal{G}^{(i)}_{\mu\nu}$ are strongly suppressed at small momenta k. [i] We consider this possibility in relation to the universality problem in next section.

6. The Goldstone graviton couplings: universality problem

In the usual General Relativity the gravitons are universally coupled to all particles and this is probably one of the main experimentally established property of the gravitation. The geometrical theory 'explains' this fact in a simple and natural way. For the Goldstone gravitons $t_{\mu\nu}$ the situation is more complicated. Since they are not geometrical objects, the universality of their coupling with other fields are not automatic at all scales and must be probably tuned to reach the agreement with the experiment. At the same time, although the local Lorentz-invariance is spontaneously broken, the translation invariance remains. Usually just this invariance leads, after localization, to diffeomorphism-invariance and universality of graviton interactions. Therefore, one can hope that in the IR limit, at the distances

---

lations like $\varepsilon^2 = \gamma_i k^2 + m^2, \ldots$, where γ_i depends from particles type. This can lead to Cherenkov radiation in vacuum and to other kinematically forbidden process. All these topics is widely discussed in last years from various angles.[6,7]

[i]Note that from experiment we have rather strong limitations on such a mean spread $|\delta G^{(i)}_{\mu\nu}|_{exper.} < 10^{-15} \div 10^{-20}$, but which can be covered by the factor k^2/Λ^2 .

much larger than $\sim \Lambda^{-1}$, at which the local Lorentz-invariance is broken, the approximate diffeomorphism invariance remains, and as a result we come to the universality of the long-range goldstone graviton mediated interactions. In fact we do not need the universality of the gravitational interactions at all scales, but only in the long-distance limit, where is fulfilled with an extreme precision as we know from experiment. But at the scales close to $\sim \Lambda^{-1}$ one may expect big nonuniversalities. These is evident for the couplings of $t_{\mu\nu}$ with different types of matter $(\Psi, \gamma, \mathrm{S})$ presented in Fig. 1.[j]

Such nonuniversality can be directly seen for the system described by the Lagrangian $\mathcal{L}_2$. The interaction of ω_{ab} goldstones with other particles is contained in terms entering the equation of motion

$$\frac{\delta\ \mathcal{L}_2(t_{\mu\nu})}{\delta\omega_{ab}(x)} = \frac{\partial\mathcal{L}_2}{\partial\omega_{ab}} - \partial_\mu \frac{\delta\ \mathcal{L}_2}{\delta(\partial_\mu\omega_{ab})} + \ ... \tag{32}$$

In the long range limit only the $\partial\mathcal{L}_2/\partial\omega_{ab}$ therm in (32) contributes to the interaction-current, connecting ω_{ab} with other fields :

$$\frac{\partial\mathcal{L}_2}{\partial\omega_{ab}} \simeq (\Gamma^{\cdots\cdots} Q^{ab\ \gamma\sigma}_{\mu\nu})(\partial_\gamma\tilde{t}_{\alpha\beta}\partial_\sigma\tilde{t}_{\delta\rho}) = \Upsilon^{\mu\nu}_{ab}(\partial_\mu\varrho^n\ \partial_\nu\varrho^n)\ , \tag{33}$$

with $\Upsilon^{\mu\nu}_{ab}$ being some constant tensors. The structure of ω_{ab} interactions, coming from the term $\partial\mathcal{L}_2/\partial\omega_{ab}$ in (32), resembles the graviton interactions in General Relativity, coming from analogous terms $T_{\mu\nu} = \delta L/\delta g_{\mu\nu}$. Note that in the week field limit ω_{ab} and $h_{\mu\nu}$ are linearly related as in (16).

The fundamental fields Ψ and physical particles S do not enter in ϱ^m symmetrically at the scale Λ. Therefore, as follows from (33), there is no universality in the goldstone-graviton coupling at such a momenta. At much larger distances $x \gg \Lambda^{-1}$ operators $\partial\mathcal{L}_2/\partial\omega_{ab}$ in (33) are renormalized and one cannot exclude that in the IR limit, for momenta $k \to 0$, the universality can be reached.

There exists a rather simple general argument[8] that in Lorentz-invariant case the massless spin two particles must be coupled universally to all particles in the limit $k \to 0$. This construction can be generalized to the case with spontaneous breaking of the LI as follows.

Consider some general N-particle amplitude with soft emission of $t_{\mu\nu}$-graviton with momentum $k \to 0$. In this limit the main contribution comes

---

[j] Let us remark, however, that we know very little whether the gravitational interaction of all types of matter (for example dark matter) is really universal – for the opposite case, see. e.g. ref.[11]

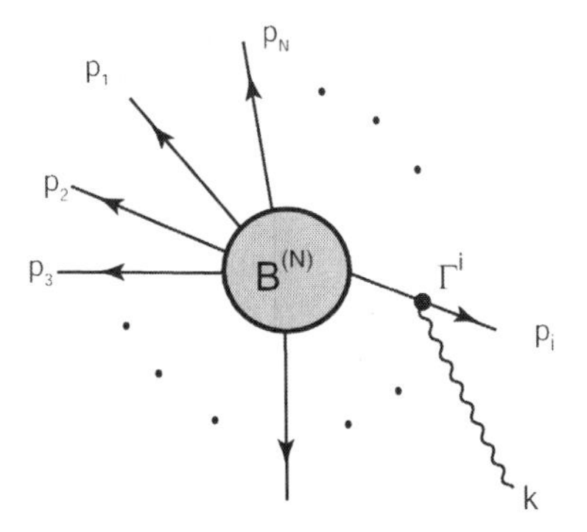

Fig. 2. A general N particle amplitude for a soft graviton emission with a momentum k.

from diagrams with the graviton emission from all external lines (See Fig. 2), and has the form:

$$A^{(N)}_{\mu\nu}(p_1, ..., p_N, \ k) \ \simeq \ B^{(N)}(p_1, ..., p_N) \sum_{i=1}^{N} \frac{\Gamma^i_{\mu\nu}(p_i)}{(2p_{i\alpha} \ k_\beta \ \mathcal{G}^i_{\alpha\beta})} \ ,$$

where we supposed that propagators of p_i-particles near mass shell have the form $2\left(p_{i\alpha} \ k_\beta \ \mathcal{G}_{\alpha\beta} - m_i^2\right)^{-1}$, and the tensors $\mathcal{G}^i_{\alpha\beta}$ can depend from $n_{\mu\nu}$. The vertexes $\Gamma^i_{\mu\nu}$ for a on mass shell graviton emission by particle of type (i) can also contain all possible combinations of p_μ and $n_{\mu\nu}$

$$\begin{aligned}\Gamma^i_{\mu\nu}(p) \ &= \ C^i p_\mu p_\nu \ + \ (H^i_1)_{\mu\nu} \ + \ p_\mu \ (H^i_2)_{\nu\lambda} \ p_\lambda \\ &+ p_\lambda \ (H^i_2)_{\lambda\mu} \ p_\nu \ + \ (H^i_3)_{\lambda\mu} \ p_\lambda \ (H^i_3)_{\nu\beta} \ p_\beta \ ,\end{aligned} \tag{34}$$

where the tensors

$$(H^i_k)_{\mu\nu} \ = \ (C^i_k n + \hat{C}^i_k nn + \tilde{C}^i_k nnn + \check{C}^i_k nnnn)_{\mu\nu} \ , \quad i, k = 1, 2, 3 \tag{35}$$

and C^i_k, $\hat{C}^i_k$, ... - some invariant functions depending at $k = 0$ only from particles type. In the usual case without LI breaking we have in (34) only the first term $p_\mu p_\nu$ in graviton particle coupling. Du to masslessness of $t_{\mu\nu}$-gravitons the amplitudes $A^{(N)}_{\mu\nu}$ should be transverse for all N and particles momenta p^i:[k]

$$2k_\mu A^{(N)}_{\mu\nu} \;=\; B(p_1,...,p_N) \sum_{i=1}^{N} \Big[\, C^i p^i_\nu \,+\, (H^i_2 p^i)_\nu$$

$$+ \frac{1}{(\mathcal{G}^i_{\alpha\beta} k_\alpha p_{i\beta})} \big[\, (H^i_1 k)_\nu \,+\, (kH_2 p^i)_\nu \,+\, (p^i H_3)_\nu (kp^i) \, \big] \Big] \;=\; 0 \; . \quad (36)$$

The only possibility that this conditions can be fulfilled identically at all N and p^i is when equation (36) is reduced to energy-momentum conservation $\sum_i p^i_\mu = 0$. And the momentum conservation take place because the translation invariance is unbroken by tensor condensate. This impose conditions on coefficients entering equation (36) which has the general solution of the form:

$$(H_1)^i_{\mu\nu} = 0 \; ; \quad (H_2)^i_{\mu\nu} = a_i \mathcal{G}^i_{\mu\nu} \; ; \quad (H_3)^i_{\mu\nu} = \; b_i \mathcal{G}^i_{\mu\nu} = \; b_i \eta_{\mu\nu} \; ;$$

$$C^i + 2a_i + (b_i)^2 \, = \, C.$$

And so we end up with the universal (at $k \to 0$) vertices for the soft $t_{\mu\nu}$-graviton interactions with other particles

$$\Gamma^i_{\mu\nu}(p,k) \;=\; C p_\mu p_\nu \,+\, \frac{k^2}{\Lambda^2} \, \hat{\Gamma}^i_{\mu\nu}(p) \,+\, ... \, , \quad (37)$$

where $\hat{\Gamma}^i_{\mu\nu}(p)$ are nonuniversal and can depend on $n_{\mu\nu}$. The details of a mechanism that can lead to the behavior depicted in (37) are not quit clear. Perhaps it can be connected to a different scale behavior of operators related to $\sim p_\mu p_\nu$ and specific dependence of vertices $\Gamma_{\mu\nu}$ on LI-breaking contributions $n_{\mu\nu}$. If we suppose that behavior as in (37) take place, the corrections to universal vertices ($\sim T_{\mu\nu}$), coming from various nonuniversal operators in $\tau_{\mu\nu}$ can decrease with energy scale as (E/Λ) or even stronger. For $E < 10^{-3}$ eV where experimental tests confirm the universality of gravity, the factor (E/Λ) is $< 10^{-31}$ if $\Lambda \sim m_p \sim 10^{19}$ GeV. This can be compared with the best experimental limits on the universality of $|1 - m_{\text{grav}}/m_{\text{inert}}| < 10^{-15}$, coming from last Etvesh-like measurements. This comparison also shows that the minimal allowed value of Λ by is ~ 1 TeV.

From a more speculative point of view, many examples are known when the maximal symmetry is restored at long distances, but it is difficult to formulate universal criteria for this. The simplest examples come from various lattice calculations (and experiments) - here not only isotropy but

[k]This point may not look as certain because LI is already broken. It corresponds to removing of the scalar part in $t_{\mu\nu}$ with respect to $\eta_{\mu\nu}$.

even translation invariance is broken at the lattice spacing scale a. But at distances $x \gg a$ the fast isotropization takes place, and the direction dependent contributions decrease as powers of (a/x). One should also mention the calculations in refs.[12] where the isotropisation in the renormalization flow to large distances is observed. But at the same time we know that there exist also such a media, whose macroscopic anisotropy reflects its the microscopic asymmetry.

From various of examples, we know, that the main reason for such an isotropization is connected with the fact that the quantities responsible for the spacial symmetry breaking are represented by a dimensional quantities (like lattice spacing a or, in our case, tensor condensate $\sim \Lambda^4 n_{\mu\nu}$). Therefore operators in which they enter have different scale dependence, in comparison with more symmetric operators represented via dimensionless quantities.

Needless to say that the questions of the local Lorentz breaking in the physical observables and the violation of the universality of the gravity must be deeply connected and this question deserves the profound study in various contexts.

It is remarkable, that the goldstone gravitons do not interact with the Lorentz invariant contribution to energy momentum tensor $T_{\mu\nu} \sim \eta_{\mu\nu}$ because local Lorentz rotations do not change it.[3] The conditions of gauge fixing reflect this fact. This can be cosnidered as an advantage towards the solution of the cosmological constant problem. On the other hand, in such a system one cannot have the de-Sitter type behavior and thus the cosmological inflation stage driven by almost constant scalar fields becomes problematic.

References

1. J.D. Bjorken, *Ann. Phys.* **24**, 174 (1963); see also arXiv:hep-th/0111196
2. P.R. Phillips, *Phys. Rev.* **146**, 966 (1966);
   H.C. Ohanian, *Phys. Rev.* **184**, 1305 (1969);
   D. Atkatz, *Phys. Rev.* D **17**, 1972 (1978);
   Y. Hosotani, *Phys. Lett.* **B319**, 332 (1993);
   V.A. Kostelecky, *Phys. Rev.* D **69**, 105009 (2004).
3. P. Kraus and E.T. Tomboulis, *Phys. Rev.* D **66**, 045015 (2002).
4. S. Deser, *Gen. Relat. Gravit.* **1**, 9 (1970).
5. Z. Berezhiani and O.V. Kancheli, arXiv:0808.3181 [hep-th].
6. S.R. Coleman and S.L. Glashow, *Phys. Rev.* D **59**, 116008 (1999).
7. T. Jacobson, S. Liberati and D. Mattingly, *Lect. Notes Phys.* **669**, 101 (2005).
8. S. Weinberg, *Phys. Rev.* **140**, 515 (1965).

9. C.J. Isham, A. Salam and J. Strathdee, *Phys. Rev.* D **3**, 867 (1970); T. Damour and I. Kogan, *Phys. Rev.* D **66**, 104024 (2002).
10. Z. Berezhiani, D. Comelli, F. Nesti and L. Pilo, *Phys. Rev. Lett.* 99, 131101 (2007); *JHEP* 07, 130 (2008).
11. Z. Berezhiani, F. Nesti, L. Pilo and N. Rossi, *JHEP* **0907**, 083 (2009); Z. Berezhiani, L. Pilo and N. Rossi, *Eur. Phys. J.* C **70**, 305 (2010).
12. H.B. Nielsen and M. Ninomiya, *Nucl. Phys.* **B141**, 153 (1978); S. Chadha and H.B. Nielsen, *Nucl. Phys.* **B217**, 124 (1983).

ON EMERGENT GAUGE AND GRAVITY THEORIES

J. L. Chkareuli[1,2]

[1] *Center for Elementary Particle Physics, ITP, Ilia State University, 0162 Tbilisi, Georgia*

[2] *Andronikashvili Institute of Physics, 0177 Tbilisi, Georgia*

j.chkareuli@iliauni.edu.ge

We present some general approach to emergent gauge theories and consider in significant detail the emergent tensor field gravity case. In essence, an arbitrary local theory of a symmetric two-tensor field $H_{\mu\nu}$ in Minkowski spacetime is considered, in which the equations of motion are required to be compatible with a nonlinear σ model type length-fixing constraint $H_{\mu\nu}^2 = \pm M^2$ leading to spontaneous Lorentz invariance violation, SLIV (M is the proposed scale for SLIV). Allowing the parameters in the Lagrangian to be adjusted so as to be consistent with this constraint, the theory turns out to correspond to linearized general relativity in the weak field approximation, while some of the massless tensor Goldstone modes appearing through SLIV are naturally collected in the physical graviton. The underlying diffeomophism invariance emerges as a necessary condition for the tensor field $H_{\mu\nu}$ not to be superfluously restricted in degrees of freedom, apart from the constraint due to which the true vacuum in the theory is chosen by SLIV. The emergent theory appears essentially nonlinear, when expressed in terms of the pure Goldstone tensor modes and contains a plethora of new Lorentz and CPT violating couplings. However, these couplings do not lead to physical Lorentz violation once this tensor field gravity is properly extended to conventional general relativity.

Keywords: Spontaneous Lorentz violation; Goldstone bosons; Emergent Gravity

1. Introduction

It is conceivable that spontaneous Lorentz invariance violation (SLIV) could provide a dynamical approach to quantum electrodynamics, gravity and Yang-Mills theories with photon, graviton and gluons appearing as massless Nambu-Goldstone bosons[1] (for some later developments see[2–4]). However, in contrast to spontaneous violation of internal symmetries, SLIV seems not to necessarily implies a physical breakdown of Lorentz invariance. Rather, when appearing in a gauge theory framework, this may eventually result in noncovariant gauge choice in an otherwise gauge invariant and Lorentz invariant theory.

Remarkably, a possible source for such a kind of the unobserved SLIV could provide the nonlinearly realized Lorentz symmetry for underlying vector field A_μ through its length-fixing constraint

$$A_\mu A^\mu = n^2 M^2 \ , \quad n^2 \equiv n_\nu n^\nu = \pm 1 \tag{1}$$

(where n_μ is a properly oriented unit Lorentz vector, while M is the proposed SLIV scale) rather than some vector field potential. This constraint in the gauge invariant QED framework was first studied by Nambu[5] a long ago, and in more detail (including the higher order corrections , extensions to spontaneously broken massive QED and non-Abelian theories etc.) in the last years.[6] The constraint (1), which in fact is very similar to the constraint appearing in the nonlinear σ-model for pions,[7] means in essence that the vector field A_μ develops some constant background value and the Lorentz symmetry $SO(1,3)$ formally breaks down to $SO(3)$ or $SO(1,2)$ depending on the time-like ($n^2 > 0$) or space-like ($n^2 < 0$) nature of SLIV. The point is, however, that, in sharp contrast to the nonlinear σ model for pions, the nonlinear QED theory, due to the starting gauge invariance involved, ensures that all physical Lorentz violating effects are proved to be strictly cancelled.

Extending the above argumentation, we consider here spontaneous Lorentz violation realized through a nonlinear length-fixing tensor field constraint of the type

$$H_{\mu\nu} H^{\mu\nu} = \mathfrak{n}^2 M^2 \ , \qquad \mathfrak{n}^2 \equiv \mathfrak{n}_{\mu\nu} \mathfrak{n}^{\mu\nu} = \pm 1 \tag{2}$$

where $\mathfrak{n}_{\mu\nu}$ is a properly oriented 'unit' Lorentz tensor, while M is the scale for Lorentz violation. Such a type of SLIV implemented into the tensor field gravity theory, which mimics linearized general relativity in Minkowski space-time, induces massless tensor Goldstone modes some of which can naturally be collected in the physical graviton.[8] Again, the theory appears essentially nonlinear and contains a plethora of new Lorentz and CPT violating couplings. However, these couplings do not lead to physical Lorentz violation once this tensor field gravity is properly extended to conventional general relativity.

2. Emergent gauge symmetries

Speaking still about vector field theories, the most important side of the nonlinear vector field constraint (1) was shown[9] to be that one does not need to specially postulate the starting gauge invariance in the framework of an arbitrary relativistically invariant Lagrangian which is proposed only to possess some global internal symmetry. Indeed, the SLIV conjecture (1) happens to be powerful enough by itself to require gauge invariance, provided that we allow the parameters in the corresponding Lagrangian density to be adjusted so as to ensure self-consistency without losing too many degrees of freedom. Namely, due to the spontaneous Lorentz violation determined by the constraint (1), the true vacuum in such a theory is chosen so that this theory acquires on its own a gauge-type invariance, which gauges the starting global symmetry of the interacting vector and matter fields involved. In essence, the gauge invariance (with a proper gauge-fixing term) appears as a necessary condition for these vector fields not to be superfluously restricted in degrees of freedom.

Let us dwell upon this point in more detail. Generally, while a conventional variation principle requires the equations of motion to be satisfied, it is possible to eliminate one component of a general 4-vector field A_μ, in order to describe a pure spin-1 particle by imposing a supplementary condition. In the massive vector field case there are three physical spin-1 states to be described by the A_μ field. Similarly in the massless vector field case, although there are only two physical (transverse) photon spin states, one cannot construct a massless 4-vector field A_μ as a linear combination of creation and annihilation operators for helicity ± 1 states in a relativistically covariant way, unless one fictitious state is added.[10] So, in both the massive and massless vector field cases, only one component of the A_μ field may be eliminated and still preserve Lorentz invariance. Once the SLIV constraint (1) is imposed, it is therefore not possible to satisfy another supplementary condition, since this would superfluously restrict the number of degrees of freedom for the vector field. In fact a further reduction in the number of independent A_μ components would make it impossible to set the required initial conditions in the appropriate Cauchy problem and, in quantum theory, to choose self-consistent equal-time commutation relations.[11]

We now turn to the question of the consistency of a constraint with the equations of motion for a general 4-vector field A_μ Actually, there are only two possible covariant constraints for such a vector field in a relativistically invariant theory - the holonomic SLIV constraint, $C(A) = A_\mu A^\mu - n^2 M^2 = 0$ (1), and the non-holonomic one, known as the Lorentz condition, $C(A) = \partial_\mu A^\mu = 0$. In the presence of the SLIV constraint $C(A) = A^\mu A_\mu - n^2 M^2 = 0$, it follows that the equations of motion can no longer be independent. The important point is that, in general, the time development would not preserve the constraint. So the parameters in the Lagrangian have to be chosen in such a way that effectively we have one less equation of motion for the vector field. This means that there should be some relationship between all the (vector and matter) field Eulerians (E_A, E_ψ, ...) involved[a]. Such a relationship can quite generally be formulated as a functional - but by locality just a function - of the Eulerians, $F(E_A, E_\psi, ...)$, being put equal to zero at each spacetime point with the configuration space restricted by the constraint $C(A) = 0$:

$$F(C = 0;\ \ E_A, E_\psi, ...) = 0\ . \tag{3}$$

This relationship must satisfy the same symmetry requirements of Lorentz and translational invariance, as well as all the global internal symmetry requirements, as the general starting Lagrangian $L(A, \psi, ...)$ does. We shall use this relationship in subsequent sections as the basis for gauge symmetry generation in the SLIV constrained vector and tensor field theories.

Let us now consider a "Taylor expansion" of the function F expressed as a linear combination of terms involving various field combinations multiplying or derivatives acting on the Eulerians[b]. The constant term in this expansion is of course zero since the relation (3) must be trivially satisfied when all the Eulerians vanish, i.e. when the equations of motion are satisfied. We now consider just the terms containing field combinations (and derivatives) with mass dimension 4, corresponding to the Lorentz invariant expressions

$$\partial_\mu (E_A)^\mu,\ A_\mu (E_A)^\mu,\ E_\psi \psi,\ \overline{\psi} E_{\overline{\psi}}. \tag{4}$$

All the other terms in the expansion contain field combinations and derivatives with higher mass dimension and must therefore have coefficients with an inverse mass dimension. We expect the mass scale associated with these coefficients should correspond to a large fundamental mass (e.g. the Planck mass M_P). Hence we conclude that such higher dimensional terms must be highly suppressed and can be neglected. A priori these neglected terms could lead to the breaking of the spontaneously generated gauge symmetry at high energy. However it could well be that a more detailed analysis would reveal that the imposed SLIV constraint requires an exact gauge symmetry. Indeed, if one uses classical equations of motion, a gauge breaking term will

[a] E_A stands for the vector-field Eulerian $(E_A)^\mu \equiv \partial L/\partial A_\mu - \partial_\nu[\partial L/\partial(\partial_\nu A_\mu)]$. We use similar notations for other field Eulerians as well.

[b] The Eulerians are of course just particular field combinations themselves and so this "expansion" at first includes higher powers and higher derivatives of the Eulerians.

typically predict the development of the "gauge" in a way that is inconsistent with our gauge fixing constraint $C(A) = 0$. Thus the theory will generically only be consistent if it has exact gauge symmetry[c].

3. Deriving diffeomorphism invariance

We now illustrate these ideas by the example of the emergent tensor field gravity case. Let us consider an arbitrary relativistically invariant Lagrangian $\mathcal{L}(H_{\mu\nu}, \phi)$ of one symmetric two-tensor field $H_{\mu\nu}$ and one real scalar field ϕ as the simplest possible matter in the theory taken in Minkowski spacetime. We restrict ourselves to the minimal interactions. In contrast to vector fields, whose basic interactions contain dimensionless coupling constants, for tensor fields the interactions with coupling constants of dimensionality sm^1 are essential. We first turn to the possible supplementary conditions which can be imposed on the tensor fields $H_{\mu\nu}$ in the Lagrangian $\mathcal{L}$, possessing still only a global Lorentz (and translational) invariance, in order to finally establish its form. The SLIV constraint (2), as it usually is when considering a system with holonomic constraints, can equivalently be presented in terms of some Lagrange multiplier term in the properly extended Lagrangian $\mathcal{L}'(H_{\mu\nu}, \phi, \lambda)$ rather than to be substituted into the starting one $\mathcal{L}(H_{\mu\nu}, \phi)$ prior to the variation of the action. Writing $\mathcal{L}'(H_{\mu\nu}, \phi, \lambda)$ as

$$\mathcal{L}'(H_{\mu\nu}, \phi, \lambda) = \mathcal{L}(H_{\mu\nu}, \phi) - \frac{1}{4}\lambda \left(H_{\mu\nu}H^{\mu\nu} - \mathfrak{n}^2 M^2\right)^2 \tag{5}$$

and varying with respect to the auxiliary field $\lambda(x)$ one has just the SLIV condition (2). Our choice of the quadratic form of the Lagrange-multiplier term[12] is only related to the fact that the equations of motion for $H_{\mu\nu}$ in this case are independent of the $\lambda(x)$ which entirely decouples from them rather than acts as some extra source of energy-momentum density, as it would be for the linear Lagrange multiplier term that could make the subsequent consideration to be more complicated. So, as soon as the constraint (2) holds

$$C(H_{\mu\nu}) = H_{\mu\nu}H^{\mu\nu} - \mathfrak{n}^2 M^2 = 0 \tag{6}$$

---

[c]The other possible Lorentz covariant constraint $\partial_\mu A^\mu = 0$, while also being sensitive to the form of the constraint-compatible Lagrangian, leads to massive QED and massive Yang-Mills theories.[11]

one has the equations of motion for $H_{\mu\nu}$ expressed through its Eulerian $(\mathcal{E}_H)^{\mu\nu}$

$$(\mathcal{E}_H)^{\mu\nu} \equiv \partial\mathcal{L}/\partial H_{\mu\nu} - \partial_\rho[\partial\mathcal{L}/\partial(\partial_\rho H_{\mu\nu})] = 0 \tag{7}$$

which is determined solely by the starting Lagrangian $\mathcal{L}(H_{\mu\nu}, \phi)$.

Despite the SLIV constraint (2) the tensor field $H_{\mu\nu}$, both massive and massless, still contains many superfluous components which are usually eliminated by imposing some supplementary conditions. In the massive tensor field case there are five physical spin-2 states to be described by $H_{\mu\nu}$. Similarly, in the massless tensor field case, though there are only two physical (transverse) spin states associated with graviton, one cannot construct a symmetric two-tensor field $H_{\mu\nu}$ as a linear combination of creation and annihilation operators for helicity ± 2 states unless three (and $2j-1$, in general, for the spin j massless field) fictitious states with other helicities are added.[7,10] So, in both massive and massless tensor field cases only five components in the 10-component tensor field $H_{\mu\nu}$ may be at most eliminated so as to preserve the Lorentz invariance. However, once the SLIV constraint (6) is already imposed, four extra supplementary conditions are only possible. Normally they should exclude the spin 1 states which are still left in the theory[d] and are described by some of the components of the tensor $H_{\mu\nu}$. Usually, they (and one of the spin 0 states) are excluded by the conventional harmonic gauge condition

$$\partial^\mu H_{\mu\nu} - \partial_\nu H_{tr}/2 = 0 \ . \tag{8}$$

or some of its analogs (see section 4). In fact, there should not be more supplementary conditions - otherwise, this would superfluously restrict the number of degrees of freedom for the spin 2 tensor field which is inadmissible.

Under this assumption of not getting too many constraints, we shall now derive gauge invariance of the Lagrangian $\mathcal{L}(H_{\mu\nu}, \phi)$. Actually, we turn to the question of the consistency of the SLIV constraint with the equations of motion for a general symmetric tensor field $H_{\mu\nu}$. For an arbitrary Lagrangian $\mathcal{L}(H_{\mu\nu}, \phi)$, the time development of the fields would not preserve the constraint (6). So the parameters in the Lagrangian must be chosen so as to give a relationship between the Eulerians for the tensor and matter fields of the type

$$\mathcal{F}^\mu(\mathcal{C}=0; \ \ \mathcal{E}_H, \mathcal{E}_\phi, ...) = 0 \quad (\mu = 0,1,2,3). \tag{9}$$

[d]These spin 1 states must necessarily be excluded as the sign of the energy for spin 1 is always opposite to that for spin 2 and 0

which, in contrast to the relationship (3), transforms in general as a Lorentz vector in the tensor field case. As a result, four additional equations for the tensor field $H_{\mu\nu}$ which appear by taking 4-divergence of the tensor field equations of motion (7)

$$\partial_\mu(\mathcal{E}_H)^{\mu\nu} = 0 \tag{10}$$

will not produce supplementary conditions at all once the SLIV condition (6) occurs. In fact, due to the relationship (9) these equations (10) are satisfied identically or as a result of the equations of motion of all the fields involved. This implies that in the absence of the equations of motion there must hold a general off-shell identity of the type

$$\partial_\mu(\mathcal{E}_H)^{\mu\nu} = P^\nu_{\alpha\beta}(\mathcal{E}_H)^{\alpha\beta} + Q^\nu\mathcal{E}_\phi \tag{11}$$

where $P^\nu_{\alpha\beta}$ and Q^ν are some operators acting on corresponding Eulerians of tensor and scalar fields (for this form of $\partial_\mu(\mathcal{E}_H)^{\mu\nu}$ the second equation in (10) is trivially satisfied). The simplest conceivable forms of these operators are

$$\begin{aligned} P^\nu_{\alpha\beta} = p_1\eta^{\nu\rho}(H_{\alpha\rho}\partial_\beta + H_{\rho\beta}\partial_\alpha + \partial_\rho H_{\alpha\beta})\ , \\ Q^\nu = q_1\eta^{\nu\rho}\partial_\rho\phi \end{aligned} \tag{12}$$

in which only terms with constants p_1 and q_1 of dimensionality cm^1 appear essential. This identity (11) implies then the invariance of $\mathcal{L}(H_{\mu\nu}, \phi)$ under the local transformations of tensor and scalar fields whose infinitesimal form is given by

$$\begin{aligned} \delta H_{\mu\nu} = \partial_\mu\xi_\nu + \partial_\nu\xi_\mu \\ +p_1(\partial_\mu\xi^\rho H_{\rho\nu} + \partial_\nu\xi^\rho H_{\mu\rho} + \xi^\rho\partial_\rho H_{\mu\nu}), \\ \delta\phi = q_1\xi^\rho\partial_\rho\phi \end{aligned} \tag{13}$$

where $\xi^\mu(x)$ is an arbitrary 4-vector function, only being required to conform with the nonlinear constraint (2). Conversely, the identity (11) in its turn follows from the invariance of the Lagrangian $\mathcal{L}(H_{\mu\nu}, \phi)$ under the transformations (13). Both direct and converse assertions are in fact particular cases of Noether's second theorem.[13]

An important point is that the operators $P^\nu_{\alpha\beta}$ and Q^ν (12) were chosen in a way that the corresponding transformations (13) could generally constitute a group (that is the Lie structure relation holds). This is why all three terms in the symmetric operator $P^\nu_{\alpha\beta}$ (12) are taken with the same constant, though the third term in it might enter with some different constant. Remarkably, though the transformations (13) were only restricted to form a group, this emergent symmetry group is proved, as one can readily confirm, to be nothing but the diff invariance. Indeed, for the quantity

$$g_{\mu\nu} = \eta_{\mu\nu} + p_1 H_{\mu\nu} \tag{14}$$

the tensor field transformation (13) may be written in a form

$$\delta g_{\mu\nu} = p_1(\partial_\mu \xi^\rho g_{\rho\nu} + \partial_\nu \xi^\rho g_{\mu\rho} + \xi^\rho \partial_\rho g_{\mu\nu}) \tag{15}$$

which shows that $g_{\mu\nu}$ transform as the metric tensors in the Riemannian geometry (the constant p_1 may be included into the transformation 4-vector parameter $\xi^\mu(x)$) with general coordinate transformations, $\delta x^\mu = \xi^\mu(x)$. So, we have shown that the imposition of the SLIV constraint (2) supplements the starting global Poincare symmetry with the local diff invariance. Otherwise, the theory would superfluously restrict the number of degrees of freedom for the tensor field $H_{\mu\nu}$, which would certainly not be allowed.

This SLIV induced gauge symmetry (13) completely determines now the Lagrangian $\mathcal{L}(H_{\mu\nu}, \phi)$. Indeed, in the weak field approximation (when $\delta H_{\mu\nu} = \partial_\mu \xi_\nu + \partial_\nu \xi_\mu$) this symmetry gives the well-known linearized gravity Lagrangian

$$\mathcal{L}(H_{\mu\nu}, \phi) = \mathcal{L}(H) + \mathcal{L}(\phi) + \mathcal{L}_{int} \tag{16}$$

consisted of the H field kinetic term of the form

$$\mathcal{L}(H) = \frac{1}{2}\partial_\lambda H^{\mu\nu}\partial^\lambda H_{\mu\nu} - \frac{1}{2}\partial_\lambda H_{tr}\partial^\lambda H_{tr} \\ - \partial_\lambda H^{\lambda\nu}\partial^\mu H_{\mu\nu} + \partial^\nu H_{tr}\partial^\mu H_{\mu\nu} \ , \tag{17}$$

(H_{tr} stands for a trace of the $H_{\mu\nu}$, $H_{tr} = \eta^{\mu\nu} h_{\mu\nu}$), and the free scalar field Lagrangian $\mathcal{L}(\phi)$ and interaction term $\mathcal{L}_{int} = (1/M_P)H_{\mu\nu}T^{\mu\nu}(\phi)$, where $T^{\mu\nu}(\phi)$ is a conventional energy-momentum tensor for scalar field. Besides, the proportionality coefficient p_1 in the metric (14) was chosen to be inverse just to the Planck mass M_P. It is clear that, in contrast to the free field terms given above by $\mathcal{L}(H)$ and $\mathcal{L}(\phi)$, the interaction term $\mathcal{L}_{int}$ is only approximately invariant under the diff transformations in the weak field limit.

To determine a complete theory, one should consider the full variation of the Lagrangian $\mathcal{L}$ as function of metric $g_{\mu\nu}$ and its derivatives (including the second order ones), and solve a general identity of the type

$$\delta\mathcal{L}(g_{\mu\nu}, g_{\mu\nu,\lambda}, g_{\mu\nu,\lambda\rho}; \phi, \phi_{,\lambda}) = \partial_\mu X^\mu \tag{18}$$

(subscripts after commas denote derivatives) which contains an unknown vector function X^μ. The latter must be constructed from the fields and local transformation parameters $\xi^\mu(x)$ taking into account the requirement of compatibility with the invariance of the $\mathcal{L}$ under transformations of the Lorentz group and translations. Following this procedure[14] for the metric and scalar field variations (15, 13) conditioned by SLIV constraint (2), one can eventually find the total Lagrangian $\mathcal{L}$ which is turned out to be properly expressed in terms of quantities similar to the basic ones in the Riemannian geometry (like as metric, connection, curvature etc.). Actually, this theory successfully mimics general relativity that allows us to conclude that the Einstein equations could be really derived in the flat Minkowski spacetime provided that Lorentz symmetry in it is spontaneously broken.

4. Graviton as a Goldstone boson

Let us turn now to spontaneous Lorentz violation in itself which is caused by the nonlinear tensor field constraint (2). This constraint means in essence that the tensor field $H_{\mu\nu}$ develops the vev configuration

$$< H_{\mu\nu}(x) > \quad = \mathfrak{n}_{\mu\nu} M \tag{19}$$

determined by the matrix $\mathfrak{n}_{\mu\nu}$, and starting Lorentz symmetry $SO(1,3)$ of the Lagrangian $\mathcal{L}(H_{\mu\nu}, \phi)$ given in (16) formally breaks down at a scale M to one of its subgroup thus producing a corresponding number of the Goldstone modes. In this connection the question about other components of a symmetric two-index tensor $H_{\mu\nu}$, aside from the pure Goldstone ones, naturally arises. Remarkably, they are turned out to be the pseudo-Goldstone modes (PGMs) in the theory. Indeed, although we only propose the Lorentz invariance of the Lagrangian $\mathcal{L}(H_{\mu\nu}, \phi)$, the SLIV constraint (2) possesses formally a much higher accidental symmetry. This is in fact $SO(7,3)$ symmetry of the length-fixing bilinear form (2). This symmetry is spontaneously broken side by side with Lorentz symmetry at scale M. Assuming a minimal vacuum configuration in the $SO(7,3)$ space with the vevs (19) developed on only one $H_{\mu\nu}$ component, we have the time-like ($SO(7,3) \rightarrow SO(6,3)$) or space-like ($SO(7,3) \rightarrow SO(7,2)$) violations of the accidental symmetry depending on the sign of $\mathfrak{n}^2 = \pm 1$ in (2), respectively. According to the number of the broken generators just nine massless NG modes appear in both of cases. Together with an effective Higgs component, on which the vev is developed, they complete the whole ten-component symmetric tensor field $H_{\mu\nu}$ of our Lorentz group. Some of them are true Goldstone modes of spontaneous Lorentz violation, the others are PGMs since, as was mentioned, an accidental $SO(7,3)$ is not shared by the whole Lagrangian $\mathcal{L}(H_{\mu\nu}, \phi)$ given in (16). Notably, in contrast to the known scalar PGM case,[7] they remain strictly massless being protected by the simultaneously generated diff invariance.

Now, one can rewrite the Lagrangian $\mathcal{L}(H_{\mu\nu}, \phi)$ in terms of the Goldstone modes explicitly using the SLIV constraint (2). For this purpose let us take the following handy parameterization for the tensor field $H_{\mu\nu}$ in the Lagrangian $\mathcal{L}(H_{\mu\nu}, \phi)$:

$$H_{\mu\nu} = h_{\mu\nu} + \frac{n_{\mu\nu}}{n^2}(\mathfrak{n}_{\alpha\beta}H^{\alpha\beta}) , \quad \mathfrak{n}_{\mu\nu}h^{\mu\nu} = 0 \tag{20}$$

where $h_{\mu\nu}$ corresponds to the pure Goldstonic modes, while the effective "Higgs" mode (or the $H_{\mu\nu}$ component in the vacuum direction) is

$$\mathfrak{n}_{\alpha\beta}H^{\alpha\beta} = (M^2 - \mathfrak{n}^2h^2)^{\frac{1}{2}} = M - \frac{\mathfrak{n}^2h^2}{2M} + O(1/M^2) \tag{21}$$

taking, for definiteness, the positive sign for the square root and expanding it in powers of h^2/M^2, $h^2 \equiv h_{\mu\nu}h^{\mu\nu}$. Putting then the parameterization (20) with the SLIV constraint (21) into Lagrangian $\mathcal{L}(H_{\mu\nu}, \phi)$ given in (16), one readily comes to the truly Goldstonic tensor field gravity Lagrangian $\mathcal{L}(h_{\mu\nu}, \phi)$ containing infinite series in powers of the $h_{\mu\nu}$ modes, which we will not display here due to its excessive length (see[8]).

Together with the Lagrangian $\mathcal{L}(h_{\mu\nu}, \phi)$ one must be also certain about the gauge fixing terms, apart from a general Goldstonic "gauge" $\mathfrak{n}_{\mu\nu}h^{\mu\nu} = 0$ given above (20). Remarkably, the simplest set of conditions being compatible with the latter is turned out to be

$$\partial^{\rho}(\partial_{\mu}h_{\nu\rho} - \partial_{\nu}h_{\mu\rho}) = 0 \tag{22}$$

(rather than harmonic gauge conditions (8)) which also automatically eliminates the (negative-energy) spin 1 states in the theory. So, with the Lagrangian $\mathcal{L}(h_{\mu\nu}, \phi)$ and the supplementary conditions (20) and (22) lumped together, one eventually comes to the working model for the Goldstonic tensor field gravity. Generally, from ten components in the symmetric-two $h_{\mu\nu}$ tensor, four components are excluded by the supplementary conditions (20) and (22). For a plane gravitational wave propagating, say, in the z direction another four components can also be eliminated. This is due to the fact that the above supplementary conditions still leave freedom in the choice of a coordinate system, $x^{\mu} \to x^{\mu} - \xi^{\mu}(t - z/c)$, much as takes place in standard GR. Depending on the form of the vev tensor $\mathfrak{n}_{\mu\nu}$, the two remaining transverse modes of the physical graviton may consist solely of Lorentz Goldstone modes or of Pseudo Goldstone modes or include both of them.

5. Summary and outlook

We presented here some approach to emergent gauge theories and considered in significant detail the emergent tensor field gravity case. Our main result can be summarized in a form of a general *Emergent Gauge Symmetry (EGS) conjecture*:

Let there be given an interacting field system $\{\boldsymbol{A}_\mu, ..., \phi, \psi, H_{\mu\nu}\}$ *containing some vector field (or vector field multiplet)* $\boldsymbol{A}_\mu$ *or/and tensor field* $H_{\mu\nu}$ *in an arbitrary Lorentz invariant Lagrangian of scalar* (ϕ), *fermion* (ψ) *and other matter field multiplets, which possesses only global Abelian or non-Abelian symmetry* G *(and only conventional global Lorentz invariance in a pure tensor field case). Suppose that one of fields in a given field system is subject to the nonlinear* σ *model type "length-fixing" constraint, say,* $\boldsymbol{A}_\mu^2 = M^2$ *(for vector fields) or* $\phi^2 = M^2$ *(for scalar fields) or* $H_{\mu\nu}^2 = M^2$ *(for tensor field). Then, since the time development would not in general preserve this constraint, the parameters in their common Lagrangian* $L(\boldsymbol{A}_\mu, ..., \phi, \psi; H_{\mu\nu})$ *will adjust themselves in such a way that effectively we have less independent equations of motion for the field system taken. This means that there should be some relationship between Eulerians of all the fields involved to which Noether's second theorem can be applied. As a result, one comes to the conversion of the global symmetry* G *into the local symmetry* G_{loc}, *being exact or spontaneously broken depending on whether vector or scalar fields are constrained, and to the conversion of global Lorentz invariance into diffeomorphism invariance for the constrained tensor field.*

Applying the EGS conjecture to tensor field theory case we found that the only possible local theory of a symmetric two-tensor field $H_{\mu\nu}$ in Minkowski spacetime which is compatible with SLIV constraint $H_{\mu\nu}^2 = \pm M^2$ is turned out to be linearized general relativity in the weak field approximation. When expressed in terms of the pure tensor Goldstone modes this theory is essentially nonlinear and contains a variety of Lorentz and CPT violating couplings. Nonetheless, as was shown in the recent calculations,[8] all the SLIV effects turn out to be strictly cancelled in the lowest order gravity processes as soon as the tensor field gravity theory is properly extended to general relativity. So, the nonlinear SLIV condition being applied both in vector and tensor field theories, due to which true vacuum is chosen and Goldstonic gauge fields are generated, may provide a dynamical setting for all underlying internal and spacetime local symmetries involved. However, this gauge theory framework, uniquely emerging for the length-fixed vector and tensor fields, makes in turn this SLIV to be physically unobservable.

From this standpoint, the only way for physical Lorentz violation to appear would be if the above local invariance were slightly broken at very small distances controlled by quantum gravity.[15] The latter could in general hinder the setting of the required initial conditions in the appropriate Cauchy problem thus admitting a superfluous restriction of vector and tensor fields in degrees of freedom through some high-order operators stemming from the quantum gravity influenced area. This may be a place where the emergent vector and tensor field theories may drastically differ from conventional gauge theories that could have some observational evidence at low energies.

Acknowledgments

This paper is largely based on the recent works[8,15,16] carried out in collaboration with Colin Froggatt, Juansher Jejelava, Zurab Kepuladze, Holger Nielsen and Giorgi Tatishvili. I would like to thank Oleg Kancheli and Archil Kobakhidze for interesting discussions and useful remarks.

References

1. J.D. Bjorken, *Ann. Phys. (N.Y.)* **24** (1963) 174; P.R. Phillips, *Phys. Rev.* **146** (1966) 966; T. Eguchi, Phys.Rev. D **14** (1976) 2755.
2. J.L. Chkareuli, C.D. Froggatt and H.B. Nielsen, *Phys. Rev. Lett.***87** (2001) 091601; *Nucl. Phys. B* **609** (2001) 46.
3. Per Kraus and E.T. Tomboulis, *Phys. Rev. D* **66** (2002) 045015.
4. V.A. Kostelecky, *Phys. Rev. D* **69** (2004) 105009; V.A. Kostelecky and R. Potting, *Phys. Rev. D* **79** (2009) 065018.
5. Y. Nambu, *Progr. Theor. Phys. Suppl. Extra* 190 (1968).
6. A.T. Azatov and J.L. Chkareuli, *Phys. Rev. D* **73,** 065026 (2006); J.L. Chkareuli and Z.R. Kepuladze, *Phys. Lett. B* **644**, 212 (2007); J.L. Chkareuli and J.G. Jejelava, *Phys. Lett. B* **659** (2008) 754.
7. S.Weinberg, *The Quantum Theory of Fields,* v.2, Cambridge University Press, 2000.
8. J.L. Chkareuli, J.G. Jejelava and G. Tatishvili, *Phys. Lett.B* **696** (2011) 124
9. J.L. Chkareuli, C.D. Froggatt, J.G. Jejelava and H.B. Nielsen, *Nucl. Phys. B* **796** (2008) 211.
10. S. Weinberg, *Phys. Rev. B* **138** (1965) 988.
11. V.I. Ogievetsky and I.V. Polubarinov, *Ann. Phys. (N.Y.)* **25** (1963) 358.
12. R. Bluhm, Shu-Hong Fung and V. A. Kostelecky, *Phys. Rev. D.* **77** (2008) 065020.
13. E. Noether, Nachrichten von der Kön. Ges. Wissenschaften zu Gettingen, *Math.-Phys. Kl.* **2,** 235 (1918)
14. V.I. Ogievetsky and I.V. Polubarinov, *Ann. Phys.* (N.Y.) **35** (1965) 167.

15. J.L. Chkareuli, Z. Kepuladze and Tatishvili, *Eur. Phys. J. C* **55** (2008) 309; J.L. Chkareuli and Z. Kepuladze, *Eur. Phys. J. C* **72** (2012) 1954.
16. J.L. Chkareuli, C.D. Froggatt and H.B. Nielsen, *Nucl. Phys.B* **848** (2011) 498.

GEODESIC MOTION IN GENERAL RELATIVITY: *LARES* IN EARTH'S GRAVITY

I.Ciufolini[1], V.G.Gurzadyan[2], R.Penrose[3] and A.Paolozzi[4]

1.Dipartimento di Ingegneria dell'Innovazione, University of Salento, Lecce, Italy
2.Center for Cosmology and Astrophysics, Alikhanian National Laboratory, Yerevan, Armenia
3.Mathematical Institute, University of Oxford, UK
4.Scuola di Ingegneria Aerospaziale and DIAEE, Sapienza University, Rome, Italy

According to General Relativity, as distinct from Newtonian gravity, motion under gravity is treated by a theory that deals, initially, only with test particles. At the same time, satellite measurements deal with extended bodies. We discuss the correspondence between geodesic motion in General Relativity and the motion of an extended body by means of the Ehlers-Geroch theorem, and in the context of the recently launched LAser RElativity Satellite (LARES). Being possibly the highest mean density orbiting body in the Solar system, this satellite provides the best realization of a test particle ever reached experimentally and provides a unique possibility for testing the predictions of General Relativity.

Keywords: General Relativity; space experiments.

1. The problem of a material body in General Relativity

The creation of a physical body that provides the best possible approximation to the free motion of a test particle space-time geodesic is a profound goal for experiments dedicated to the study of space-time geometry in the vicinity of external gravitating bodies, this enabling high precision tests of General Relativity and alternative theories of gravity to be carried out.

There are important issues regarding the approximation to a geodesic that are being addressed by the motion of an actual extended body. On the one hand, in General Relativity,[1–3] the problem of an extended body is subtle due, not only to the non-linearity of the equations of motion, but also because of the need to deal with the internal structure of a compact body, constructed of continuous media, where kinetic variables and thermodynamic potentials, are involved, and where there may be intrinsically non-local effects arising from the internal structure of an extended body, such as tidal ones. Moreover, there are problems concerning approximations that need to be made in order to describe a given extended body as a test particle moving along a geodesic, these being related with the fact that many of the common Newtonian gravitational concepts such as centre of mass, total mass, or size of an extended material body do not have well defined counterparts in General Relativity.[4]

The Ehlers-Geroch theorem,[5] which generalized an earlier result by Geroch and Jang,[6] provides a justification, under appropriate conditions, for the trajectory of an extended body, having a limitingly small gravitational field of its own, to be a geodesic:

A timelike curve γ on a 4-manifold of Lorenzian metric g must necessarily be a geodesic, if for each closed neighborhood U of γ there exists, within each neighborhood of the metric g (with accompanying connection) in U, a metric $\tilde{g}$ whose the Einstein tensor

$$\tilde{\mathbf{G}} = \tilde{\mathbf{R}} - \frac{1}{2}\tilde{g}\tilde{R}$$

($\tilde{\mathbf{R}}$ being the Ricci tensor and $\tilde{R}$ the scalar curvature of $\tilde{g}$) satisfies the dominant energy condition in U, is non-zero in U, and vanishes on ∂U.

The conditions of the theorem are not specific to any equations of state that the material of a test body might be subject to, beyond its satisfaction of the dominant energy condition.

This theorem, asserting that *small massive bodies move on near-geodesics*, thus achieves a rigorous bridge from General Relativity to satellite experiments. This suggests a high level of suppression of non-gravitational and self-gravitational effects from the satellite's own small gravitational field, enabling us to consider that it has a very nearly geodesic motion and, hence, providing a genuine testing ground for General Relativity effects.

2. Space Experiment LARES

LAser RElativity Satellite (LARES)[7] was launched on February 13, 2012 from the European Space Agency's spaceport in Kourou, French Guiana. That launch also marked the first qualification flight of the new vehicle-rocket VEGA of the European Space Agency. The satellite is a tungsten alloy sphere of 18.2 cm radius and mass 386.8 kg (Fig. 1), covered by 92 retro-reflectors to reflect the laser signals from the International Laser Ranging Service stations. The orbit is circular to high accuracy, the eccentricity is 0.0007, and inclination 69°.5 at semi-major axis 7820 km.

The efficiency of LARES is in the simplicity of its design, aimed at the suppression of non-gravitational orbital perturbations including atmospheric drag and the anisotropic thermal radiation from its surface due to the anisotropic heating of the sphere due to the solar and Earth's radiations (Yarkovsky effect): it has the smallest ratio of cross-sectional area to mass of any other artificial satellite, including the previous two LAser GEOdynamics Satellites (LAGEOS) and a single piece structure. (LAGEOS had a heterogeneous structure.) The experience with the LAGEOS satellites had proved the efficiency of laser ranging passive satellites for testing the frame-dragging effect predicted by General Relativity.[8] An important fact was the availability of the improved Earth's gravity models of GRACE, providing the possibility of the accurate evaluation of their spherical harmonic coefficients. The obtained behavior for the residuals of the nodes of the LAGEOS 1 and 2 confirmed the Lense-Thirring effect,[9] i.e. the rate of drag of the longitude of the nodes The efficiency of LARES is in the simplicity of its design, aimed at the suppression of non-gravitational orbital perturbations including atmospheric drag and the anisotropic thermal radiation from its surface due to the anisotropic heating of the sphere due to the solar and Earth's radiations (Yarkovsky effect): it has the smallest ratio of cross-sectional area to mass of any other artificial satellite, including the previous two LAser GEOdynamics Satellites (LAGEOS) and a single piece structure. (LAGEOS had a heterogeneous structure.) The experience with the LAGEOS satellites had proved the efficiency of laser ranging passive satellites for testing the frame-dragging effect predicted by General Relativity.[8] An important fact was the availability of the improved Earth's gravity models of GRACE, providing the possibility of the accurate evaluation of their spherical harmonic coefficients. The obtained behavior for the residuals of the nodes of the LAGEOS 1 and 2 confirmed the Lense-Thirring effect,[9] i.e. the rate of drag of the longitude of the nodes

$$\dot{\Omega}(Lense - Thirring) = \frac{2\mathbf{J}}{a^3(1-e^2)^{3/2}},$$

to an accuracy of the order of 10%,[8] where **J** is the angular momentum of the central mass, a and e are the semi-major axis and the eccentricity of the orbit of the test particle, respectively.

The analysis of the first months of measurements[10] confirmed the efficiency of the design of LARES, even though its orbit is lower than that of both LAGEOS: the acceleration due to the non-gravitational perturbations for LARES achieves 2-3-times improvement with respect to LAGEOS. It has to be stressed that until the launch of LARES, the two LAGEOS satel-

Fig. 1. The LARES satellite in the laboratory. Photo by Italian Space Agency.

lites had the smallest residual acceleration of any other artificial satellite.

The creation of a General Relativistic test particle moving along a geodesic in the Earth's space-time gravitational field has to be considered as a principal achievement for high accuracy testing of any relativistic effect. These studies are able to constrain possible extensions of General Relativity such as the Chern-Simons modified gravity (e.g.[11]), thus bridging Earth-vicinity measurements to issues of dark energy and cosmology.

References

1. Penrose R., Structure of Spacetime, in: Battelle Rencontres, (C.DeWitt and J.A.Wheeler, Eds.), New York, 1968.
2. Ciufolini I., Wheeler J.A., Gravitation and Inertia, Princeton University Press, Princeton, 1995.
3. Rindler, W., Relativity: Special, General, and Cosmological, Oxford University Press, Oxford, 2001.
4. Ehlers J. in: Relativity, Astrophysics and Cosmology, (W.Israel Ed.), 1-125, Reidel Publ. 1973.
5. Ehlers J., Geroch R., Annals of Phys. 309 (2004) 232.
6. Geroch R., Jang P.S., J.Math.Phys. 16 (1975) 65
7. Ciufolini I., Paolozzi A., Pavlis E.C., Ries J.C., Koenig R. and Matzner R., in: General Relativity and John Archibald Wheeler, (I.Ciufolini and R.Matzner Eds.), 467-492, Springer, 2010.
8. Ciufolini I., Nature, 449 (2007) 41.
9. Lense J., Thirring H., Phys.Zeitschr. 19 (1918) 156.
10. Ciufolini I., Paolozzi A., Pavlis E., Ries J., Gurzadyan V.G., Koenig R., Matzner R., Penrose R., Sindoni G., LARES, the test-particle for gravitational physics and General Relativity (submitted)
11. Yagi K., Yunes N., Tanaka T. arXiv:1206.6130.

QUANTUM DOTS AND QUANTUM RINGS: REAL LOW DIMENSIONAL SYSTEMS*

I. Filikhin, S.G. Matinyan and B. Vlahovic

North Carolina Central University, 1801 Fayetteville St. Durham, NC 27707, USA

We consider the several phenomena which are taking place in Quantum Dots (QD) and Quantum Rings (QR): The connection of the Quantum Chaos (QC) with the reflection symmetry of the QD, Disappearance of the QC in the tunnel coupled chaotic QD, electron localization and transition between Double Concentric QR in the transverse magnetic field, transition of electron from QR to the QD located in the center of QR. Basis of this consideration is the effective Schrödinger equation for the corresponding systems.

1. Introduction

The progress of semiconductor physics in the decade 1970-1980 is connected with gradual deviation from the electronic band structure of ideal crystal of Bloch picture[1] where, unlike atomic world with its discrete and precisely defined, in the limits of uncertainty relation, energy levels, energy of bound electron is a multivalued function of momentum in the energy band and density of states are continuous (For the earlier short but comprehensive survey see[2]).

In principle, Bloch theory deals with infinite extension of lattice, with the understandable (and important) surface effects. The decreasing of the size of the object to a few micrometers principally does not change the picture of the extended crystal qualitatively. It takes a place until one reaches the scale where the size quantization essentially enters the game and we can speak about microscopic limit of matter. What generally divides macro-

*This paper is the basis of two talks of Sergei Matinyan ("Quantum Chaos and its disappearance in the Coupled Quantum Dots" (Nor-Amberd, 21 September 2011) and "Quantum Mechanics of Electron Transition between concentric Quantum Ring" (Tbilisi, 28 September 2011)

This work is supported by NSF CREST award; HRD-0833184 and NASA award NNX09AV07A

scopic limit of the solid state from the microscopic one? It is defined by some correlation length (or, more generally, all such relevant lengths)): for carriers it is mean free path length l or Broglie length $l_B = h/p$ (p - momentum), which is smaller. One may say that the quantum mechanical properties of matter clearly reveal if $l/a \geq 1$, where a is the size of the lattice constant. In the opposite limit $l/a < 1$, matter is considered macroscopically.

In this light, it is worthy to remind that as long as 1962, L. V. Keldysh[3] ([3] as cited in[4]) considered electron motion in a crystal with periodic potential with the period that is much larger than the lattice constant. In this limit he discovered so called minizones and negative resistance. Just in this limit $l/a \geq 1$ we expect the size quantization with its discrete levels and coherence in the sense that electron can propagate across the whole system without scattering, its wave function maintains a definite phase. In this limit, mesoscopic relates to the intermediate scale dividing the macro and micro limits of matter and nanoscopic objects (Quantum Wells (QW), Wires and Dots (QD)) shown very interesting quantum mechanical effects. In this limit many usual rules of macroscopic physics may not hold. For only one example, rules of addition of resistance both in series and parallel are quite different and more complicated.[5–7]

Closing this brief introduction concerning some aspects of genuine quantum objects (QW, QWires, QD) we would like to emphasize the conditional sense of the notion of dimensions in this world: in the limit $l/a \geq 1$ dimensions are defined as difference between real spatial dimension (in our world D=3) and numbers of the confined directions: Quantum Well: D=2, Quantum Wire: D=1, Quantum Dot: D=0. However, for example, QD which will be one of our subject for study, has very rich structure with many discrete levels, their structure define the presence or absence of Chaos, as we will see below, inside QD. Minimal size of QD is defined by the condition to have at least one energy level of electron (hole) or both: $a_{\min} = \pi\hbar/2m * \Delta E \sim 4$ nm, where ΔE is average distance between neighboring energy levels. Maximal size of QD is defined by the conditions that all three dimensions are still confined. It depends, of course, on temperature: at room temperature it is 12 nm (GaAs), 20 nm (InAs) ($\Delta E \approx$ 3kT). The lower temperature, the wider QD is left as quantum object with D=0 and the number of energy levels will be higher.

2. Schrödinger equation and effective mass approximation

In the present review a semiconductor 3D heterostructure (QD or QR) is modelled utilizing a ***kp***-perturbation single sub-band approach with quasi-particle effective mass.[8–10] The energies and wave functions of a single carrier in a semiconductor structure are solutions the Schrödinger equation:

$$(H_{kp} + V_c(\vec{r}))\Psi(\vec{r}) = E\Psi(\vec{r}) \tag{1}$$

Here H_{kp} is the single band kp-Hamiltonian operator, $H_{kp} = -\nabla\frac{\hbar^2}{2m*(\vec{r})}\nabla$, $m*$ is the electron/hole effective mass for the bulk, which may depend on coordinate, and $V_c(\vec{r})$ is the confinement potential. The confinement of the single carrier is formed by the energy misalignment of the conduction (valence) band edges of the QD material (index 1) and the substrate material (index 2) in the bulk. $V_c(\vec{r})$ is so called "band gap potential". The magnitude of the potential is proportional to the energy misalignment. The band structure of the single band approximation one can be found in many textbooks (see, for example,[8–10]). We consider here the model in which the band gap potential is defined as follows:

$$V_c(\vec{r}) = \begin{cases} 0, & \in QD, \\ E_c, & \vec{r} \notin QD, \end{cases}$$

where $E_c = \kappa(E_{g,2} - E_{g,1})$, E_g is the band gap and the coefficient $\kappa < 1$ can be different for the conduction and valence bands gap potential. The BenDaniel-Duke boundary conditions are used on interface of the materials:[11] The single electron Schrödinger equation for wave function $\Psi(\vec{r})$ and its derivative $1/m * (\vec{n}, \nabla)\Psi(\vec{r})$ on interface of QD and the substrate are continues.

3. The non-parabolicity of the conduction band. The Kane formula

Traditionally applied in the macroscopic scale studies parabolic electron spectrum needs to be replaced by the non-parabolic approach, which is more appropriate to nano-sized quantum objects.[12,13] The Kane formula[14] is implemented in the model to take into account the non-parabolicity of the conduction band. The energy dependence of the electron effective mass is defined by the following formula:

$$\frac{m_0}{m^*} = \frac{2m_0P^2}{3\hbar^2}\left(\frac{2}{E_g + E} + \frac{1}{E_g + \Delta + E}\right). \tag{2}$$

Here m_0 is free electron mass, P is Kane's momentum matrix element, E_g is the band gap, and Δ is the spin-orbit splitting of the valence band.

Taking into account the relation (2) the Schrödinger equation (1) is expressed as follow

$$(H_{kp}(E) + V_c(\vec{r}))\Psi(\vec{r}) = E\Psi(\vec{r}). \quad (3)$$

Here $H_{kp}(E)$ is the single band kp-Hamiltonian operator $H_{kp}(E) = -\nabla\frac{\hbar^2}{2m*(E,\vec{r})}\nabla$, $m*(E,\vec{r})$ is the electron (or hole) effective mass, and $V_c(\vec{r})$ is the band gap potential. As a result, we obtain a non-linear eigenvalue problem.

Solution of the problem (3)-(2) results that the electron/hole effective mass in QD (or QR) varies between the bulk values for effective mass of the QD and substrate materials. The same is given for the effective mass of carriers in the substrate. The energy of confinement states of carries is rearranged by the magnitude of the band gap potential V_c.

The Schrödinger equation (1) with the energy dependence of effective mass can be solved by the iteration procedure.[15–18]

$$\begin{aligned} &H_{kp}(m*_i^{k-1})\Psi^k(\vec{r}) = E^k\Psi^k(\vec{r}), \\ &m*_i^k = f_i(E^k), \end{aligned} \quad (4)$$

where k is the iteration number, i refers to the subdomain of the system; $i = 1$ for the QD, $i = 2$ for the substrate. $H_{kp}(m*_i^k)$ is the Hamiltonian in which the effective mass does not depend on energy and is equal to the value of $m*_i^k$, f_i is the function defined by the relation (2). For each step of the iterations the equation (1) is reduced to Schrödinger equatio with the effective mass of the current step which does not depend on energy. At the beginning of iterations the bulk value of the effective mass is employed. Obtained eigenvalue problem can be solved numerically (by the finite element method, for example). After that, a new value for effective mass is taken by using Eq. (2) and procedure is repeated. The convergence of the effective mass during the procedure has a place after 3-5 steps. As an example, the typical convergences for election effective mass and confinement energy of single electron are displayed in Fig. 1 for the InAs/GaAs QR.[18] Description of other methods for the solution of the problem (3)-(2) can be found in.[19]

Remarks: at the first, in the present review the consideration was restricted by the electron and heavy hole carriers, and, the second, the Coulomb interaction was excluded. Often the linear approximation for the function$m*_i/m_0 = f(E,r)$ is used. We also will be applied the linear fit in the present paper.

4. Effective approach for strained InAs/GaAs quantum structures: effective potential

Here we propose the effective potential method to calculate the properties of realistic semiconductor quantum dot/ring nanostructures with the explicit consideration of quantum dot size, shape, and material composition. The method is based on the single sub-band approach with the energy dependent electron effective mass (Eq. (2)). In this approach, the confined states of carriers are formed by the band gap offset potential. Additional effective potential is introduced to simulate the cumulative band gap deformations due to strain and piezoelectric effects inside the quantum dot nanostructure. The magnitude of the effective potential is selected in such a way that it reproduces experimental data for a given nanomaterial.

We rewrite the Schrödinger equation (3) in the following form:

$$(H_{kp}(E) + V_c(\vec{r}) + V_s(\vec{r}))\Psi(\vec{r}) = E\Psi(\vec{r}). \tag{5}$$

Here $H_{kp}(E)$ is the single band kp-Hamiltonian operator $H_{kp}(E) = -\nabla\frac{\hbar^2}{2m^*(E,\vec{r})}\nabla$. As previously, $m^*(E,\vec{r})$ is the electron (or hole) effective mass, and $V_c(\vec{r})$ is the band gap potential, $V_s(\vec{r})$ is the effective potential. $V_c(\vec{r})$ is equal zero inside the QD and is equal to V_c outside the QD, where V_c is defined by the conduction band offset for the bulk. The effective potential $V_c(\vec{r})$ has an attractive character and acts inside the volume of the QD. This definition for the effective potential is schematically illustrated by Fig. 2 for the conduction band structure of InAs/GaAs QD. In the figure, the confinement potential of the simulation model with effective potential V_s is denoted as "strained".[20] The band gap potential for the conduction band (valence band) can be determin as V_c=0.594 eV (V_c=0.506 eV). The magnitude of the effective potential can be chosen to reproduce experimental data. For example, the magnitude of V_s for the conduction (valence) band chosen in[21] is 0.21 eV (0.28 eV). This value was obtained to reproduce results of the 8-th band ***kp***-calculations of[22] Schliwafor InAs/GaAs QD. To reproduce the experimental data from,[23] the V_s value of 0.31 eV was used in[20] for the conduction band.

Possibility for the substitution of the function describing the strain distribution in QD and the substrate was firstly proposed in.[24] Recent works[25,26] in which the strain effect taken into account rigorously applying the analytical method of continuum mechanics allow us to say that the approximation of the effective potential is appropriate.

In the next sub-section of the section 2 we will review the results ob-

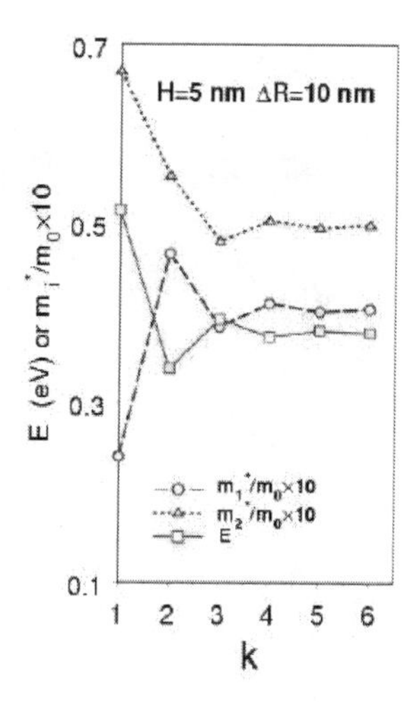

Fig. 1. Convergence of the iterative procedure (4) for the confinement energy$E$ (solid line) and electron effective mass $m*_i/m_0$ calculated for InAs/GaAs QR (dashed line) and GaAs substrate (dotted line). Here the height of QR is H, radial width is ΔR and inner radius is R_1(R_1=17 nm), V_c=0.77 eV.

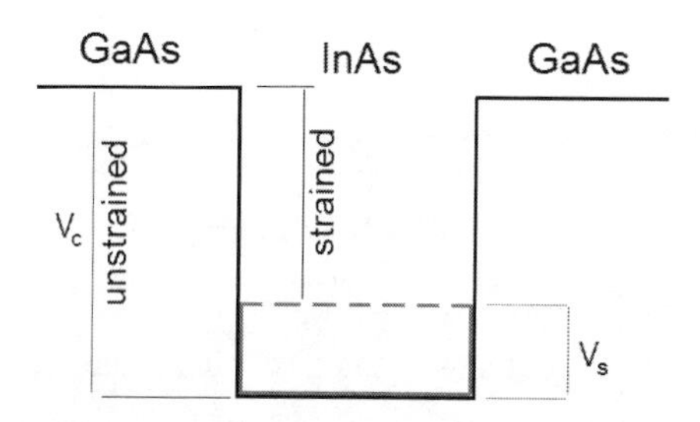

Fig. 2. Effective potential V_s and band gap structure of the conductive band of InAs/GaAs QD.

tained in both these approximations as the non-parabolic one as well as the effective potential method.

5. Electron energy in quantum rings with varieties of geometry: effect of non-parabolicity

In this section a model of the InAs/GaAs quantum ring with the energy dispersion defined by the Kane formula (2) (non-parabolic approximation) based on single sub-band approach is considered. This model leads to the confinement energy problem with three-dimensional Schrödinger equation in which electron effective mass depens on the electron energy. This problem can be solved using the iterative procedure (4). The ground state energy of confined electron was calculated of[17,18,27]where the effect of geometry on the electron confinement states of QR was studied and the non-parabolic contribution to the electron energy was estimated. The size dependence of the electron energy of QR and QD was subject of several theoretical studies.[15,28] We present here, unlike the previous papers, a general relation for the size dependence of the QR energy.

Consider semiconductor quantum ring located on the substrate. Geometrical parameters of the semi-ellipsoidal shaped QR are the height H, radial width ΔR and inner radius R_1. It is assumed that $H/\Delta R \ll 1$ which is appropriate technologically. QR cross section is schematically shown in Fig. 3. The discontinuity of conduction band edge of the QR and the substrate forms a band gap potential, which leads to the confinement of electron.

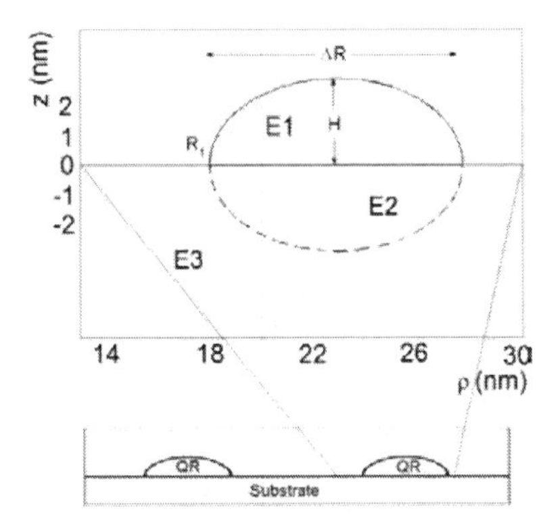

Fig. 3. Profile of cross section of quantum ring (E1) and substrate (E2 and E3). Cylindrical coordinates ρ and z shown on axis.

The band gap potential $V_c(\vec{r})$ is equal to zero inside the QR ($V_c(\vec{r})$=0) and it is equal to the confinement potential E_c outside of the QR: The spatial dependence of the electron effective mass is given as $m^*(E,\vec{r}) = m_i^*(E)$, i=1, 2, 3, where m_1^* is the effective mass in the material of QR ($\vec{r} \in$ E1), and $m_2^*(E)$, m_3^* are the effective mass of the substrate material ($\vec{r} \in$ E2 and E3). Within each of the regions E1, E2 and E3 m_i^* does not depend on the coordinates. The effective mass m_3^* is equal to a constant bulk value. The energy dependence of the electron effective mass from the E1 and E2 subdomains is defined by the formula (2). The equation (1) satisfies the asymptotical boundary conditions: $\Psi(\vec{r})|_{|\vec{r}|\to\infty} \to 0$, $\vec{r}$ ∈substrate and $\Psi(\vec{r})|_{|\vec{r}|\in S} = 0$, where S is free surface of QR. On the surface of boundaries the wave function and the first order derivative$(\vec{n}, \vec{\nabla}\Psi)/m_i^*$ are continuous with different materials (the surface normal $\vec{n}$).

The Schrödinger equation (3) was numerically solved by the finite element method and iterative procedure (4). The following typical QR/substrate structures with experimental parameters were chosen: InAs/GaAs and CdTe/CdS. The parameters of the model are given in Table 1 for the each hetero-structure.

Table 1. Parameters of the QR and substrate materials

QR/Substrate	$m*_1/m*_2$	$m*_1/m*_2$	$\frac{2m_0 P_1^2}{\hbar^2}/\frac{2m_0 P_2^2}{\hbar^2}$	Δ_1/Δ_2
InAs/GaAs	0.024/0.067	0.77	22.4/24.6	0.34/0.49
CdTe/CdS	0.11/0.20	0.66	15.8/12.0	0.80/0.07

It has to be noted that the effective mass substrate calculated for the InAs/GaAs and CdTe/CdS QRs is slightly differ from the bulk values within area E2. One can consider a simpler model when the properties of the area E2 and E3 are similar. It means that the wave function of electron does not penetrate through surface of QR (area E1) essentially. The simpler model does not change qualitative results of these calculations.

Analysis of the results of numerical calculations shows that the ground state energy of QR can be best approximated as a power function of the inverse values of the height and the radial width:

$$E \approx a(\Delta R)^{-\gamma} + bH^{-\beta}, \tag{6}$$

where the coefficients γ=3/2 and β=1 were obtained numerically by the least square method. An example of this relation is illustrated in Fig. 4 for InAs/GaAs QR. Parameters a and b are remain constant except for extremely low values of H and ΔR. Our analysis also reveals a significant numerical difference between the energy of QR electron ground states, calculated in non-parabolic and parabolic approximations. The results of the calculation with parabolic approximation are represented in the Fig. 4 by the dashed lines.

Computation of the electron confinement energy of QRs for different materials show that the non-parabolic contribution is quite significant when

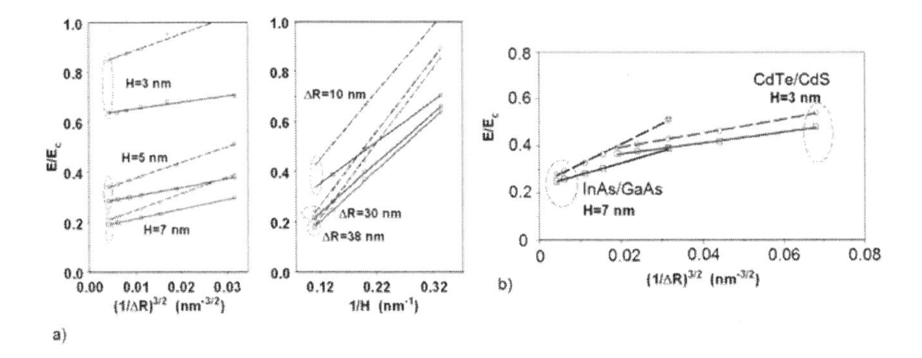

Fig. 4. a) Normalized electron ground state energy of semi-ellipsoidal shape InAs/GaAs QR with parabolic (dashed line) and non-parabolic (solid lines) approximation as function of the QR size (R_1=17 nm). b) Normalized electron confinement energy of QRs of various materials in the parabolic (dashed line) and non-parabolic (solid lines) approximation.

chosen QR geometrical parameters are close to those of the QRs produced experimentally: $H < 7$ nm, $R < 30$ nm for InAs/GaAs, $H < 5$ nm, $R < 20$ nm for CdTe/CdS. Magnitude of this effect for InAs/GaAs can be greater than 30%. According with this fact, the coefficients a and b in Eq. (6) also depend on the approximation used: a/b=3.4/1.9 for the non-parabolic and a/b=6.2/3.0 for parabolic approximation.

As can be seen from the Fig. 4b), coefficients γ and β in the relation (6) do not depend on QR/substrate materials. Their values are defined by geometry and by the boundary conditions of the applied model. The model described above corresponds to the boundary condition as "hard wall at one side" (top side of the QR). For the model without the walls when the QR embedded into the substrate one can obtain γ=1, and β=1/3. In contrast with it, the coefficients $a$and b depend on the QR/substrate material set essentially.

Concluding, we have shown that for wide QR sizes the non-parabolicity effect does considerably alter the energy of the electron states, especially when the height or width of QR is relatively small.

6. The C-V measurements and the effective model: choosing the parameters

The well-established process of QDs formation by epitaxial growth and consecutive transformation of QDs into InAs/GaAs quantum rings (QR)[29] allows the production of 3D structures with a lateral size of about 40-60 nm

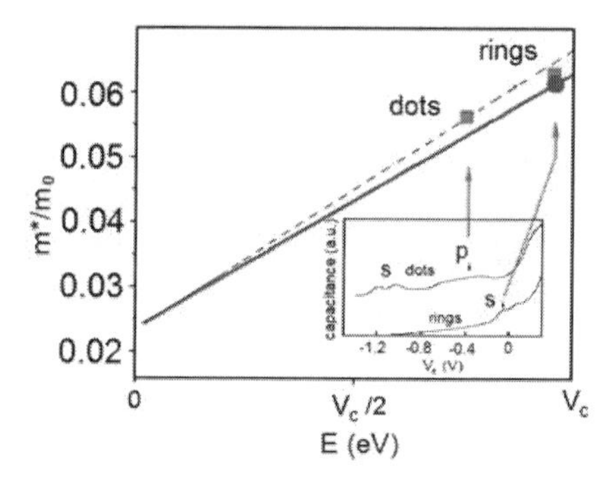

Fig. 5. Calculated (circle) and experimentally obtained by[29,30] (squares) values for the electron effective mass and the confinement energies of the electron s- and p-levels of QD and QR. The solid line is obtained by the Kane formula (2), and the dashed line connects the bulk values of the effective mass. The inset: the capacitance-gate voltage traces.[29]

and a height of 2-8 nm. In produced QDs and QRs it is possible directly to observe discrete energy spectra by applying capacitance-gate-voltage (CV) and far-infrared spectroscopy (FIR). In this section we will show how the effective model works using as an example the CV data. We use results of the CV experiment from[29–31] for QD and QR.

The effective mass of an electron in QD and QR changes from the initial bulk value to the value corresponding to the energy given by the Kane formula (2). Results of the effective model calculations for the InAs/GaAs QR are shown in Fig.5. The effective mass of an electron in the InAs QR is close to that of the bulk value for the GaAs substrate. Since the effective mass in the QD is relatively smaller, as it is clear from Fig. 5, for QD the electron confinement is stronger; the s-shell peak of the CV trace is lower relative upper edge of conduction band of GaAs. The lower s-shell peak corresponds to the tunneling single electron into the QD. The picture is a starting point for the choosing the parameters of the effective potential model. In this section we follow the paper[20] where the semi-ellipsoidal InAs/GaAs QD has been considered. The average sizes of InAs/GaAs QD reported in[29] were: H=7 nm (the height) and R=10 nm (the radius). A cross section of the quantum dot is shown in Fig. 6a). The quantum dot has rotation symmetry. Thus the cylindrical coordinate was chosen in Eq. (5) which defines the effective model. For each step of iterative procedure (4) the problem (3-2) is reduced to a solution of the linear eigenvalue problem for the Schrödinger equation.

Taking into account the axial symmetry of the quantum dot (ring) considered, this equation may be written in the cylindrical coordinates (ρ, z, ϕ) as follows:

$$(-\frac{\hbar^2}{2m^*}(\frac{\partial^2}{\partial\rho^2} + \frac{1}{\rho}\frac{\partial}{\partial\rho} - \frac{l^2}{\rho^2} + \frac{\partial^2}{\partial z^2}) + V_c(\rho, z) + V_s(\rho, z) - E)\Phi(\rho, z) = 0. \quad (7)$$

The wave function is of the form: $\Psi(r) = \Phi(\rho, z)exp(il\phi)$, where l=0, ±1, $\pm2\ldots$ is the electron orbital quantum number. For each value of the orbital quantum number l, the radial quantum numbers $n = 0, 1, 2, \ldots$are defined corresponding to the numbers of the eigenvalues of (4) which are ordered in increasing. The effective mass $m*$ must be the mass of electron for QD or for the substrate depending on the domain of the Eq. (2) is considered. The wave function $\Phi(\rho, z)$, and its first derivative in the form $\frac{\hbar^2}{2m^*}(\vec{n}, \nabla)\Phi$, have to be continuous throughout the QD/substrate interface, where $\vec{n}$ is the normal vector to the interface curve. The Neumann boundary condition $\frac{\partial}{\partial\rho}\Phi(\rho, z) = 0$ is established for $\rho = 0$ (for case of QD). The asymptotical boundary conditions is $\Phi(\rho, z) \to 0$, when $\rho \to \infty$, $|z| \to \infty$ (QD is located near the origin of z-axes). When quantum dots are in an external perpendicular magnetic field, as it will be considered below, the magnetic potential term must be added to the potentials of Eq. (7)[32] in the form $V_m(\rho) = \frac{1}{2m^*}(\beta\hbar l + \frac{\beta^2}{4}\rho^2)$, where $\beta = eB$, B is the magnetic field strength, and e is the electron charge. We consider the case of a magnetic field normal to the plane of the QD and do not take into account the spin of electron because the observed Zeeman spin-splitting is small.

The confinement potential in Eq. (7) was defined as follow: $V_c = 0.7(E_g^{\mathrm{S}} - E_g^{\mathrm{QD}})$; $V_c = 0.77eV$. The parameters of the QD and substrate materials were $m^*_{bulk,1}/m^*_{bulk,2}$=0.024/0.067, E_g^{QD}/E_g^S=0.42/1.52, $\frac{2m_0P_1^2}{\hbar^2}/\frac{2m_0P_2^2}{\hbar^2}$=20.5/24.6, Δ_1/Δ_2=0.34/0.49. The magnitude of the effective potential V_s was chosen as 0.482 eV. There are three electron confinement states: the s, p, and d, as shown in the Fig. 6b). The energy of the s single electron level measured from the top of the GaAs conduction band can be obtained from CV experimental data. To explain it, in Fig. 6c) the capacitance-gate-voltage trace from[33] is shown. The peaks correspond to the occupation of the s and p energy shells by tunneled electrons. The Coulomb interaction between electrons results to the s-shell splits into two levels and the p-shell splits into four levels if one takes into account the spin of electron and the Pauli blocking for fermions. The gate voltage-to-energy conversion coefficient f =7 ($\Delta E = e\Delta V_g/f$) was applied to recalculate the gate voltage to the electron energy. The value of the effective potential V_s was chosen in order to accurately reproduce the observed s-wave level position with respect to the bottom of GaAs conduction band. The approximate size of this energy region is 180 meV.

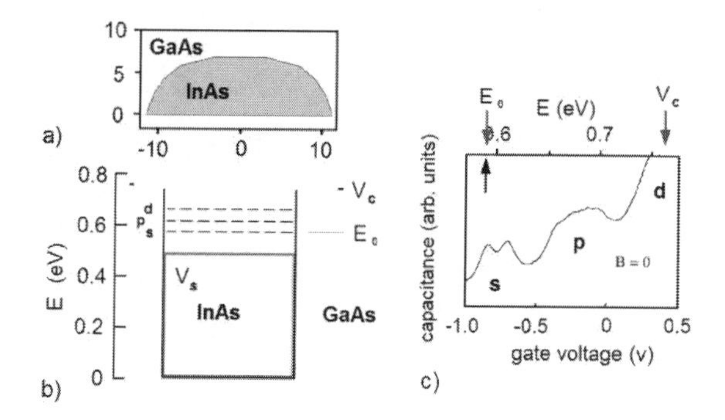

Fig. 6. a) A cross section of the quantum dot. The dimensions are given in nm. b) Localization of the s, p and d single electron levels relatively to the bottom of the GaAs conduction band. V_c is the band-gap potential, V_s is the effective potential simulating the sum of the band-gap deformation potential, the strain-induced potential and the piezoelectric potential. c) The capacitance-gate-voltage trace.[33] The peaks correspond to the occupation of the s and p energy shells by tunneled electrons. The arrows denote the s level (E_0) and the bottom of the GaAs conduction band.

The non-parabolic effect causes a change in the electron effective mass of QD with respect to the bulk value. According to the relation Eq. (2), the effective electron mass for InAs is sufficiently increased from the initial value of $0.024m_0$ to $0.054m_0$, whereas for GaAs substrate it is slightly decreased from $0.067m_0$ to $0.065m_0$ within the region of transmission of the wave function out of the quantum dot. The obtained value of the electron effective mass of InAs is close to the one ($0.057m_0 \pm 0.007$) extracted in[33] from the CV measurements of orbital Zeeman splitting of the p level.

Applying the obtained effective model, one can take into account the effect the Coulomb interaction between electrons (the Coulomb blockade). The goal is to reproduce the C-V data presented in Fig. 6 for the InAs QD. The calculations[34] have been carried out using the perturbation procedure, proposed in.[35] The Coulomb energy matrix elements were calculated by applying single electron wave functions obtained from the numerical solution of Eq. (7). Both the direct terms of E^c_{ij} and the exchange terms E^x_{ij} of the Coulomb energy between electron orbitals with angular momentum projection of $\pm i$ and $\pm j$ were calculated (notation is given in[35]). The results of calculations of the electron energies of the s, p and d levels are shown in Fig. 3($Cal.2$). The s shell Coulomb energy was found to be close to the experimental value which is about 20 meV.

Returning to the Fig. 5 we have to note that the effective potential obtained for InAs/GaAs QD has to be corrected for the case of the InAs/GaAs quantum rings. The reason is the topological, geometrical dependence of the depth of the effective potential. This dependence is weak for the considered QD and QR. The corresponding V_s potentials have the magnitude of 0.482 eV and 0.55 eV for QD and QR, respectively. Accordingly to the experimental data the electron effective mass in quantum dots and rings changes from $0.024m_0$ to $(0.057\pm0.007)m_0$[33] and $0.063m_0$,[29] respectively. The Kane's formula describes these variations well as it is shown in Fig. 5. The calculated values for the effective masses for quantum dots and rings are $0.0543m_0$ and $0.0615m_0$, respectively.[34]

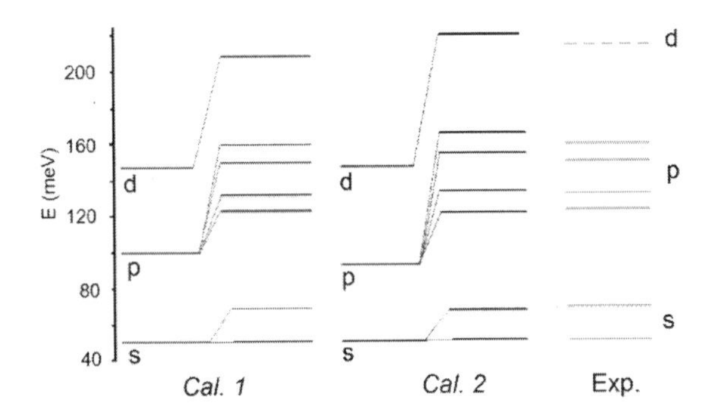

Fig. 7. Energies of the electrons occupying a few first levels of the quantum dot at zero magnetic field. The calculations *Cal.*1 are that of parabolic model.[35] Our calculations are denoted by *Cal.*2. The splitting of the single electron levels of a corresponding energy shell is presented. CV experimental data are taken from.[35]

Correct choice of the average QD profile is important for an analysis of the C-V data. It was shown in,[36] where the calculation of the energy shifts due to the Coulomb interaction between electrons tunneling into the QD was performed for comparison with the C-V experiments.

One can see in Fig. 7 that the agreement between our results and the experimental data is satisfactory. Slight disagreement can be explained by uncertainty in the QD geometry which has not been excluded by available experimental data. In[36] it was shown that small variations of the QD cross section lead to significant changes in the levels presented in Fig. 7. The variations of the QD profile we considered are shown in Figure 8a, and the results of calculations for the electron energies are presented in Fig. 8b) for

s and p – shell levels. The results of the calculations shown in Fig. 8 reveal rather high sensitivity to these variations of the QD profile. In particular, the spectral levels shift is noticeable due to a small deformation of the QD profile. Thus, we have seen that the average QD profile is important when we are comparing the result of the calculations and the experimental data.

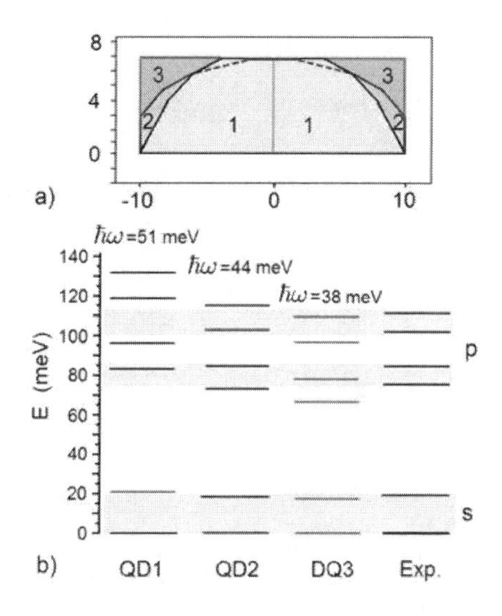

Fig. 8. a) Cross sections of the QD. The dimensions are given in nm. b) Excitation energies of the electrons occupying s and p -energy shells of the InAs/GaAs quantum dot for various QD profiles are shown in Figure 7a). CV experimental data are taken from.[35] Here $\hbar\omega$ is the excitation energy$\hbar\omega = E_{(0,0)} - E_{(0,1)}$, where $E_{(n,l)}$ is a single electron energy of the (n, l) state.

Finally, we may conclude that the effective model of QD/substrate semiconductor structure with the energy dependent effective mass and realistic 3D geometry taken into account, can quantitatively well interpret the CV spectroscopy measurements.

7. Experimental data for InAs/GaAs QR and the effective model

In this section we continue the description of the effective model use as an example the InAs/GaAs quantum ring. The geometry of the self-assembled QRs, reported in,[29] is shown in Fig. 9 (Geometry 1). The InGaAs QRs have a height of about 2 nm, an outer diameter of about 49 nm, and an inner diameter of about 20 nm. Also, three-dimensional QR geometry (Geometry

2), which follows from the oscillator model[31] is used. The confinement of this model is given by the parabolic potential: $U(r) = \frac{1}{2}m * \omega(r - r_0)^2$, where ω, r_0 are parameters.[37] The QR geometry constructed by the relation between the adopted oscillator energy and a length l as follows:[38]

$$l = \sqrt{2\hbar/m * \omega}. \tag{8}$$

Here the width d for the considered rings is defined by $d = 2l$. The obtained geometry with the parameters m^* and ω from[31] is shown in Fig. 9 (Geometry 2); m^*=0.067m_0 and ω=15 meV. The averaged radius of QR is 20 nm.

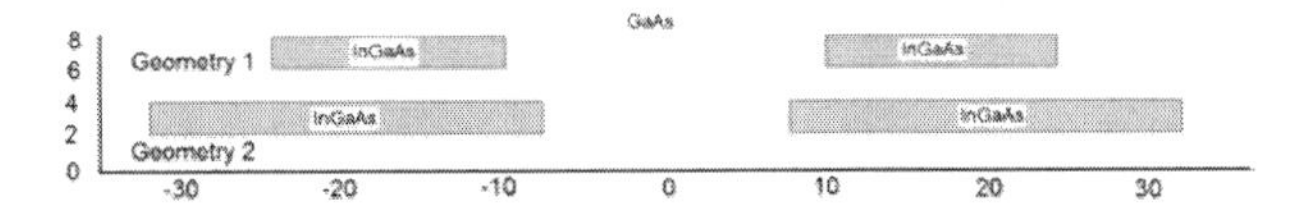

Fig. 9. QR cross section profile corresponding to Geometry 1 and Geometry 2; sizes are in nm.

Results of the effective model calculations for the ground state energy of electron in a magnetic field are shown in Fig. 10.[39] The picture of the change of the orbital quantum number of the ground state is similar to that obtained in[31] with the oscillator model. The change occurred at 2.2 T and 6.7 T. The obtained energy fits the experimental data rather well.

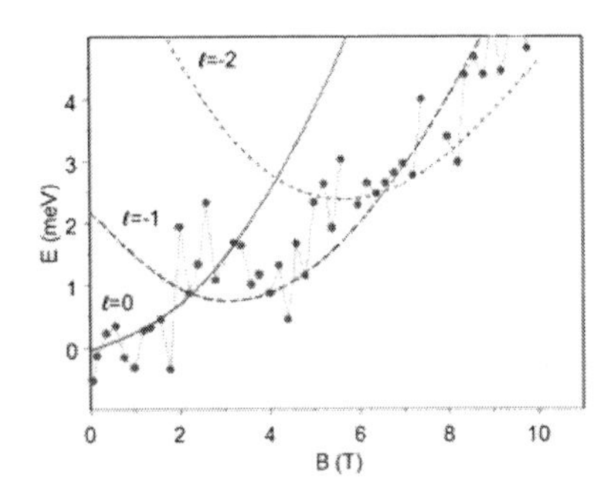

Fig. 10. Additional energy of an electron in QR in a magnetic field B. The C-V experimental energies (circles) were obtained in[31] by using a linear approximation $\Delta E = e\Delta V_g/f$, with the lever arm $f = 7.84$. The curves $l = 0, -1, -2$ are the results of our calculations multiplied by a factor of 1.18.[31]

It has to be noted here that one cannot reproduce this result using the geometry proposed in[31] (Geometry 1) for this QR. The correspondence between the confinement potential parameters of the oscillator model and the real sizes of quantum objects has to be formalized by Eq. (8). Only using the geometry followed from Eq. (8) can we reproduce result of,[31] as is shown in Fig. 10. The strength parameter of the effective potential, in the case of the Geometry 2, was chosen to be 0.382 eV, which is close to that for QD from,[36] where V_s=0.31 eV. The difference is explained by the topology dependence of the effective potentials (see section above and also[20]).

Note that the considered QRs are the plane quantum rings with the condition $H \ll D$, which enhances the role of the lateral size confinement effect. To qualitatively represent the situation shown in Fig. 12, one can used an approximation for the 3D QR based on the formalism of one dimensional ideal quantum ring. Additional electron energy, due to the magnetic field, can be calculated by the relation: $E = \hbar^2/(2m{*}R^2)(l + \Phi/\Phi_0)^2$(see for instance[30]), where$\Phi = \pi R^2 B$, $\Phi_0 = h/e$ ($\Phi_0 = 4135.7$ T nm2); R is radius of the ideal ring. The Aharonov-Bohm (AB)[40] period ΔB[41] is estimated by the relation: $\Delta B = \Phi_0/\pi/R^2$. Using the root mean square (rms) radius for R (R=20.5 nm), one can obtain $\Delta B/2$=1.56 T and $\Delta B/2 + \Delta B$=4.68 T for the ideal ring. This result is far from the result of 3D calculations shown in Fig. 12 where $\Delta B/2 \approx 2.2$T and $\Delta B/2 + \Delta B \approx$6.7 T are determined. Note here that the electron root mean square radius $R_{n,l}$ is defined by the relation: $R^2_{n,l} = \int |\Phi^N_{n,l}(\rho, z)|^2 \rho^3 d\rho dz$, where $\Phi^N_{n,l}(\rho, z)$is the normalized wave function of electron state described by the quantum numbers (n, l).

One can obtain better agreement by using the radius for the most probable localization of the electron $R_{loc.}$, defined at the maximum of the square of the wave function. The electron is mostly localized near 17.1 nm, for B=0. With this value, the ideal ring estimation leads to the values for $\Delta B/2$and $\Delta B/2+\Delta B$ as 2.25 T and 6.75 T, respectively. That agrees with the result of the 3D calculations (see Fig. 10). Obviously, the reason for this agreement is the condition $H \ll D$, for the considered QR geometry as it was mentioned above. The mostly localized position of the electron in QR depends weakly on the magnetic field. We present $R_{loc.}$ as a function of the magnetic field B in Fig. 11. $R_{loc.}(B)$ is changed in an interval of ± 1 nm around the mean value $R_{loc.}(0)$ of 17 nm. It is interesting to note that the magnetization of a single electron QR demonstrates the same behavior as it does for $R_{loc.}(B)$ if the one dimensional ring (circle) is used in[30] (see[30] for details).

Additionally we compare the results of calculations for the QR geometry

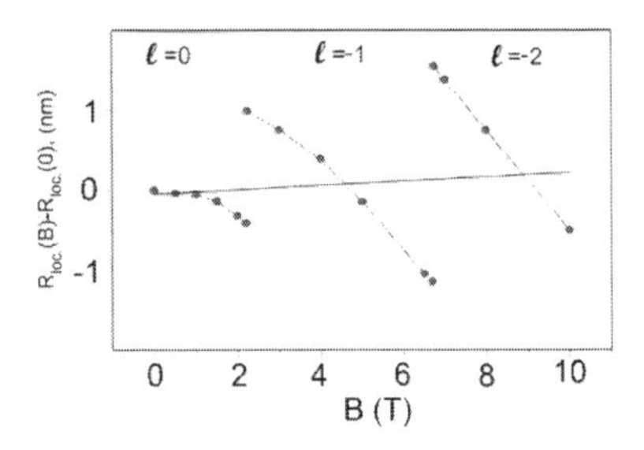

Fig. 11. The radius ($R_{loc.}$) of the most localized position of an electron as a function of a magnetic field B. The electron of the ground state is considered. The circles indicate the calculated values and the solid line indicates the result of the least squares fitting of the calculated values. The orbital quantum number of the ground state is shown.

parameters corresponding to Geometry 1 and Geometry 2 in Fig. 9 with the far-infrared (FIR) data, reported in.[35] The results are presented in Fig. 12. One can see that the QR geometry proposed in[31]] leads to a significant difference between the FIR data and the effective model calculations (see Fig. 12a), whereas the results obtained with Geometry 2 are in satisfactory agreement with the data (Fig. 12b). Again we conclude that the QR geometry of[31] does not provide an adequate description of electron properties of the InAs/GaAs QRs measured in.[29,31]

To summarize, we wish to point out that the problem of reliable theoretical interpretation of the C-V (and FIR) data for InAs/GaAs quantum rings is far from resolved. Obtained geometry can be considered as a possible version of geometry for experimentally fabricated QR.

8. Quantum Chaos in Single Quantum Dots

8.1. *Quantum Chaos*

Quantum Chaos concerns with the behavior of quantum systems whose classical counterpart displays chaos. It is quantum manifestation of chaos of classical mechanics.

The problem of quantum chaos in meso - and nano-structures has a relatively long history just since these structures entered science and technology. The importance of this problem is related to wide spectrum of the transport phenomena and it was actively studied in the last two decades.[42–44] One of the main results of these studies, based mainly on the classical and semi-classical approaches, is that these phenomena sensitively depend on the geometry of these quantum objects and, first of all, on their

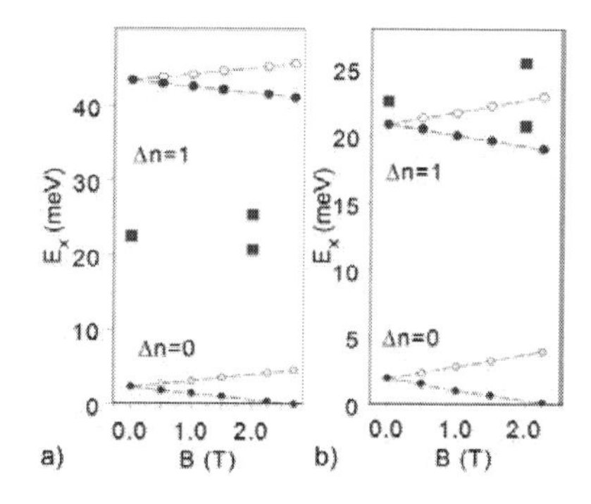

Fig. 12. Solid squares represent the observed resonance positions[30] of the FIR transmissions at various magnetic field B. Calculated energies of the excited states with $|\Delta l|=1$ are marked by the circles. a) QR with shape given by Geometry 1, b) QR with shape given by Geometry 2. The orbital quantum number of the ground state is $l = 0$. The quantum number n is changed as it shown.

symmetry: Right - Left (RL) mirror symmetry, up-down symmetry and preserving the loop orientation inversion symmetry important in the presence of the magnetic field.[45,46]

These results are well-known and discussed widely. There is another, actively studied in numerous fields of physics, aspect which ,in essence, is complimentary to the above mentioned semi classical investigations: Quantum Chaos with its inalienable quantum character, including, first of all, Nearest Neighbor level Statistics (NNS) which is one of the standard quantum-chaos test.

Mathematical basis of the Quantum Chaos is a Random Matrix Theory (RMT) developed by Wigner, Dyson, Mehta and Goudin (for comprehensive review see book[42]). RMT shows that the level repulsion of quantum systems (expressed by one of the Wigner-Dyson -like distributions of RMT) corresponds to the chaotic behavior and, contrary, level attraction described by Poisson distribution tells about the absence of chaos in the classical counterpart of the quantum system. This theorem-like statement checked by numerous studies in many fields of science. For the completeness, we add that there are other tests of Quantum Chaos based on the properties of the level statistics: Δ_3 statistics (spectral rigidity$\Delta_3(L)$), Number variance$\Sigma_2(L)$), spectral form-factor, two- and multipoint correlation functions, two level cluster function $Y_2(E)$ etc. They play an important subsidiary role to enhance and refine the conclusions emerging from the NNS.

The present review surveys the study of the NNS of nanosize quantum objects - quantum dots (QD) which demonstrate atom-like electronic structure under the regime of the size confinement. To use effectively NNS, we have to consider so called weak confinement regime where the number of levels can be of the order of several hundred. QD of various shape embedded into substrate are considered here under the effective model.[47] We use the sets of QD/substrate materials (Si/SiO_2, GaAs/Al0.7Ga0.25As, GaAs/InAs).

8.2. *The nearest neighbor spacing statistics*

For the weak confinement regime (for the Si/SiO_2 QD, the diameter $D \geq 10$ nm), when the number of confinement levels is of the order of several hundred,[47] we studied NNS statistics of the electron spectrum. The low-lying single electron levels are marked by E_i, $i = 0, 1, 2, ...N$. One can obtain the set $\Delta E_i = E_i - E_{i-1}$, $i = 1, 2, 3, ...N$ of energy differences between neighboring levels. An example of the energy spectrum and set of the neighbor spacings for Si/SiO_2 QD are in Fig. 13. We need to evaluate the distribution function$R(\Delta E)$, distribution of the differences of the neighboring levels. The function is normalized by $\int R(\Delta E) d\Delta E = 1$. For numerical calculation, a finite-difference analog of the distribution function is defined by following relation:

$$R_j = N_j / H_{\Delta E} / N, \quad j = 1, ...M,$$

where $\sum N_j = N$ represents total number of levels considered, $H_{\Delta E} = ((\Delta E)_1 - (\Delta E)_N)/M$ is the energy interval which we obtained by dividing the total region of energy differences by M bins. N_j ($j = 1, 2, ...M$) is the number of energy differences which are located in the j-th bin.

The distribution function $R(\Delta E)$is constructed using the smoothing spline method. If R_j, $j = 1, 2, ..M$ are calculated values of the distribution functions corresponding to ΔE_j, the smoothing spline is constructed by giving the minimum of the form $\sum_M^{\sum} j = 1$. The parameter $\lambda > 0$ is controlling the concurrence between fidelity to the data and roughness of the function sought for. For $\lambda \to \infty$ one obtains an interpolating spline. For $\lambda \to 0$ one has a linear least squares approximation.

We studied neighbouring level statistics of the electron/hole spectrum treated by way described above. The Si quantum dots having strong difference of electron effective mass in two directions is considered as appropriate example for the study of role of the effective mass asymmetry. In this study

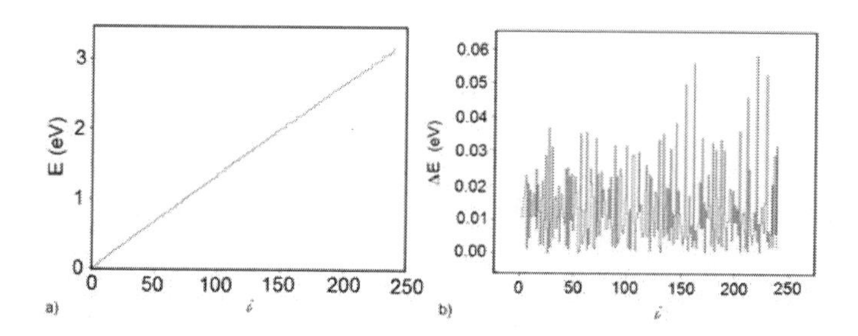

Fig. 13. a) The energy levels of the spherical Si/SiO2 QD with diameter D=17 nm. The parameters N=245, M=9 were used. b) The neighbor spacings $\Delta E_i = E_i - E_{i-1}$, $i = 1, 2, ..N$ for of the spherical Si/SiO_2 QD with diameter D=17 nm. The parameters N=245, M=9 were used.

we do not include the Coulomb potential between electrons and holes. The shape geometry role is studied for two and three dimensions.

8.3. *Violation of symmetry of the QD shape and nearest neighbor spacing statistics*

Distribution functions for the nearest neighboring levels are calculated for various QD shapes.[47] Our goal here to investigate the role of violation of the QD shape symmetries on the chaos. The two and three dimensional models are considered. It is important to note that existing of any above mentioned discrete symmetry of QD shape leads to the Poisson distribution of the electron levels.

In Fig. 14 the numerical results for the distribution functions of Si/SiO_2 QD are presented. The QD has three dimensional spherical shape. We considered the two versions of the shape. The first is fully symmetrical sphere, and the second shape is a sphere with the cavity damaged the QD shape. The cavity is represented by semispherical form; the axis of symmetry for this form does not coincide with the axis of symmetry of the QD. In the first case, the distribution function is the Poisson-like distribution. The violation symmetry in the second case leads to non-Poisson distribution.

We fit the distribution function $R(\Delta E)$ using the Brody distribution:[48]

$$R(s) = (1 + \beta)bs^{\beta} \exp(-bs^{1+\beta}), \tag{9}$$

with the parameter β=1.0 and $b = (\Gamma[(2 + \beta)/(1 + \beta)]/D)^{1+\beta}$, D is the average level spacing. Note that for the Poisson distribution the Brody parameter is zero.

If the QD shape represents a figure of rotation (cylindrical, ellipsoidal and others) then the 3D Schrödinger equation is separable. In cylindrical coordinates the wave function is written by the following form$\psi(\vec{r}) = \Phi(\rho, z)\exp(il\varphi)$, where $l = 0, \pm 1, \pm 3, ...$is the electron orbital quantum number. The function $\Phi(\rho, z)$ is a solution of the two dimensional equation for cylindrical coordinates ρ and z.

Our results for the distribution function for the ellipsoidal shaped Si/SiO_2 QD are presented in Fig. 15a) (left). In the inset we show the cross section of the QD. The fitting of the calculated values for $R(\Delta E)$ gives the Poisson-like distribution. For the case of QD shape with the break of the ellipsoidal symmetry (Fig. 15b) (left)) by the cut below the major axis we obtained a non-Poisson distribution.

Fig. 15 (right) It is shown that slightly deformed rhombus-like shape leads to the NNS with Brody parameter β=1 (9). It is obvious why systems with different discrete symmetries reveal Poisson statistics: the different levels of the mixed symmetry classes of the spectrum of the quantum system are uncorrelated.

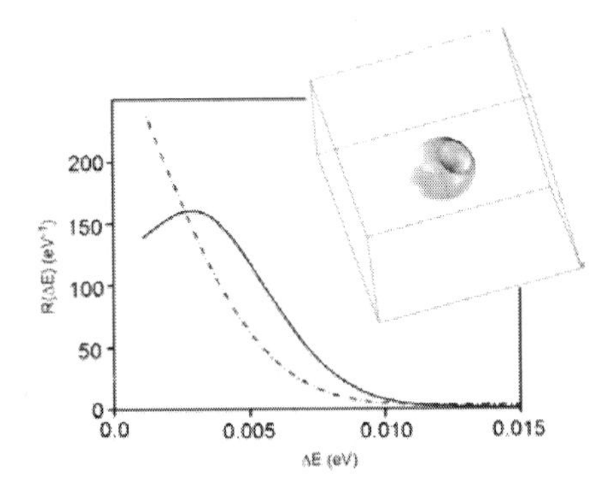

Fig. 14. Distribution functions for electron neighboring levels in Si/SiO2 QD for spherical-like shape with cut. The Brody parameter β=1.0. The geometry of this QD is shown in 3D. The QD diameter is 17 nm (see the inset).

In Schrödinger equation (7) in the asymptotical region of large ρ one can neglect the two terms $\frac{1}{\rho}\frac{\partial}{\partial\rho}$ and $-\frac{l^2}{\rho^2}$of this equation. The solution of Eq. (7) can demonstrate the same properties of the solution of the Schrödinger equation for 2D planar problem in Cartesian coordinates with the same geometry of QD shape in the asymptotical region. We illustrate this fact by Fig. 16. In this figure the violation of the shape Up-Down symmetry for 2D Si/SiO2 QD is clarified. We compare the distribution functions for

QD with "regular" semi-ellipsoidal shape (dashed curve in Fig. 16a) and for QD with the semi-ellipsoidal shape having the cut (solid curve) as it are shown in Fig. 16b). In the first case there is Up-Down symmetry of the QD shape. Corresponding distribution functions has the Poissonian type. In second case the symmetry is broken by cut. The level statistics become to non Poissonian. We have qualitative the same situation as for QD having rotation symmetry in 3D, presented in Fig. 15 (left) for the QD shape with rotation symmetry in cylindrical coordinates. The relation between the symmetry of QD shape and NNS is presented visually by Fig. 17 where we show the results of calculation of NNS for the 2D InAs/GaAs quantum well (QW). The two types of the statistics are presented in Fig. 16(left). The Poisonian distribution corresponds to shapes shown in Fig. 17 (b)-(d)(left) with different type of symmetry. The non-Poissonian distribution has been obtained for the QW shape with cut (a) which violates symmetry of initial shape (b), which is square having left-right symmetry, up-down symmetry, and diagonal reflection symmetry. The shape of the Fig. 17c) has only diagonal reflection symmetry. In Fig. 17d) the left-right symmetry of the shape exists only. The electron wave function of the high excited state, which contour plot is shown with the shape contour in Fig. 17(left), reflects the symmetry properties of the shapes.

9. Double Quantum Dots and Rings: new features

9.1. *Disappearance of Quantum Chaos in Coupled Chaotic Quantum Dots*

In the previous section, we investigated the NNS for various shape of the single quantum dots (SQD) in the regime of the weak confinement when the number of the levels allows to use quite sufficient statistics. Referring for details to,[47] we briefly sum up the main conclusions of previous section: SQDs with at least one mirror (or rotation) symmetry have a Poisson type NNS whereas a violation of this symmetry leads to the Quantum Chaos type NNS.

In this section we study quantum chaotic properties of the double QD (DQD). By QD here we mean the three dimensionally (3D) confined quantum object, as well its 2D analogue - quantum well (QW). In three dimensional case we use an assumption of the rotational symmetry of QD shape. The presented effective approach is in good agreement with the experimental data and previous calculations in the strong confinement regime.[47] Here, in the regime of weak confinement, as in,[47] we also do not consider

Coulomb interaction between electron and hole: Coulomb effects are weak when the barrier between dots is thin leading to the strong interdot tunneling and dot sizes are large enough. In these circumstances, studied in detail in[49] (see also for short review a monograph citebimberg, one may justify disregard of the Coulomb effects. The physical effect, we are looking for, has place just for thin barriers; to have sufficient level statistics, we need large enough QDs ($\geq$100 nm for InAs/GaAs QW).

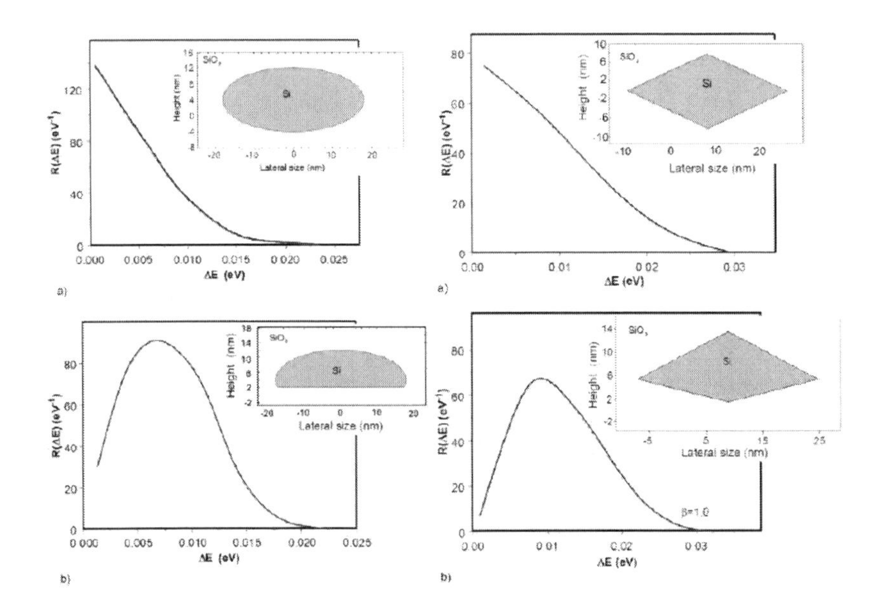

Fig. 15. (Left) Distribution functions for electron neighboring levels in Si/SiO_2 QD for different shapes: a) ellipsoidal shape, b) ellipsoidal like shape with cut. Brody parameter β is defined to be equal 1.02 for the fitting of this distribution. The 3D QD shape has rotation symmetry. Cross section of the shapes is shown in the inset. (Right). Violation of the shape Up-Down symmetry for Si/SiO_2 QD. Distribution functions for electron neighboring levels in Si/SiO_2 QD for different shapes: a) with rhombus cross section, b) with slightly deformed rhombus cross section. The 3D QD shape has rotation symmetry. The Brody parameter β for the curve fitting this distribution is shown. Cross section of the shapes is shown in the inset.

Thus, we consider tunnel coupled two QDs with substrate between, which serves as barrier with electronic properties distinct from QD. Boundary conditions for the single electron Schrödinger equation are standard. We take into account the mass asymmetry inside as well outside of QDs.[47] To avoid the complications connected with spin-orbit coupling, *s*-levels of elec-

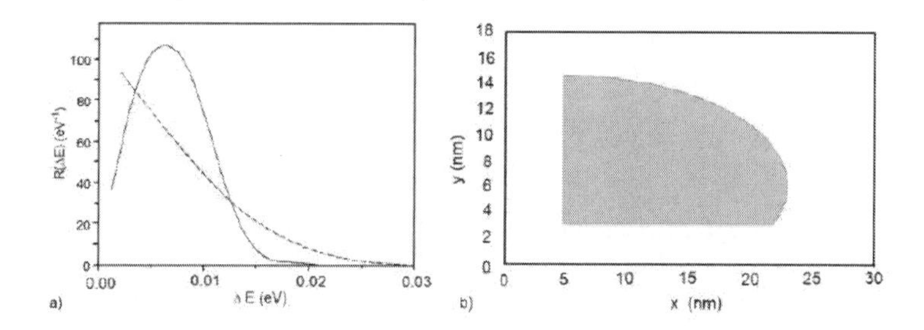

Fig. 16. Violation of the shape Up-Down symmetry for two dimensional Si/SiO_2 QD. a) Distribution functions for electron neighboring levels for the "regular" semi-ellipsoidal shape (dashed curve), for the semi-ellipsoidal shape with the cut (solid curve). b) The shape of the QD with cut (in Cartesian coordinates).

tron are only considered in the following. We would like to remind that the selection of levels with the same quantum numbers is requisite for study of NNS and other types of level statistics.

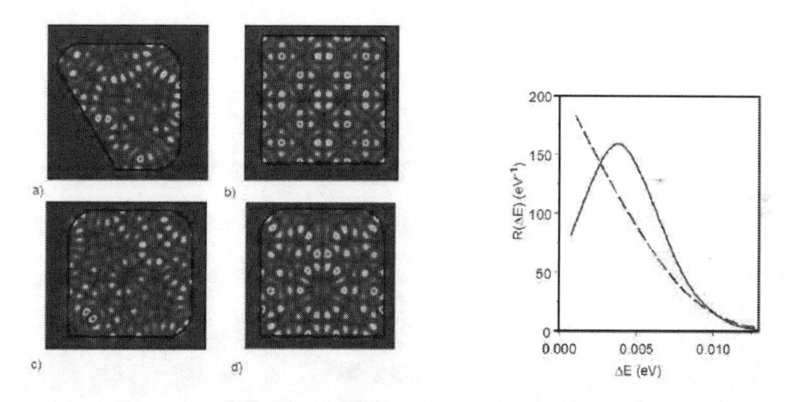

Fig. 17. Shape of the 2D InAs/GaAs quantum dots (Left). The black curves mean the perimeters. The electron wave function contour plots of the excited state (with energy about 0.5 eV are shown). The corresponding types of the level statistics are shown (Right). The shape a) leads to non-Poissonian statistics (solid curve). The shapes b)-d) result to the Poissonian statistics (dashed curve).

Whereas at the large distances between dots each dot is independent and electron levels are twofold degenerate, expressing the fact that electron can be found either in one or in the other isolated dot, at the smaller inter-dot distances the single electron wave function begins to delocalize and extends to the whole DQD system. Each twofold degenerated level of the SQD splits by two, difference of energies is determined by the overlap, shift and transfer integrals.[50] Actually, due to the electron spin, there is fourfold degeneracy, however that does not change our results and below we consider electron as spinless. Note that the distance of removing degeneracy is different for different electron levels. This distance is larger for levels with higher energy measured relative to the bottom quantum well (see Fig. 21 below). By the proper choice of materials of dots and substrate one can amplify the "penetration" effects of the wave function.

Below we display some of our results for semiconductor DQDs. The band gap models are given in.[47] Fig. 18 shows distribution function for two Si/SiO_2 QDs of the shape of the 3D ellipsoids with a cut below the major axis. Isolated QD of this shape, as we saw in the previous section, is strongly chaotic. It means that distribution function of this QD can be well fitted by Brody formula with the parameter which is close to unity.[47] We see that the corresponding up-down mirror symmetric DQD shows Poisson-like NNS. Note that these statistics data involved 300 confined electron levels, which filled the quantum well from bottom to upper edges. We considered the electron levels with the orbital momentum $l = 0$,as was mentioned above. The orbital momentum of electron can be defined due to rotational symmetry of the QD shape.

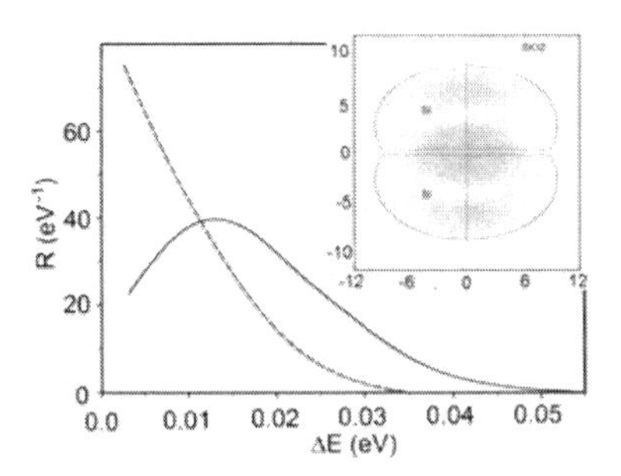

Fig. 18. The electron wave function of the ground state is shown by the contour plot. (The lower figure) Distribution functions for energy differences of the electron neighboring levels in Si/SiO_2 single QD (solid line) and DQD (dashed line). The coefficient of the spline smoothing is equal 6. The cross section of DQD shape is shown in inset (sizes are given in nm).

In Fig. 19, SQD (2D quantum well) without both type of symmetry reveals level repulsion, two tunnel coupled dots show the level attraction.

From the mirror symmetry point of view, the chaotic character of such single object is due to the lack of the R-L and up-down mirror symmetries. The symmetry requirements in this case, for the coupled dots are less restrictive: presence of one of the mirror symmetry types is sufficient for the absence of quantum chaos.

Dependence NNS on the interdot distance shows a gradual transition to the regular behaviour with intermediate situation when Poisson-like behavior coexists with chaotic one: they combine but the level attraction is not precisely Poisson-like. Further decreasing distance restores usual Poisson character (see Fig. 18). Fig. 19 shows how the degeneracy gradually disappears with the distance b between QDs in InAs/GaAs DQD.

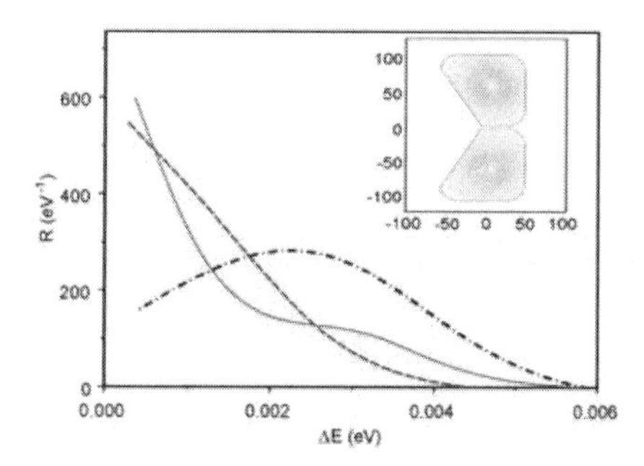

Fig. 19. Distribution functions for energy differences of the electron neighboring levels in the 2D InAs/GaAs DQW calculated for various distances b between QWs. Dashed (solid) line corresponds to b=4 nm (b=2 nm). Distribution functions of single QW is also shown by the dot-dashed line. The DQW shape is shown in inset (sizes are in nm).

Finally, we would like to show the disappearance of the Quantum Chaos when chaotic QW is involved in the "butterfly double dot"[46] giving huge conductance peak in the semi-classical approach. Fig. 20 shows the NNS for chaotic single QW of[46] by dashed line. Mirror (up-down and L-R) symmetry is violated. The NNS for an L-R mirror symmetric DQW is displayed by solid line in Fig. 20. It is clear that Quantum Chaos disappears.

We conjecture that the above mentioned peak in conductance of[46] and observed here a disappearance of Quantum Chaos in the same array are the expression of the two faces of the Quantum Mechanics with its semi-classics and genuine quantum problem of the energy levels of the confined objects, despite the different scales (what seems quite natural) in these two phenomena (several micrometers and 10–100 nm, wide barrier in the first case and narrow one in the second). We have to emphasize here that the transport

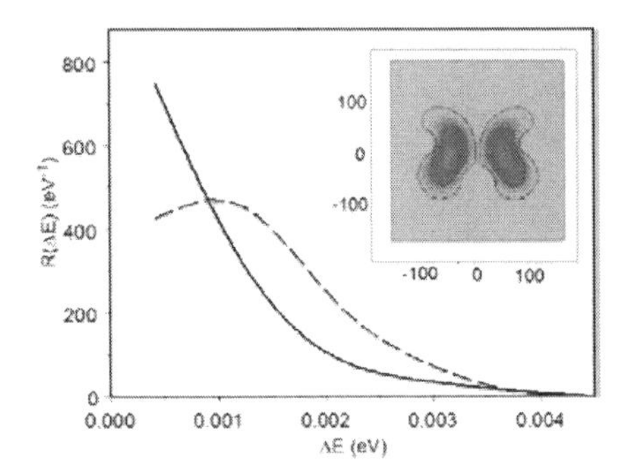

Fig. 20. Distribution functions for electron neighboring levels in InAs/GaAs single QW (dashed line) and DQW (solid line). Shape of DQW is shown in the inset. The electron wave function of the ground state is shown by the contour plot in the inset. Data of the statistics include 200 first electron levels.

properties are mainly the problem of the wave function whereas the NNS is mainly the problem of eigenvalues. Similar phenomena are expected for the several properly arranged coupled multiple QDs and QD superlattices. In the last case, having in mind, for simplicity, a linear array, arranging the tunnel coupling between QDs strong enough, we will have wide mini-bands containing sufficient amount of energy levels and the gap between successive mini-bands will be narrow. Since the levels in the different mini-bands are uncorrelated, the overall NNS will be Poissonian independently of the chaotic properties of single QD. We would like to remark also that our results have place for 3D as well as for 2D quantum objects. It is important to notice that the effect of reduction of the chaos in a system of DQD could appear for interdot distances larger than considered, for instance see Fig. 18, if an external electrical field is applied. By properly designed bias the electric field will amplify wave function "penetration" effectively reducing a barrier between QDs.

Thus, we have shown that the tunnel coupled chaotic QDs in the mirror symmetric arrangement have no quantum chaotic properties, NNS shows energy level attraction as should to be for regular, non-chaotic systems. These results are confronted with the huge conductance peak found by the semi-classical method in.[46] We think that our results have more general applicability for other confined quantum objects, not only for the quantum nanostructures, and may be technologically interesting. Concerning the last issue, problem is what easer: try to achieve regular, symmetric shape of SQDs, or, not paying attention to their irregular, chaotic shape arranges more or less symmetric mutual location.[51]

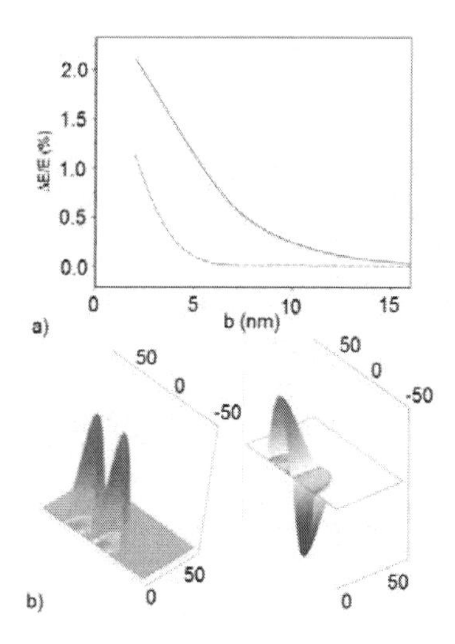

Fig. 21. (The upper figure) Doublet splitting ΔE of single electron levels dependence on the distance $b$between QDs in InAs/GaAs DQD. The ground state (E=0.23 eV) level splitting is expressed by dashed line. The solid line corresponds to doublet splitting of a level which is close to upper edge of the quantum well (E=0.56 eV). The shape of DQD is the same as in Fig. 18 (The lower figure). The electron wave functions of the doublet state: the ground state (left) and first excited state (right), are shown.

9.2. *Electron transfer between rings of Double Concentric Quantum Ring in magnetic field*

Quantum rings are remarkable meso- and nanostructures due to their non-simply connected topology and attracted much attention last decade. This interest supported essentially by the progress in the fabrication of the structures with wide range of geometries including single and double rings. This interest rose tremendously in the connection to the problem of the persistent current in mesoscopic rings.[52] Transition from meso - to nano -scale makes more favorable the coherence conditions and permits to reduce the problem to the few or even to single electron.

Application of the transverse magnetic field B leads to the novel effects: Whereas the quantum dots (QDs) of the corresponding shape (circular for two dimensional (2D), cylindrical or spherical for 3D) has degeneracy in the radial n and orbital l quantum numbers, QR due to the double connectedness in the absence of the magnetic field B has degeneracy only in l, and at the nonzero B lifts the degeneracy in l, thus making possible the energy level crossing at some value of B, potentially providing the single electron transition from one state to the another.

Use the configurations with double concentric QR (DCQR) reveals a new pattern: one can observe the transition between different rings in the analogy with atomic phenomena. For the DCQR, the 3D treatment is especially important when one includes the inter-ring coupling due to the tunneling. The dependence on the geometries of the rings (size, shape and etc.) becomes essential.

We investigate the electron wave function localization in double concentric quantum rings (DCQRs) when a perpendicular magnetic field is applied. In weakly coupled DCQRs can be arisen the situation, when the single electron energy levels associated with different rings may be crossed. To avoid degeneracy, the anti-crossing of these levels has a place. In this DCQR the electron spatial transition between the rings occurs due to the electron level anti-crossing. The anti-crossing of the levels with different radial quantum numbers (and equal orbital quantum numbers) provides the conditions when the electron tunneling between rings becomes possible. To study electronic structure of the semiconductor DCQR, the single sub-band effective mass approach with energy dependence was used (see section 2). Realistic 3D geometry relevant to the experimental DCQR fabrication was employed taken from.[53,54] The GaAs QRs and DQRs rings, embedded into the Al0.3Ga0.7As substrate, are considered.[55] The strain effect between the QR and the substrate materials was ignored here because the lattice mismatch between the rings and the substrate is small. Due to the non-parabolic effect taken into account by energy dependence effective mass of electron in QR, the effective mass of the electron ground state is calculated to be the value of $0.074m_0$ that is larger than the bulk value of $0.067m_0$. For the excited states, the effective mass will increase to the bulk value of the Al0.3Ga0.7As substrate. Details of this calculation one can find in.[55]

Electron transfer in the DCQR considered is induced by external factor like a magnetic or electric fields. Probability for this transfer strongly depends on the geometry of DCQR. The geometry has to allow the existing the weakly coupled electron states. To explain it, we note that DCQR can be described as a system of double quantum well. It means that there is duplication of two sub-bands of energy spectrum (see[9] for instance) relative the one for single quantum object. In the case of non-interacting wells (no electron tunneling between wells) the each sub-band is related with left or right quantum well. The wave function of the electron is localized in the left or right quantum well. When the tunneling is possible (strong coupling state of the system), the wave function is spread out over whole volume of the system. In a magnetic field, it is allowed an intermediate situation (weak coupled states) when the tunneling is possible due to anti-crossing of the levels. Anti-crossing, of course, is consequence of the impossibility to cross of levels with the same space symmetry.[56,57]

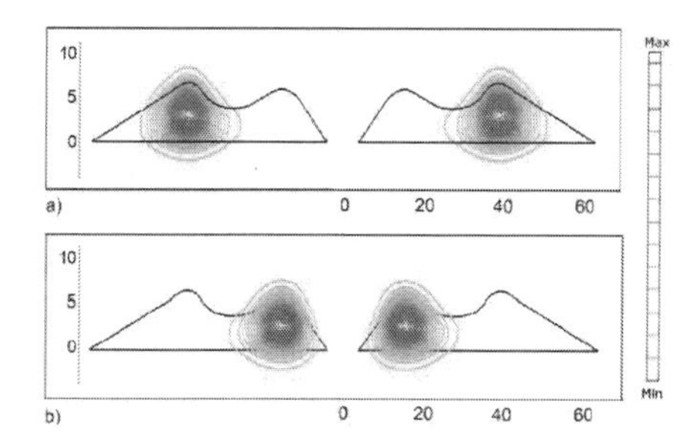

Fig. 22. The squares of wave functions for the a)$(1,0)$,outer ($E = 0.072$ eV) and b)$(2,0)$,inner ($E = 0.080$ eV) states are shown by contour plots. The contour of the DCQR cross-section is given. The sizes are in nm.

There is a problem of notation for states for DCQR. If we consider single QR (SQR) then for each value of the orbital quantum number $|l| = 0, 1, 2...$ in Eq. (2) we can definite radial quantum number n=1,2,3,... corresponding to the numbers of the eigenvalues of the problem (7) in order of increasing. One can organize the spectrum by sub-bands defined by different n. When we consider the weakly coupled DCQR, in contrast of SQR, the number of these sub-bands is doubled due to the splitting the spectrum of double quantum object.[50] Electron in the weakly coupled DCQR can be localized in the inner or outer ring. In principle, in this two ring problem one should introduce a pair of separate sets of quantum numbers (n_i, l) where index i=1,2 denoted the rings where electron is localized. However, it is more convenient, due to the symmetry of the problem, to have one pair (n, l) numbers ascribed to both rings (inner or outer), in other words, we use a set of quantum numbers $(n, l), p$ where p is dichotomic parameter attributed to the electron localization ("inner" or "outer").

Since we are interested here in the electron transition between rings and, as we will see below, this transition can occur due to the electron levels anti-crossing followed a tunneling, we concentrate on the changing of the quantum numbers n. The orbital quantum numbers must be equal providing the anti-crossing of the levels with the same symmetry (see[57]). Thus, the anti-crossing is accompanied by changing the quantum numbers n and p of the $(n, l), p$ set.

Strongly localized states exist in the DCQR with the geometry motivated by the fabricated DCQR in.[53,54] The wave functions of the two s-states of the single electron with n=1,2 are shown in Fig. 22, where the electron state n=1 is localized in outer ring, and the electron state n=2 is localized in inner ring. Moreover all states of the sub-bands with n=1,2, and $|l|$=1,2,3... are well localized in the DCQR. The electron localization is outer ring for n=1, $|l|$=0,1,2,..., and inner ring for n=2, $|l|$=0,1,2....

The difference between spectra of the two sub-bands can be explained by competition of two terms of the Hamiltonian of Eq. (7) and geometry factor. The first term includes first derivative of wave function over ρ in kinetic energy; the second is the centrifugal term. For $|l| \neq 0$ the centrifugal force pushes the electron into outer ring. One can see that the density of the levels is higher in the outer ring. Obviously, the geometry plays a role also. In particular, one can regulate density of levels of the rings by changing a ratio of the lateral sizes of the rings.

Summarizing, one can say that for B=0 the well separated states are only the states $(1,l),p$ and $(2,l),p$. Thus, used notation is proper only for these states. The wave functions of the rest states $(n > 2, l)$ are distributed between inner and outer rings. These states are strongly coupled states.

Crossing of electron levels in the magnetic field $B$are presented in Fig. 23 There are crossings of the levels without electron transfer between the rings. This situation is like when we have crossing levels of two independent rings. There are two crossings when the orbital quantum number of the lower state is changed due to the Aharonov-Bohm effect. It occurs at about 0.42 T and 2.5 T. There are two anti-crossings: the first is at 4.8 T, another is at 5.2 T. These anti-crossings are for the states with different n; the first are states (1,0) and (2,0) and the second are states (1,-1) and (2,-1). In these anti-crossings the possibility for electron tunneling between rings are realized. In Fig. 24 we show how the root mean square (rms) of the electron radius is changed due to the tunneling at anti-crossing. One can see from Fig. 23 that the electron transition between rings is only possible when the anti-crossed levels have different radial quantum numbers and equal orbital quantum numbers, in accordance of see.[56]

Transformation of the profile of the electron wave function during the process of anti-crossing with increasing B is given in Fig. 25. The electron state (1,-1), outer is considered as "initial" state of an electron (B=0). The electron is localized in outer ring. Rms radius is calculated to be R=39.6 nm. For B=5.2 T the second state is the tunneling state corresponding to the anti-crossing with the state (0,-1). The wave function is spreaded out in

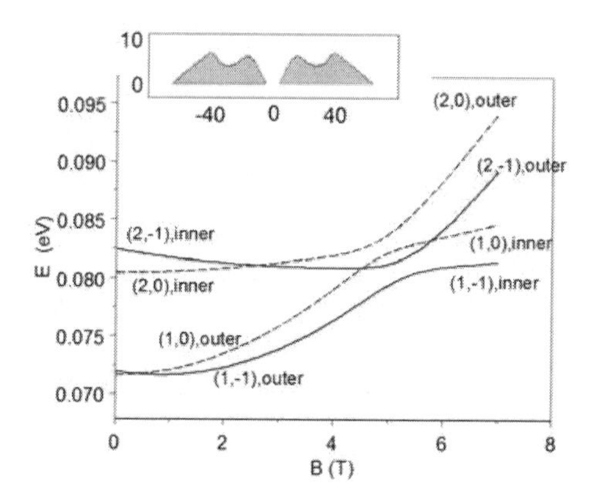

Fig. 23. Single electron energies of DCQR as a function of magnetic field magnitude$B$. Notation for the curves: the double dashed (solid) lines mean states with l=0 (l=-1) with n=1,2. The quantum numbers of the states and positions of the electron in DCQR are shown. The cross section of the DCQR is given in the inset.

both rings with R=32.7 nm. The parameter p has no definite value for this state. The "final" state is considered at B=7 T. In this state the electron was localized in inner ring with R=17.6 nm. Consequently connecting these three states of the electron, we come to an electron trapping, when the electron of outer ring ("initial" state) is transferred to the inner ring ("final" state). The transfer process is governed by the magnetic field.

Note that the energy gap between anti-crossed levels which one can see in Fig. 26 can be explained by the general theory for double interacting quantum well [50]. The value of the gap depends on separation distance between the rings, governed by the overlapping wave functions corresponding to the each ring, and their spatial spread which mainly depends on radial quantum number of the states.[55]

Other interesting quantum system is one representing QR with QD located in center of QR. The cross section of such heterostructure (GaAs/Al0.3Ga0.7As) is shown in Fig. 26a. In Fig. 26b we present the results of calculations for electron energies of the (1,0) and (3,0) states in the magnetic field B.[55] Once more we can the level anti-crossing (for about of 12.5 T). This anti-crossing is accompanied by exchange of electron localization between the QD and the QR. It means that if initial state (for $B < 12.5$ T) of electron was the state (1,0),outer, then the "final" state (for $B > 12.5$ T) will be (1,0),inner. It can be considered as one of possibilities for trapping of electron in QD.

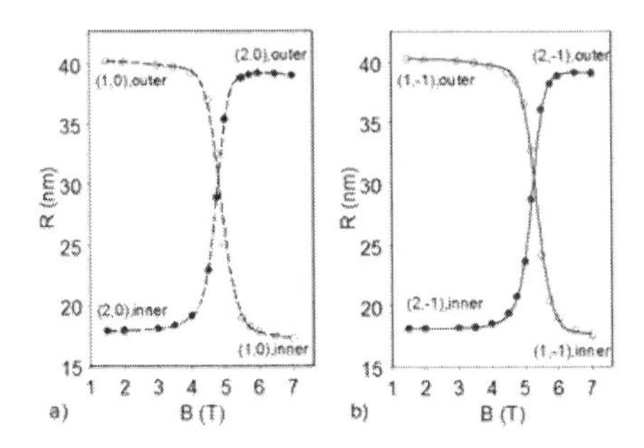

Fig. 24. Rms radius of an electron in DCQR as a function of magnetic field for the states a) $((n=1,2), l=0)$ and b) $((n=1,2), l=-1)$ near point of the anti-crossing. The calculated values are shown by solid and open circles. The dashed (solid) line, associated with states of $l=0$ ($l=-1$), fits the calculated points.

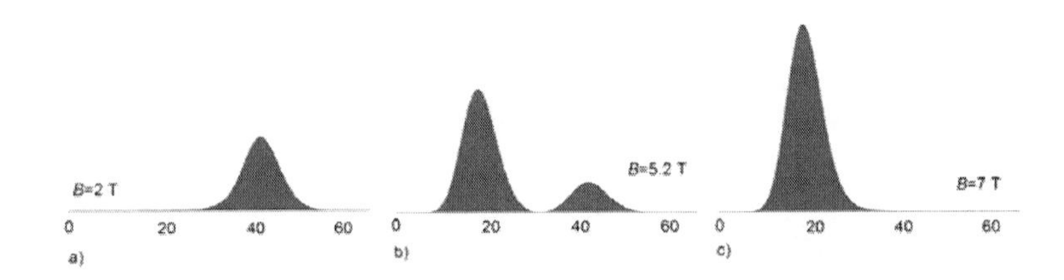

Fig. 25. Profiles of the normalized square wave function of electron in the states a) $(1,-1)$,outer; b) $(1,-1)$,n/a and c) $(1,-1)$,inner for different magnetic field B. The a) is the "initial" state (B=0) with R=39.6 nm, the b) is the state of electron transfer (B=5.2 T) with R=32.7 nm, the c) is the "final" state (B=7 T) with R=17.6 nm. The radial coordinate ρ is given in nm (see Fig. 22 for the DCQR cross section).

One can see from Fig. 26b that the energy of the dot-localized state grows more slowly than the envelope ring-localized state. At the enough large B the dot-localized state becomes the ground state.[58] In other words, when the Landau orbit of electron becomes smaller then dot size, electron can enter the dot without an extra increase of kinetic energy.

Concluding, we made visible main properties of this weakly coupled DCQD established by several level anti-crossings that occurred for the states with different radial quantum number n (n=1,2) and equal orbital quantum number l. One may conclude that the fate of the single electron in DCQRs is governed by the structure of the energy levels with their crossing and anti-crossing and changing with magnetic field. The above described behavior is the result of the nontrivial excitation characteristic of the DC-QRs. Effect of the trapping of electron in inner QR (or QD) of DCQR may be interesting from the point of view of quantum computing.

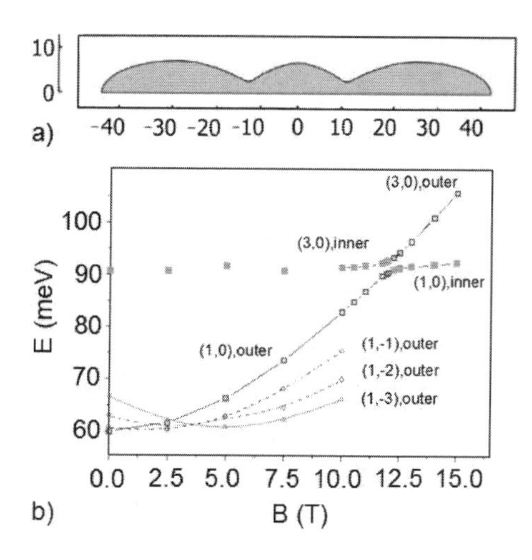

Fig. 26. a) Cross section of the QR with QD system. Sizes are given in nm. b) Energies of the (1,0) and (3,0) states in the magnetic field B for the QR with QD system. The open symbols show that the electron is localized in the ring. The solid squares show that the electron localized in QD.

References

1. F. Bloch, *Zeitschrift fur Physik,* **52**, 555 (1928).
2. Zh. I. Alferov, *Semiconductors* **32**, 1 (1998).
3. L.V. Keldysh, *Soviet Physics Solid State* **4**, 2265 (1962).
4. D. Bimberg, M. Grundmann,; N.N. Ledentsov, Quantum Dot Heterostructure, John Wiley and Sons Inc. (1999), ISBN 0-471-97388-2, Chichester, England
5. R. Landauer, *Philosophical Magazine,* **21**, 863 (1970).
6. P.W. Anderson; D.J. Thouless; E. Abrahams, D.S. Fisher, *Phys. Rev.* **B22**, 3519 (1980).
7. Y.Gefen; Y. Imry, M.Ya. Azbel, *Phys. Rev. Let.* **52**, 129 (1984).
8. P. Harrison, Quantum Wells, Wires And Dots: Theoretical And Computational Physics of Semiconductor Nanostructures, Wiley-Interscience, (2005), pp. 1-497.
9. O. Manasreh, Semiconductor Heterojunctions and Nanostructures (Nanoscience & Technology), McGraw-Hill, (2005), p. 554, ISBN 0-07-145228-1, New York.
10. Yu, P. & Cardona, M. (2005). Fundamentals of Semiconductors: Physics and Materials Properties (3rd ed.). Springer. Section 2.6, (2005), pp. 68, ISBN 3-540-25470-6.

11. D.J. BenDaniel, C.B. Duke, *Phys. Rev.* **152**, 683 (1966).
12. C. Wetzel; R. Winkler, M. Drechsler, B. K. Meyer, U. Rössler, J. Scriba, J. P. Kotthaus, V. Härle, F. Scholz, *Phys. Rev.* **B53**, 1038 (1996).
13. H. Fu, L.-W. Wang, A. Zunger, (1998). *Phys. Rev.* **B57**, 9971 (1998)
14. E.O. Kane, *Journal of Physics and Chemistry of Solids* **1**, 249 (1957).
15. Y. Li, O. Voskoboynikov, C.P. Lee. Computer simulation of electron energy states for three-dimensional InAs/GaAs semiconductor quantum ring, Proceedings of the International Conference on Modeling and Simulating of Microsystems (San Juan, Puerto Rico, USA), p. 543, Computational Publication, (2002), Cambridge
16. H. Voss, Electron energy level calculation for a three dimensional quantum dot, Advances in Computational Methods in Sciences and Engineering 2005, selected papers from the International Conference of Computational Methods in Sciences and Engineering 2005 (ICCMSE 2005) pp. 586 - 589, Leiden, The Netherlands; Editors: Theodore Simos and George Maroulis.
17. I.N. Filikhin, E. Deyneka, B. Vlahovic, *Modelling and Simulation in Materials Science and Engineering*, **12**, 1121 (2004).
18. I. Filikhin, E. Deyneka, H. Melikyan, B. Vlahovic, *Journal Molecular Simulation*, **31** 779 (2005).
19. M.M. Betcke, H. Voss, Analysis and efficient solution of stationary Schrödinger equation governing electronic states of quantum dots and rings in magnetic field, https://www.mat.tu-harburg.de/ins/forschung/rep/rep143.pdf; to appear in CiCT (2011).
20. I. Filikhin, V.M. Suslov, B. Vlahovic, *Phys. Rev.* **B73**, 205332 (2006).
21. I. Filikhin, V.M. Suslov, B. Vlahovic, *Physica E: Low-dimensional Systems and Nanostructures*, **41**, 1358 (2009).
22. A. Schliwa, M. Winkelnkemper, D. Bimberg, *Phys. Rev.* **B76**, 205324 (2007).
23. J.M. Garsia, G. Medeiros-Riberiro, K. Schmidt, T. Ngo, J.L. Feng, A. Lorke, J.P. Kotthaus, P.M. Petroff, *Appl. Phys. Lett.* **71**, 2014 (1997).
24. M. Califano, P. Harrison, *Phys. Rev.* **B61**, 10959 (2000).
25. Q. Zhao, T. Mei, *Journal of Applied Physics*, **109**, 063101 (2011).
26. Li Bin, F. M. Peeters *Phys. Rev.* **B83**, 115448-13 (2011).
27. I. Filikhin E. Deyneka, B. Vlahovic, *Molecular Simulation* **33**, 589 (2007).
28. S.S. Li, J.B. Xia, (2001). *Journal Applied Physics*, **89**, 3434 (2001).
29. A. Lorke, R.J. Luyken, A.O. Govorov, J.P. Kotthaus,; J.M. Garcia, P. M. Petroff, (2000). *Phys. Rev. Let.* **84**, 2223 (2000).
30. A. Emperador, M. Pi, M. Barranco, A. Lorke,. *Phys. Rev.* **B62**, 4573 (2000).
31. W. Lei, C. Notthoff, A. Lorke, D. Reuter, A. D. Wieck, (2010). *Applied Physics Letters* **96**, 033111 (2010).
32. O. Voskoboynikov, Yiming Li, Hsiao-Mei Lu, Cheng-Feng Shih, C.P. Lee *Phys. Rev.* **B66**, 155306 (2002).
33. B.T. Miller, W.Hansen, S. Manus, R.J. Luyken, A. Lorke, J.P. Kotthaus, S. Huant, G. Medeiros-Ribeiro, P.M. Petroff, *Phys. Rev.* **B56**, 6764 (1997).)
34. I. Filikhin, E. Deyneka, B. Vlahovic, *Solid State Communications* **140**, 483 (2006).
35. R.J. Warburton; B.T. Miller; C.S. Durr, C. Bodefeld, K. Karrai,; J.P. Kot-

thaus, G. Medeiros-Ribeiro, P.M. Petroff, S. Huant, *Phys. Rev.* **B58**, 16221 (1998).

36. I. Filikhin, V.M. Suslov, M. Wu, & B. Vlahovic. *Physica E: Low-dimensional Systems and Nanostructures*, **40**, 715 (2008).
37. T. Chakraborty, P. Pietiläinen, *Phys. Rev.* **B50**, 8460 (1994).
38. B. Szafran, F.M. Peeters, *Phys. Rev.* **B72**, 155316 (2005).
39. I. Filikhin, V.M. Suslov, & B. Vlahovic, *Journal of Computational and Theoretical Nanoscience*, **9**, 1 (2012).
40. Y. Aharonov. and D. Bohm, *Phys. Rev.* **115**, 485 (1959).
41. A.G. Aronov, Yu.V. Sharvin, *Review Modern Physics* **59**, 755 (1987).
42. C.W.J. Beenakker, H. van Houten, Solid State Physics: Advances in Research and Applications, Ed. H. Ehrenreich and D. Turnbull, Academic Press (April 2, 1991), Vol. 44, (1991), p.1-454, N-Y, USA.
43. H.U. Baranger, A.D. Stone, *Phys. Rev. Let.* **63**, 414 (1989).
44. H.U. Baranger, D.P DiVincenzo, R.A. Jalabert, A.D. Stone, *Phys. Rev.* **B44**, 10637 (1991).
45. R.S. Whitney, H. Schomerus, M. Kopp, *Phys. Rev.* **E80**, 056209 (2009).
46. R.S. Whitney, P. Marconcini, M. Macucci, *Phys. Rev. Let.*, **102**, 186802 (2009).
47. I. Filikhin, S.G. Matinyan, B.K. Schmid, & B. Vlahovic, *Physica E: Low-dimensional Systems and Nanostructures*, **42**, 1979 (2010).
48. T.A. Brody, J. Flores, J.B. French, P.A. Mello, A. Pandey, S.S.M. Wong, *Reviews of Modern Physics* **53**, 385 (1981).
49. G.W. Bryant, *Phys. Rev.* **B48**, 8024 (1993).
50. G. Bastard, Wave Mechanics applied to Semiconductor Heterostructures, John Wiley and Sons Inc. (Jan 01, 1990), p. 357, ISBN 0-471-21708-1, New York, NY (USA)
51. L.A. Ponomarenko, F. Schedin, M.I. Katsnelson, R. Yang, E.W. Hill, K.S. Novoselov, A. K. Geim, *Science* **320**, 356 (2008).
52. M. Buttiker, Y. Imry, R. Landauer. *Phys. Let.* **A96**, 365 (1983).
53. T. Mano, T. Kuroda, S. Sanguinetti, T. Ochiai, T. Tateno, J. Kim, T. Noda, M. Kawabe, K. Sakoda, G. Kido, and N. Koguchi, *Nano Letters*, **5**, 425 (2005).
54. T. Kuroda, T. Mano, T. Ochiai, S. Sanguinetti, K. Sakoda, G. Kido and N. Koguchi, *Phys. Rev.* **B72**, 205301 (2005).
55. I. Filikhin, S. Matinyan, J. Nimmo, B. Vlahovic, *Physica E: Low-dimensional Systems and Nanostructures*, **43**, 1669 (2011).
56. J. v.Neumann, E. Wigner, *Physikalische Zeitschrift*, **30**, 467 (1929).
57. L.D. Landau, E.M. Lifschitz, Quantum Mechanics (Non-Relativistic Theory), 3rd ed. Pergamon Press, (1977), ISBN 0080209408, Oxford, England.
58. B. Szafran, F.M. Peeters, S. Bednarek, *Phys. Rev.* **B70**, 205318 (2004).

COLLECTIVE STATES OF $D(D_3)$ NON-ABELIAN ANYONS

P. E. Finch and H. Frahm*

Institut für Theoretische Physik, Leibniz Universität Hannover,
30655 Hannover, Germany
** E-mail: frahm@itp.uni-hannover.de*
www.itp.uni-hannover.de

We study an exactly solvable model of non-Abelian anyons symmetric under the quantum double of the dihedral group D_3 on a one-dimensional lattice. Bethe ansatz methods are employed to compute the ground states of this model in different regions of the parameter space. The finite size spectrum is studied and the corresponding low energy field theories are identified.

Keywords: anyons, Bethe ansatz, conformal field theory

1. Introduction

In recent years the properties of two-dimensional topological quantum liquids have been the subject of great attention in condensed matter physics. These systems which are realized e.g. in fractional quantum Hall systems[1] and certain frustrated quantum magnets[2–4] are characterized by the existence of topological order without a local order parameter. Within a topologically ordered phase they have gapped quasi-particle excitations which exhibit non-trivial *anyonic* statistics. Put into a more general context they can be seen as topological defects appearing in planar gauge theories in a spontaneously broken symmetry (Higgs) phase with a finite residual gauge group.[5] The anyonic statistics can be revealed by interchanging two of these particles: in the case of Abelian anyons the many-body wave function picks up a fractional exchange phase $e^{i\theta}$ instead of the usual factor ± 1 for bosons or fermions. In a system where the many-particle ground state is k-fold degenerate the wave-function can be changed by a rotation described by non-commuting $k \times k$ unitary matrix.

2. Integrable many anyon models

The investigation of collective anyon states can be based on the implementation of the two-particle operations *braiding*, i.e. their properties under exchange, and *fusion* determining the decomposition of product states into irreducible multiplets of the underlying symmetry. This approach has been used to construct microscopic lattice models which can then be analyzed by established many-body techniques: starting from the non-Abelian degrees of freedom in $su(2)_k$ Chern Simons theories one obtains a class of chains and ladder systems of so-called Ising or Fibonacci anyons which have been studied using both numerical field theoretical methods.[4,6–8] Here, we propose a different route to uncovering the collective properties of anyons: the algebraic structure underlying the (discrete) gauge theory with D_3, the dihedral group of order six, as its gauge group is the Drinfeld double $D(D_3)$. The explicit representations of $D(D_3)$ have been constructed in various papers, here we shall use the notation from Ref. 9. In short, one has two one-dimensional irreducible representations $\pi_1^{\pm}$, four two-dimensional ones $\pi_2^{(a,b)}$, $(a,b) = (0,1)$ or $(1,b)$ with $b = 0,1,2$, and two three-dimensional ones $\pi_3^{\pm}$. The fact that $D(D_3)$ forms a quasi-triangular Hopf algebra provides it with a natural tensor product structure: the so-called quantum dimensions of these anyons are integers. This allows to build up the Hilbert space of a many-particle system using products of single particle states. Therefore we can consider a chain of length L with a $D(D_3)$ anyon in one of three possible internal states represented by a copy of π_3^+ on each site.

Equally important in the present context is that the quantum double shares much of the algebraic structure of an integrable model. Baxterization of the braid operator leads to a two-parameter R-matrix $R(z_1, z_2)$ satisfying a Yang-Baxter equation.[9] It is related to that of the spin-1 Fateev-Zamolodchikov model[10] in a particular singular limit. Within the framework of the Quantum Inverse Scattering Method[11] a family of commuting transfer matrices $\tilde{t}(z_1, z_2)$ is obtained, from which an integrable quantum spin chain with a one parametric hamiltonian coupling spins on neighbouring sites can be derived[12]

$$H_\theta = \sum_{i=1}^{L} \cos\theta\, h_{i,i+1}^{(1)} + \sin\theta\, h_{i,i+1}^{(2)} \,. \tag{1}$$

In terms of the projection operators on irreps[9] appearing in the tensor product $\pi_3^+ \otimes \pi_3^+$ the local hamiltonian can be expressed as

$$h^{(1)} = \frac{2\sqrt{3}}{3} p_1^+ - \frac{\sqrt{3}}{3} p_2^{(0,1)} - \frac{\sqrt{3}}{3} p_2^{(1,0)} - \frac{\sqrt{3}}{3} p_2^{(1,1)} + \frac{2\sqrt{3}}{3} p_2^{(1,2)} \,. \tag{2}$$

The operator $h^{(2)} = Ph^{(1)}P = \left(h^{(1)}\right)^*$ can be obtained either by a permutation P of the spins on the neighbouring sites or by complex conjugation: with the exception of $p_2^{(1,1)} = \left(p_2^{(1,2)}\right)^*$ the projection operators are invariant under these operations. Therefore the eigenstates of (1) are independent of the parameter θ, i.e. the commutator $[H_\theta, H_{\theta'}]$ vanishes. Furthermore, the operators H_0 and $H_{\pi/2}$ are related by a spatial inversion have the same eigenvalues. These properties can be traced back to the underlying two parameter transfer matrix of this model and its symmetries.[12,13]

To study the spectrum of (1) we have therefore to address two separate problems: first, the energy spectrum of H_0 (or, equivalently $H_{\pi/2}$) has to be determined. In a second step we need to identify so-called *pairing rules* determining which eigenvalues of H_0 and $H_{\pi/2}$ can be combined to give an eigenvalue of H_θ. Variation of the parameter θ allows to tune the interactions in (1) to favour different channels for the fusion of spins on neighbouring sites, e.g. the vacuum channel π_1^+ for $\pi < \theta < 3\pi/2$. As a consequence one has to expect phase transitions due to level crossings appearing at θ being a multiple of $\pi/2$.

3. Bethe ansatz

The integrability of the model (1) can be used to study the spectrum of $H_0 = -H_\pi$ analytically by means of Bethe ansatz methods. Since the model considered here has no $U(1)$ symmetry the solution of the eigenvalue problem has to rely on functional methods where the analytical properties of the transfer matrix and functional equations, in particular *fusion relations*, are exploited. Within this approach it has been found that the eigenvalues of H_θ can be parameterized by two sets of L complex rapidities x_j, each satisfying the Bethe equations[12]

$$(-1)^{L+1}\left(\frac{1+(i/\omega)\mathrm{e}^{x_j}}{1-i\omega\,\mathrm{e}^{x_j}}\right)^L = \prod_{k=1}^{L}\frac{\mathrm{e}^{x_k}-(1/\omega)\mathrm{e}^{x_j}}{\mathrm{e}^{x_k}-\omega\,\mathrm{e}^{x_j}} \qquad (3)$$

($\omega = \exp 2\pi i/3$). A root configuration of these algebraic equations contributes

$$E\left(\{x_j\}\right) = i\sum_{j=1}^{L}\frac{1}{1-i\omega\,\mathrm{e}^{x_j}} - \frac{\sqrt{3}}{3}\omega L \qquad (4)$$

to the energy of the corresponding state as

$$E_\theta\left(\{x_j^{(\alpha)}\},\{x_j^{(\beta)}\}\right) = \cos\theta\, E\left(\{x_j^{(\alpha)}\}\right) + \sin\theta\, E\left(\{x_j^{(\beta)}\}\right), \qquad (5)$$

provided that the combination (α, β) satisfies the pairing rules discussed below. Similarly, the partial momentum of a state due to a given root configuration is $p(\{x_j\}) = -i\sum_{j=1}^{L} \ln(1 - i\omega\, e^{x_j})$.

To identify the root configurations $\{x_j\}$ solving Eqs. (3) that are relevant to description of the ground state and low energy excitations of H_0 and H_π we have performed a complete numerical diagonalization of the transfer matrix $\tilde{t}(z_1, z_2)$ for systems of up to $L = 10$ sites. From these numerical data we have determined the Bethe roots $\{x_j\}$ for *all* states. It turns out that for L sufficiently large the configurations group into three different 'string' types,[12,14] i.e.

- $+$-strings: $\mathrm{Im}(x_j) = 0$,
- $-$-strings: $\mathrm{Im}(x_j) = \pi$,
- 2-strings: pairs of roots $x_j^+ = (x_j^-)^*$ with $\mathrm{Im}(x_j^\pm) = \pm\frac{2\pi}{3}$.

plus up to two roots at $\pm\infty$. Computing the corresponding energies, see Figure 1, we can identify the root configurations for the low and high energy states of the quantum spin chain, in particular the ground states of $\pm H_0 = \mp H_\pi$:

The ground state of H_π is composed of $N_2 = L/2$ 2-strings, which is realized for chains of even length. Using a combination of exact results for the thermodynamic limit $L \to \infty$ and numerical analysis its energy and partial momentum has been determined to be[12]

$$E = -\left[\frac{1}{\pi} + \frac{2\sqrt{3}}{9}\right] L - \frac{12}{5} \times \frac{\pi}{6L} + o(L^{-1}) \quad \text{and} \quad p = \frac{\pi}{12} L\,. \tag{6}$$

Similarly, the ground state of H_0 is found to be given by a solution of the Bethe equations (3) with $N_+ = L/4$ $+$-strings and $N_- = 3L/4$ $-$-strings for systems sizes being multiples of 4. In the thermodynamic limit the ground state energy and partial momentum are[12]

$$E = -\left[\frac{1}{2\pi} - \frac{2\sqrt{3}}{9} + \frac{3}{4}\right] L - \frac{3}{2} \times \frac{\pi}{6L} + o(L^{-1}) \quad \text{and} \quad p = -\frac{7\pi}{24} L\,. \tag{7}$$

We note here that the total number of distinct sets $\{x_j\}$ solving the Bethe equations (3) which are found to appear in the decomposition of the eigenvalues of H_θ according to Eq. (5) is $4 \cdot 3^{(L/2-1)}$ for chains of even length L and $2 \cdot 3^{(L-1)/2}$ for L odd. This implies the existence of massive degeneracies in the spectrum of $H_{\theta=k\pi/2}$ for integer k!

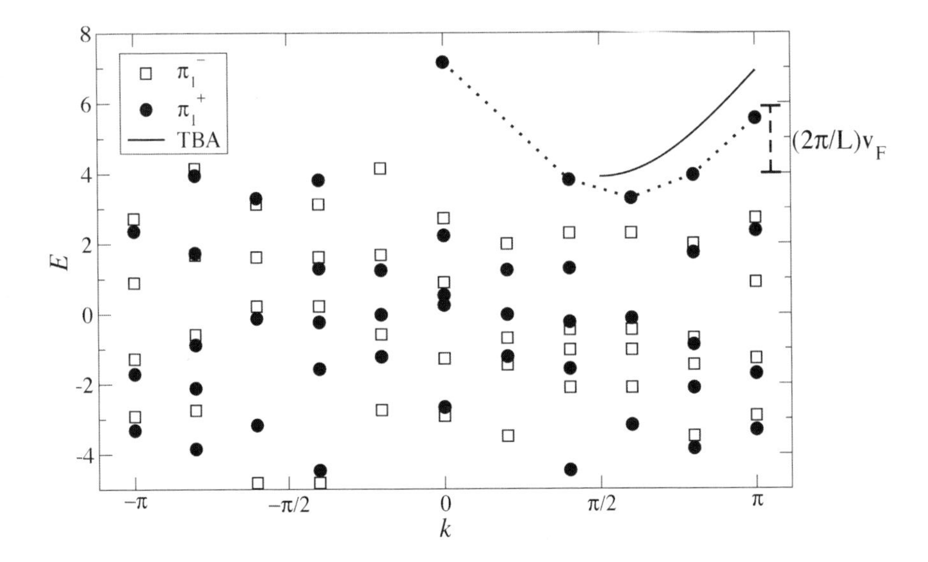

Fig. 1. Complete set of energies and momenta of the eigenstates of $H_{\theta=0}$ for $L = 10$ in the symmetry sectors corresponding to the one-dimensional representations $\pi_1^{\pm}$ of $D(D_3)$ from numerical diagonalization. Also shown is the dispersion (9) of elementary excitations of H_π as obtained from the thermodynamic Bethe ansatz and the finite size spacing $(2\pi/L)v_F$ of low energy levels of H_π. The dotted line connecting the ground state and one branch of excitations of H_π is just a guide to the eye. Note that the lowest energy state (ground state of H_0) is doubly denegerate for this system size.

4. Low energy spectrum of H_π

To characterize the low energy properties of this many-anyon system further we need to study the low lying excitations over these ground states. Conformal field theory (CFT) predicts the finite size scaling behaviour of these states for periodic chains to be ($X = h + \bar{h}$, $s = h - \bar{h}$)[15,16]

$$\begin{aligned} E_0 &= L\epsilon_\infty - \frac{\pi v_F}{6L} c\,, \\ \Delta E &= \frac{2\pi v_F}{L}(X + n + \bar{n})\,, \quad \Delta p = \frac{2\pi}{L}(s + n - \bar{n}) \end{aligned} \tag{8}$$

with positive integers n, $\bar{n}$. Comparing with Eq. (6) we have $v_F\,c = \frac{12}{5}$ for the product of the conformal anomaly c and the Fermi velocity v_F of the gapless excitations. Usually, the Fermi velocity can be determined within the root density approach from the thermodynamic Bethe ansatz (TBA).[17,18] In the present case we have to take into account the constraint that the number of Bethe roots is fixed to L for all states. Combining the TBA results with our numerical findings we obtain the dispersion relation of elementary excitations to be

$$\epsilon(p) = 3|\sin p|\,, \quad \frac{\pi}{2} < p < \frac{3\pi}{2}\,. \tag{9}$$

Hence, we obtain for the Fermi velocity $v_F = \partial\epsilon/\partial p|_{p=\pi}$ and therefore the central charge of the corresponding CFT is $c = \frac{4}{5}$.[19] This allows to identify the low energy effective theory for H_π as the unitary minimal model $\mathcal{M}_{5,6}$. The possible conformal weights determining the finite size gaps of the excitations in (8) are given by the Kac table

$$h \in \left\{0, \frac{1}{40}, \frac{1}{15}, \frac{1}{8}, \frac{2}{5}, \frac{21}{40}, \frac{2}{3}, \frac{7}{5}, \frac{13}{8}, 3\right\} \tag{10}$$

The actual operator content of a specific realization of a given CFT is given by a subset of (10) due to the requirement of modular invariance of the partition function and depending on the boundary conditions.[20,21] For the minimal model $\mathcal{M}_{5,6}$ the possible modular invariants allowing for a representation in terms of local quantum fields are classified.

For the $D(D_3)$ chain (1) the conformal weights can be determined from the exact finite size spectrum using the CFT prediction (8) for excitated states: based on our numerical analysis of small systems we have been able to identify many of the root configurations related to levels in the conformal spectrum of H_π, see e.g. Figure 2. As mentioned above, the configuration extrapolating to the thermodynamic ground state is realized for even number of lattice sites only. As a consequence the finite size spectrum is different for even and odd L. Solving the Bethe equations for even length lattices up to $L \lesssim 200$ and extrapolating the finite size gaps using (8) we have found the excitations shown in Table 1. As a consequence of the periodic boundary conditions imposed in (1) the full $D(D_3)$ invariance is broken, but the levels can be grouped according to a residual symmetry. According to Table 1 we observe the Virasoro characters

$$\chi_0 + \chi_3\,, \quad \chi_{\frac{2}{5}} + \chi_{\frac{7}{5}}\,, \quad \chi_{\frac{1}{15}}\,, \quad \chi_{\frac{2}{3}} \tag{11}$$

in the spectrum of H_π for even L. These combinations also characterize the (ferromagnetic) 3-states Potts model subject to cyclic boundary conditions and a Z_3 charge depending on the symmetry sector.[20,22,23] Note that spin $\frac{1}{3}$ parafermions are present in the symmetry sectors $\pi_2^{(0,1)}$ and $\pi_2^{(1,0)}$.

The finite size scaling results from a analogous study of the spectrum of H_π for odd length lattices are shown in Table 2. In this case the symmetry of the states can only be characterized by their transformation properties under the D_3-rotation σ. Here we observe the appearance of twist operators with conformal weight $h \in \{\frac{1}{40}, \frac{1}{8}, \frac{21}{40}, \frac{13}{8}\}$. To describe the low energy

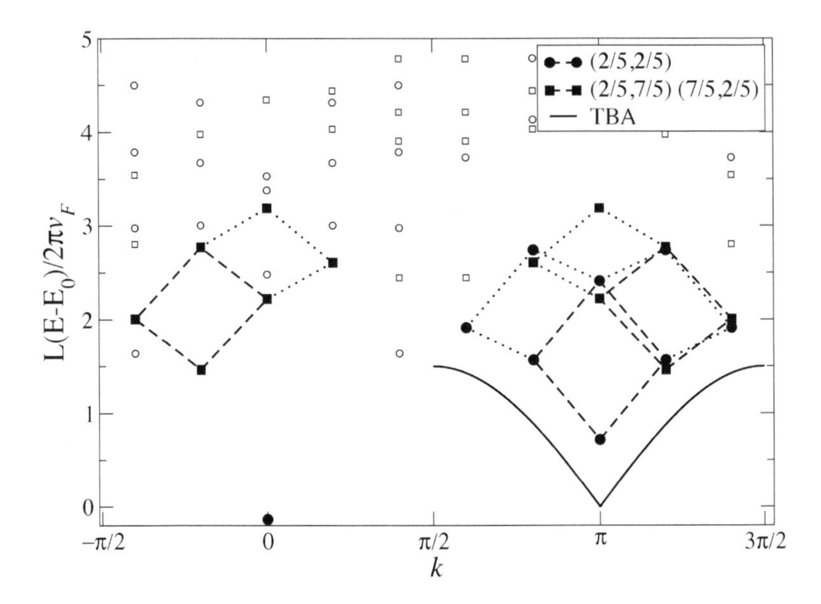

Fig. 2. Conformal states in the low energy spectrum of H_π for $L = 10$. Levels in the π_1^+ (π_1^-) symmetry sector are shown as circles (Squares), the ground state and the blocks corresponding to the $(h, \bar{h}) = (\frac{2}{5}, \frac{2}{5})$, $(\frac{2}{5}, \frac{7}{5})$ and $(\frac{7}{5}, \frac{2}{5})$ are shown as filled symbols. The full line is the dispersion (9) of elementary excitations as obtained from the thermodynamic Bethe ansatz with Fermi velocity $v_F = 3$.

Table 1. Extrapolation of scaling dimensions and conformal weights from low energy excitations in the spectrum of H_π for L even. $(h, \bar{h})$ are the predictions from the $\mathcal{M}_{5,6}$ minimal model. The operators appearing in the sector with $\pi_2^{(1,0)}$ residual $D(D_3)$-symmetry is obtained from that of $\pi_2^{(0,1)}$ by interchange of h and $\bar{h}$.

$D(D_3)$	X_π^{num}	$(h, \bar{h})$	spin	pairing mult.
$\pi_1^+ \oplus \pi_1^-$	0.000000(1)	$(0, 0)$	0	1
	0.801(3)	$(\frac{2}{5}, \frac{2}{5})$	0	1
	1.80(1)	$(\frac{2}{5}, \frac{7}{5}), (\frac{7}{5}, \frac{2}{5})$	± 1	1
$\pi_2^{(0,1)}$	0.4668(2)	$(\frac{1}{15}, \frac{2}{5})$	$-\frac{1}{3}$	2
	0.666666(1)	$(\frac{2}{3}, 0)$	$\frac{2}{3}$	2
$\pi_2^{(1,1)} \oplus \pi_2^{(1,2)}$	0.13334(6)	$(\frac{1}{15}, \frac{1}{15})$	0	4
	1.33333(3)	$(\frac{2}{3}, \frac{2}{3})$	0	4

spectrum in terms of Virasoro characters one has to use more general combinations than have been considered in the literature before. This is also reflected in the appearance of operators with non-integer conformal spins.

Table 2. Extrapolation of scaling dimensions and conformal weights from low energy excitations in the spectrum of H_π for L odd.

σ	$X_\pi^{\rm num}$	$(h, \bar{h})$	spin	pairing mult.
1	0.125000(5)	$(0, \frac{1}{8})$	$-\frac{1}{8}$	1
	0.42502(2)	$(\frac{2}{5}, \frac{1}{40})$	$\frac{3}{8}$	1
	0.92490(6)	$(\frac{2}{5}, \frac{21}{40})$	$-\frac{1}{8}$	1
	1.625000(1)	$(0, \frac{13}{8})$	$-\frac{13}{8}$	1
ω, ω^{-1}	0.091665(2)	$(\frac{1}{15}, \frac{1}{40})$	$\frac{1}{24}$	2
	0.59168(7)	$(\frac{1}{15}, \frac{21}{40})$	$-\frac{11}{24}$	2
	0.791667(1)	$(\frac{2}{3}, \frac{1}{8})$	$\frac{13}{24}$	2

Note that *physical* operators are direct products of the primary fields appearing in the two commuting components of the theory with local hamiltonians $h^{(1,2)}$ (2). Taking into account the allowed pairing between these sectors the total spin $s = s^{(1)} - s^{(2)}$ of the physical fields takes integer values as it should. From our finite size data we conclude that pairing is determined by the (residual) symmetry sectors: all states of a given symmetry can pair with each other, states of different symmetry do not pair. The resulting fields appear in the spectrum of the model (1) with the pairing multiplicity indicated in the Tables.

5. Low energy excitations of H_0

The low energy effective theory for H_0 can be identified in a similar fashion as that for H_π above: using the root density formalism within the thermodynamic Bethe ansatz approach together with data from the numerical solution of the Bethe equations (3) we find that the Fermi velocity of low excitations of H_0 is $v_F = \frac{3}{2}$, hence the Virasoro central charge is $c = 1$ according to (7).

Since the root configuration corresponding to the thermodynamic ground state can be realized only for L being a multiple of 4 the operator content found from the finite size analysis also depends on the number of lattice sites $\bmod 4$.[19] In Table 3 we present the excitations observed for even $L = 2 \bmod 4$, again grouped by their residual $D(D_3)$ symmetry of the even length chain. The conformal field theory describing the low energy properties of H_0 can be identified with that of Z_4 parafermions[24,25] (or anti-ferromagnetic 3-states Potts model). The lowest states for a given length L are the Z_4 spin fields σ_ℓ, $\ell = L \bmod 4$, with conformal dimensions $h_\ell = 0, \frac{1}{16}, \frac{1}{12}, \frac{1}{16}$. As for the operator content of H_π we find that again fields with non-integer or non-parafermionic spins are present in the

Table 3. Extrapolation of scaling dimensions and conformal weights from low energy excitations in the spectrum of H_0 for $L = 2 \mod 4$.

$D(D_3)$	X_0^{num}	$(h, \bar{h})$	spin	pairing mult.
$\pi_1^+ \oplus \pi_1^-$	0.750000	$(0, \frac{3}{4}), (\frac{3}{4}, 0)$	$\pm\frac{3}{4}$	1
$\pi_2^{(0,1)}$	0.083333	$(0, \frac{1}{12})$	$-\frac{1}{12}$	2
	1.083333	$(\frac{3}{4}, \frac{1}{3})$	$\frac{5}{12}$	2
$\pi_2^{(1,1)} \oplus \pi_2^{(1,2)}$	0.416667	$(\frac{1}{12}, \frac{1}{3}), (\frac{1}{3}, \frac{1}{12})$	$\pm\frac{1}{4}$	4

effective theory. This is resolved once pairing is taken into account. The corresponding rules for the allowed combinations of two primary fields from two Z_4 theories or between fields from a Z_4 theory and the minimal model $\mathcal{M}_{5,6}$ are based on the residual symmetry as stated above.

6. Phase diagram

As pointed out before the Bethe ansatz study performed so far just gives half of the spectrum. To complete the analysis we have to implement the pairing rules and add the θ-dependence resulting from the second spectral parameter in the transfer matrix

$$H_\theta = \cos\theta\, H^{(1)} + \sin\theta\, H^{(2)}$$

Depending on θ the two possible low energy effective theories identified above have to be combined in different ways to describe the collective behaviour of the interacting $D(D_3)$ anyons:

- $0 < \theta < \pi/2$:
  both terms of the hamiltonian enter a positive pre-factor: the critical theory is the direct product of two Z_4 parafermionic theories.
- $\pi/2 < \theta < \pi$ and $3\pi/2 < \theta < 2\pi$:
  $H^{(1)}$ and $H^{(2)}$ enter with different signs: the low energy excitations are described by the direct product of a Z_4 parafermion and an $\mathcal{M}_{(5,6)}$ theory.
- $\pi < \theta < 3\pi/2$:
  both terms of the hamiltonian enter with a negative pre-factor: the critical theory is the direct product of two minimal $\mathcal{M}_{(5,6)}$ models.

When $\theta = m\pi/2$, $m \in \mathbb{Z}$ one of the sectors does not contribute the energy. This leads to a huge degeneracy of the spectrum scaling exponentially

($\propto 3^{L/2}$) with the system size L. The corresponding level crossings accompany the first order transitions between the four phases listed above.

7. Summary

We have constructed a class of integrable quantum chains based on the algebraic structure of a discrete (in this case D_3) gauge theory resulting in a model for non-Abelian interacting $D(D_3)$ anyons. To identify the properties of the collective states formed by these objects we have performed a finite-size study of this system. Using the exact Bethe ansatz solution we have identified the low energy behaviour to be described by a direct product of two conformal field theories corresponding to the ferro- and anti-ferromagnetic 3-state Potts model. In the context of interfaces between different topological quantum liquids[26–29] this can serve as a realization of a boundary supporting a pair of gapless modes. The observed pairing in the spectrum of the quantum chain implies that the interaction between these edge modes has to be of purely topological nature.

For a complete understanding of this interaction the boundary conditions imposed on the quantum chain can be varied: in addition to the periodic closure of the spin chain used here one can consider closed braided boundary conditions and free ends[12] (both of which will restore the full $D(D_3)$ symmetry of the chain) or include boundary fields, see Ref. 30. A different class of quantum chains based on this symmetry can be constructed in the anyon picture where the state space is spanned not by tensoring local states but by the so-called fusion paths allowed by underlying algebra.[31] The bulk hamiltonian of these models is equivalent to the one given by Eqs. (1) and (2) here, therefore they are expected to be in the same universality class although the low energy spectrum will be governed by different pairing rules. Based on these rules the possible modular invariants appearing in the partition function of the low energy effective theory can be identified. How they are obtained by combination of the Virasoro characters of the individual CFTs is the origin of the topological interaction between the gapless edge modes and will provide support for the CFT description of interfaces between different topological quantum liquids.

Another possible direction for generalization of the model studied in this work is to consider the anyons arising as quasi-particle excitations in a D_n gauge theory. Just as (1) can be seen as a descendent of the 6-vertex model with anisotropy ω, these models would be related to the 6-vertex model at n^{th} root of unity.

Acknowledgments

This work has been supported by a grant from the Deutsche Forschungsgemeinschaft.

References

1. R. B. Laughlin, *Phys. Rev. Lett.* **50**, 1395 (1983).
2. R. Moessner and S. L. Sondhi, *Phys. Rev. Lett.* **86**, 1881 (2001).
3. L. Balents, M. P. A. Fisher and S. M. Girvin, *Phys. Rev. B* **65**, p. 224412 (2002).
4. A. Kitaev, *Ann. Phys. (NY)* **321**, 2 (2006).
5. M. de Wild Propitius and F. A. Bais, Discrete gauge theories, in *Particles and Fields*, eds. G. W. Semenoff and L. VinetCRM Series in Mathematical Physics (Springer, 1999).
6. M. A. Levin and X.-G. Wen, *Phys. Rev. B* **71**, p. 045110 (2005).
7. A. Feiguin, S. Trebst, A. W. W. Ludwig, M. Troyer, A. Kitaev, Z. Wang and M. H. Freedman, *Phys. Rev. Lett.* **98**, p. 160409 (2007).
8. S. Trebst, E. Ardonne, A. Feiguin, D. A. Huse, A. W. W. Ludwig and M. Troyer, *Phys. Rev. Lett.* **101**, p. 050401 (2008).
9. P. E. Finch, K. A. Dancer, P. S. Isaac and J. Links, *Nucl. Phys. B* **847**, 387 (2011).
10. V. A. Fateev and A. B. Zamolodchikov, *Phys. Lett. A* **92**, 37 (1982).
11. V. E. Korepin, N. M. Bogoliubov and A. G. Izergin, *Quantum Inverse Scattering Method and Correlation Functions* (Cambridge University Press, Cambridge, 1993).
12. P. E. Finch, H. Frahm and J. Links, *Nucl. Phys. B* **844**, 129 (2011).
13. P. E. Finch, *J. Stat. Mech.* , p. P04012 (2011).
14. M. Takahashi and M. Suzuki, *Prog. Theor. Phys.* **48**, 2187 (1972).
15. H. W. J. Blöte, J. L. Cardy and M. P. Nightingale, *Phys. Rev. Lett.* **56**, 742 (1986).
16. I. Affleck, *Phys. Rev. Lett.* **56**, 746 (1986).
17. C. N. Yang and C. P. Yang, *J. Math. Phys.* **10**, 1115 (1969).
18. F. H. L. Essler, H. Frahm, F. Göhmann, A. Klümper and V. E. Korepin, *The One-Dimensional Hubbard Model* (Cambridge University Press, Cambridge (UK), 2005).
19. P. E. Finch and H. Frahm, *J. Stat. Mech.* , p. L05001 (2012).
20. J. L. Cardy, *Nucl. Phys. B* **275**, 200 (1986).
21. A. Cappelli, C. Itzykson and J.-B. Zuber, *Nucl. Phys. B* **280** **[FS18]**, 445 (1987).
22. G. von Gehlen and V. Rittenberg, *J. Phys. A* **19**, L655 (1986).
23. J.-B. Zuber, *Phys. Lett. B* **176**, 127 (1986).
24. A. B. Zamolodchikov and V. A. Fateev, *Sov. Phys. JETP* **62**, 215 (1985).
25. D. Gepner and Z. Qiu, *Nucl. Phys. B* **285**, 423 (1987).
26. E. Grosfeld and K. Schoutens, *Phys. Rev. Lett.* **103**, p. 076803 (2009).
27. C. Gils, E. Ardonne, S. Trebst, A. W. W. Ludwig, M. Troyer and Z. Wang, *Phys. Rev. Lett.* **103**, p. 070401 (2009).

28. P. Fendley, M. P. A. Fisher and C. Nayak, *Ann. Phys. (N.Y.)* **324**, 1547 (2009).
29. F. J. Burnell, S. H. Simon and J. K. Slingerland, *Phys. Rev. B* **84**, p. 125434 (2011).
30. K. A. Dancer, P. E. Finch, P. S. Isaac and J. Links, *Nucl. Phys. B* **812**, 456 (2009).
31. P. E. Finch (2012), arXiv:1201.4470.

EFFECTIVE HAMILTONIAN FOR THE HALF-FILLED SPIN-ASYMMETRIC IONIC HUBBARD CHAIN WITH STRONG ON-SITE REPULSION

I. Grusha and G.I. Japaridze

Ilia State University, Cholokashvili Ave. 3-5,
Tbilisi, 0162, Georgia
E-mail: gia_japaridze@iliauni.edu.ge
www.iliauni.du.ge

We derive an effective spin Hamiltonian for the one-dimensional half-filled spin asymmetric ionic-Hubbard model in the limit of strong on-site repulsion. We have shown that the effective Hamiltonian, which describes spin degrees of freedom of the initial lattice electron system is given by the Hamiltonian of the frustrated XXZ Heisenberg chain with an additional alternating three spin coupling term in the presence of alternating magnetic field.

Keywords: Low-dimensional spin systems; strongly correlated electrons.

1. Introduction

During last two decades the correlation induced metal-insulator transition is one of the challenging problems in condensed matter physics.[1–3] In main cases breaking of spatial symmetry is a prerequisite for the insulator.[4] However, the one-dimensional Hubbard model[5]

$$\mathcal{H} = t \sum_{i,j,\alpha} N_{i,j} c^{\dagger}_{i,\alpha} c_{j,\alpha} + U \sum_{i} n_{i,\uparrow} n_{i,\downarrow} \tag{1}$$

at half-filling represents the case, where dynamical generation of a charge gap is not connected with the breaking of a discrete symmetry.[6] In Eq. (1) where we have used standard notation, $n_{i,\alpha} = c^{\dagger}_{i,\alpha} c_{i,\alpha}$, and U is the Hubbard on-site repulsive interaction and $N_{i,j} = 1$ if i and j are labels for neighboring sites and equals zero otherwise. Thus the kinetic part presents hops between neighboring sites and the interaction part gives contribution only from electrons on the same site. At half-filling the exact solution shows a uniform ground state with exponentially suppressed density correlations[7] and gapless $SU(2)$ symmetric spin degrees of freedom.[8] This is in agreement with the large U expansion result, that at $U \gg |t|$ the model (1) is equivalent to the effective spin $S = 1/2$ Heisenberg antiferromagnet Hamiltonian,[9]

$$\mathcal{H} = J \sum_i \mathbf{S}_i \cdot \mathbf{S}_{i+1} + J' \sum_i \mathbf{S}_i \cdot \mathbf{S}_{i+2}\,, \tag{2}$$

where $J = 4t^2/U\,(1 - 4t^2/U^2)$, $J' = 4t^4/U^3$ in the forth-order perturbation. Several very elegant mathematical tools have been developed for calculation of the effective spin-chain Hamiltonian in higher orders.[10–12] In agreement with the exact solution, all higher order exchange terms are strongly irrelevant and the ground state remains featureless.

The transition from a band-insulating phase with manifestly displayed broken translation symmetry to the featureless Mott insulating phase in electron system has been the subject of intensive studies in past two decades.[13–25] In many cases these transition has been studied within the framework of the $SU(2)$-symmetric ionic-Hubbard model[26]

$$\mathcal{H}_{IHM} = t \sum_{i,j,\alpha} N_{i,j} c^{\dagger}_{i,\alpha} c_{j,\alpha} + \frac{\Delta}{2} \sum_{i,\alpha} (-1)^i n_{i,\alpha} + U \sum_i n_{i,\uparrow} n_{i,\downarrow} \tag{3}$$

where the translational symmetry is implicitly broken due to presence of two types of atoms with different on-site energy given by Δ. At $U = 0$ the IHM is a regular band insulator with long-range ordered charge-density-wave (CDW), while in the strong-coupling limit, at $U \gg t, \Delta$ is the Mott insulator with a charge gap and gapless spin sector. The scrupulous analysis of the model (3) based on the continuum-limit bosonization approach[17] and the numerical (DMRG) studies[21,23,25] have reveal the two-step nature of the CDW-Mott insulator transition: with increasing U first there is the *Ising type transition* from a CDW band phase into a LRO dimerized phase at U^c_{ch} and, with further increase of the Hubbard repulsion, at $U^c_{sp} > U^c_{ch}$, a continuous *Kosterlitz-Thouless transition* from the dimerized into a MI phase $U > U^c_{sp}$ takes place. The charge gap vanishes *only* at $U = U^c_{ch}$, while the spin sector is gapless for $U > U^c_{sp}$ and its infrared behavior is again described by the Heisenberg Hamiltonian (2), with slightly modified by Δ spin exchange parameters J and J'.[13]

The spin-asymmetric Hubbard model

$$\mathcal{H} = \sum_{i,j,\alpha} N_{i,j} t_{\alpha} c^{\dagger}_{i,\alpha} c_{\,j,\alpha} + U \sum_i n_{i,\uparrow} n_{i,\downarrow}\,, \tag{4}$$

is the other extended version of the Hubbard model which describes light and heavy electrons ($t_{\uparrow} > t_{\downarrow}$)interacting with an on-site repulsion. Initially this model has been introduced[27,29] to describe a fermion system which shows an interpolative behavior between the regime described by the standard Hubbard model with equal masses of electrons ($t_\uparrow = t_\downarrow$) and the regime described by the Falicov-Kimball model[31] with mass of one spin component equal to infinity ($t_\uparrow > 0, t_\downarrow = 0$). An increased recent interest in theoretical studies of the fermion systems with spin-dependent hopping[32–38] is connected with great progress in studies of the ultra-cold atoms trapped in the optical lattice, where the corresponding behavior of the fermion system can be experimentally realized.[39]

At ($t_\uparrow \neq t_\downarrow$) the model (4) is characterized by the $U(1)$ spin symmetry. The reduced spin symmetry manifests itself in the strong coupling limit ($U \gg t_\uparrow, t_\downarrow$), where at half-filling the spin sector is described by the XXZ Heisenberg chain[29]

$$\mathcal{H} = J \sum_i \left(S_i^x S_{i+1}^x + S_i^y S_{i+1}^y + \gamma S_i^z S_{i+1}^z \right) , \tag{5}$$

with $J = 4t_\uparrow t_\downarrow / U$ and $\gamma = (t_\uparrow^2 + t_\downarrow^2)/2t_\uparrow t_\downarrow$. Since the spin anisotropy parameter $\gamma > 1$ for arbitrary $t_\uparrow \neq t_\downarrow$, at $t_\uparrow \neq t_\downarrow$, $|t_\uparrow - t_\downarrow| \ll |t_\uparrow|, |t_\downarrow|$ the system is characterized by an exponentially small but finite spin gap and the long-range antiferromagnetic ordered in the ground state.[40] However, due to the translational invariance of the effective Hamiltonian (9) the spin degrees of freedom in the ground state are degenerate with respect of the translation transformations and the broken translation symmetry of the initial lattice model manifess itself only in the charge degrees of freedom.

In this paper we derive the effective spin Hamiltonian in the strong on-site repulsion limit $U \gg |t_\uparrow|, |t_\uparrow|$ for the half-filled spin asymmetric ionic-Habbard model given by the Hamiltonian $\mathcal{H}_{AIH} = \hat{T} + \hat{V}$, where

$$\hat{T} = \sum_{i,j,\alpha} N_{i,j} t_\alpha c_{i,\alpha}^\dagger c_{\ j,\alpha} \tag{6}$$

$$\hat{V} = \frac{\Delta}{2} \sum_{i,\alpha} (-1)^i n_{i,\alpha} + U \sum_i n_{i,\uparrow} n_{i,\downarrow} . \tag{7}$$

As we show below, in marked contrast with the ordinary spin-asymmetric Hubbard model (4), the universal infrared behavior of the spin asymmetric ionic-Hubbard model in the strong repulsion limit of the spin asymmetric ionic-Hubbard model is described by the effective XXZ Heisenberg chain in the presence of alternating magnetic field

$$\mathcal{H} = J \sum_i \left(S_i^x S_{i+1}^x + S_i^y S_{i+1}^y + \gamma S_i^z S_{i+1}^z \right) + h_0 \sum_i (-1)^i S_i^z , \tag{8}$$

with $J = 4t_\uparrow t_\downarrow/U(1-\Delta^2/U^2)$, $\gamma = (t_\uparrow^2 + t_\downarrow^2)/2t_\uparrow t_\downarrow$ and

$$h_0 \simeq \frac{2\Delta(t_\uparrow^2 - t_\downarrow^2)}{U^2-\Delta^2}\,. \tag{9}$$

Thus, at $t_\uparrow \neq t_\downarrow$ and finite Δ the translation symmetry is broken already on the level of the effective spin Hamiltonian via the presence of the alternating magnetic field. Since the alternating magnetic represents strongly relevant perturbation to the spin spin-chain system, in the ground state the system shows a long-range antiferromagnetic order with implicitly broken translation symmetry. The excitation spectrum is gapped and the gap shows almost linear dependence on the parameter h_0.[41]

2. The strong-coupling expansion approach

Below in this paper we apply the method developed by MacDonald, Girvin and Yoshioka in the case of ordinary Hubbard chain[11] to obtain the effective spin Hamiltonian for the one-dimensional *spin-asymmetric* ionic Hubbard model (7). We consider the strong coupling limit, assuming $U \gg |t_\alpha|$. Contrary to the case of the ordinary Hubbard model where subbands can be classified by the total number of double-occupied states (doublons) N_d, in the case of IH model we have to deal with the system, where each band is characterized by two different numbers: by the number of doubly occupied sites in even and odd sublattices, denoted by N_{de} and N_{do} respectively. The hopping term mixes states from these subbands. The "unmixing" of the AIH model subbands can be achieved by introducing suitable linear combinations of the uncorrelated basic states. The $\mathcal{S}$ matrix for this transformation, and the transformed Hamiltonian,

$$\mathcal{H}_{eff} = e^{i\mathcal{S}}\mathcal{H}_{AIH}e^{-i\mathcal{S}}\,,$$

are generated by an iterative procedure, which results in expansion in powers of the hopping integrals t_α divided by energies $U \pm \Delta$.

This expansion is based on separation of the kinetic part of the Hamiltonian into three terms: T_1 which increases the number of doubly occupied sites by one, T_{-1} which decreases the number of doubly occupied sites by one and T_0 which leaves the number of doubly occupied sites unchanged. In addition, in the case of spin-asymmetric Hubbard model each of these terms splits into several different terms, depending on whether corresponding hopping process of un "up" or "down" spin electron takes place from even on odd site or vice versa.

In particular, we split the T_0 term into three separate hopping processes:

$$T_0 = T_0^0 + T_0^{de} + T_0^{do}\,, \tag{10}$$

where

$$T_0^0 = \sum_\alpha \sum_{i,j} t_\alpha\, N_{i,j} h_{i,-\alpha} c_{i,\alpha}^\dagger c_{\ j,\alpha} h_{j,-\alpha}\,, \tag{11}$$

is a "hole" hopping term (as a remark here in the case of asmmetric ionic Hubbard this term is to be separated into *even* and *odd* parts for "up" and "down" spins respectively),

$$T_0^{de} = \sum_\alpha \sum_{i,j} t_\alpha\, N_{2i,j} n_{2i,-\alpha} c_{2i,\alpha}^\dagger c_{j,\alpha} n_{j,-\alpha}\,, \tag{12}$$

is a "pair" hopping term, which hops the pair to even site and

$$T_0^{do} = \sum_\alpha \sum_{i,j} t_\alpha\, N_{2i+1,j} n_{2i+1,-\alpha} c_{2i+1,\alpha}^\dagger c_{\ j,\alpha} n_{j,-\alpha}\,, \tag{13}$$

is a "pair" hopping term, which hops the pair to odd site. Here $h_{i,\alpha} = 1 - n_{i,\alpha}$.

Respectively, the term which increases the number of doubly occupied sites by one, is also separated into two terms $T_1 = T_1^e + T_1^o$, where

$$T_1^e = \sum_\alpha \sum_{i,j} t_\alpha\, N_{2i,j} n_{2i,-\alpha} c_{2i,\alpha}^\dagger c_{j,\alpha} h_{j,-\alpha}\,, \tag{14}$$

is the term which increase the number of doubly occupied sites by one on the sublattice of even sites and

$$T_1^o = \sum_\alpha \sum_{i,j} t_\alpha\, N_{2i+1,j} n_{2i+1,-\alpha} c_{2i+1,\alpha}^\dagger c_{j,\alpha} h_{j,-\alpha}\,, \tag{15}$$

is the term which increase the number of the doubly occupied sites by one on the sublattice of odd sites.

Similarly, the term which decreases the number of doubly occupied sites by one is also separated into two terms $T_{-1} = T_{-1}^e + T_{-1}^o$, where

$$T_{-1}^e = \sum_\alpha \sum_{i,j} t_\alpha\, N_{i,2j} h_{i,-\alpha} c_{i,\alpha}^\dagger c_{2j,\alpha} n_{2j,-\alpha} \tag{16}$$

and

$$T_{-1}^o = \sum_{i,j,\alpha} N_{i,2j+1} t_\alpha h_{i,-\alpha} c_{i,\alpha}^\dagger c_{2j+1,\alpha} n_{2j+1,-\alpha}\,, \tag{17}$$

are the terms, which decrease the number of doubly occupied sites by one on the sublattice of even and odd sites, respectively.

One can easily check the following commutation relations:

$$[\hat{V}, T^q_m] = (m + \delta_{m,0})\Lambda_q T^q_m \,, \tag{18}$$

where

$$\Lambda_q = \begin{cases} U+\Delta, & q=e \\ U-\Delta, & q=o \\ \Delta, & q=de \\ -\Delta, & q=do \\ 0, & q=0 \end{cases} \,. \tag{19}$$

We must emphasize that the Eq. (18) is true in all cases where allowed hopping processes connect only sublattices with different on-site repulsion.

Let us now start to search for such a unitary transformation $\mathcal{S}$, which eliminates hops between states with different numbers of doubly occupied sites

$$\mathcal{H}' = e^{i\mathcal{S}}\mathcal{H}e^{-i\mathcal{S}} = \mathcal{H} + [i\mathcal{S}, \mathcal{H}] + \frac{1}{2}[i\mathcal{S}, [i\mathcal{S}, \mathcal{H}]] + \dots \,. \tag{20}$$

We follow the recursive scheme[11] which allows to determine a transformation which has the requested property to any desired order in $t/(U \pm \Delta)$. To proceed further we define:

$$T^{(k)}\left[\{a\}, \{m\}\right] = T^{a_1}_{m_1} T^{a_2}_{m2} \dots T^{a_k}_{m_k} \,. \tag{21}$$

Using Eq. (18) we can write

$$\left[\hat{V}, T^{(k)}[\{a\}, \{m\}]\right] = \sum_{i=1}^{k} \Lambda_{a_i}(m_i + \delta_{m_i,0}) T^{(k)}[\{a\}, \{m\}] \,. \tag{22}$$

$\mathcal{H}'^{(k)}$ contains terms of order t^k_α, denoted by $\mathcal{H}'^{[k]}$, which couple the states in different subspaces. By definition $[V, \mathcal{H}'^{[k]}] \neq 0$ and $\mathcal{H}'^{[k]}$ can be expressed in the following way

$$\mathcal{H}'^{[k]} = \sum_{\{a\}} \sum_{\{m\}} C^{(k)}_{\{a\}}(\{m\}) T^{(k)}[\{a\}, \{m\}], \qquad \sum_{i=1}^{k} m_i \neq 0 \,. \tag{23}$$

If in the each 'k'-th order step, we choose $\mathcal{S}^{(k)}$ in the following way $\mathcal{S}^{(k)} = \mathcal{S}^{(k-1)} + \mathcal{S}^{[k]}$, where $\mathcal{S}^{[k]}$ is the solution of the following equation

$$[i\mathcal{S}^{[k]}, V] = -\mathcal{H}'^{[k]} \tag{24}$$

and therefore equals

$$\mathcal{S}^{[k]} = -i \sum_{a,\{m\}} \frac{C^{(k)}_{\{a\}}(\{m\})}{\sum_{i=1}^{k} \Lambda_{a_i}(m_i + \delta_{m_i,0})} T^{(k)}[\{a\}, \{m\}], \qquad \sum_{i=1}^{k} m_i \neq 0 \,, \tag{25}$$

then the transformed Hamiltonian

$$\mathcal{H}'^{(k+1)} = e^{i\mathcal{S}^{(k)}}\mathcal{H}e^{-i\mathcal{S}^{(k)}} \tag{26}$$

contains terms of the order t^k/U^{k-1} which commute with the unperturbed Hamiltonian and mix states within each subspace only.

3. The Hubbard operators

To treat correlations properly, it is important to know, whether at the beginning or at the end of hopping process, a particular site is doubly occupied or not. The introduction of the so-called Hubbard operators[42] provides us with the tool necessary for such a full description of the local environment. The X_j^{ab}-operator is determined on each site of the lattice and describes all possible transitions between the local basis states: unoccupied $\mid 0\rangle$, single occupied with "up"-spin $\mid +\rangle$ and "down"-spin $\mid -\rangle$ and double occupied $\mid 2\rangle$. The original electron creation (annihilation) operators can be expressed in terms of the Hubbard operators in the following way:

$$c_{i,\alpha}^{\dagger} = X_i^{\alpha 0} + \alpha X_i^{2-\alpha} \qquad c_{i,\alpha} = X_i^{0\alpha} + \alpha X_i^{-\alpha 2}\,. \tag{27}$$

Correspondingly, in terms of creation (annihilation) operators the Hubbard operators have the form:

$$\begin{aligned} X_i^{\alpha 0} &= c_{i,\alpha}^{\dagger}(1 - n_{i,-\alpha})\,, & X_i^{2\alpha} &= -\alpha c_{i,-\alpha}^{\dagger} n_{i,\alpha}\,, \\ X_i^{\alpha-\alpha} &= c_{i,\alpha}^{\dagger} c_{i,-\alpha}\,, & X_i^{20} &= -\alpha c_{i,-\alpha}^{\dagger} c_{i,\alpha}^{\dagger}\,, \\ X_i^{00} &= (1 - n_{i,\uparrow})(1 - n_{i,\downarrow}), & X_i^{22} &= n_{i,\uparrow} n_{i,\downarrow}\,, \\ X_i^{\alpha\alpha} &= n_{i,\alpha}(1 - n_{i,-\alpha})\,. \end{aligned} \tag{28}$$

The Hubbard operators which contain even (odd) number of electron creation and annihilation operators are the Bose-like (Fermi-like) operators. They obey following on-site multiplication rules $X_i^{pq} X_i^{rs} = \delta_{q,r} X_i^{ps}$ and commutation relations:

$$[X_i^{pq}, X_j^{rs}]_{\pm} = \delta_{i,j}(\delta_{qr} X_j^{ps} \pm \delta_{ps} X_j^{rq}), \tag{29}$$

where the upper sign stands for the case when both operators are Fermi-like, otherwise the lower sign should be adopted.

It is straightforward to obtain, that

$$T_0^0 = \sum_{i,j} \sum_{\alpha} N_{i,j} t_\alpha X_i^{\alpha 0} X_j^{0\alpha}\,, \tag{30}$$

$$T_0^{do} = \sum_{i,j} \sum_{\alpha} N_{2i+1,j} t_\alpha X_{2i+1}^{2-\alpha} X_j^{-\alpha 2}\,, \quad T_0^{de} = \sum_{i,j} \sum_{\alpha} N_{2i,j} t_\alpha X_{2i}^{2-\alpha} X_j^{-\alpha 2}\,, \tag{31}$$

$$T_1^{o} = \sum_{i,j} \sum_{\alpha} \alpha N_{2i+1,j} t_\alpha X_{2i+1}^{2-\alpha} X_j^{0\alpha}\,, \quad T_1^{e} = \sum_{i,j} \sum_{\alpha} \alpha N_{2i,j} t_\alpha X_{2i}^{2-\alpha} X_j^{0\alpha}\,, \tag{32}$$

$$T_{-1}^{o} = \sum_{i,j} \sum_{\alpha} \alpha N_{i,2j+1} t_\alpha X_i^{\alpha 0} X_{2j+1}^{-\alpha 2}\,, \quad T_{-1}^{e} = \sum_{i,j} \sum_{\alpha} \alpha N_{i,2j} t_\alpha X_i^{\alpha 0} X_{2j}^{-\alpha 2}\,. \tag{33}$$

One can easily find that the spin $S = 1/2$ operators can be rewritten in terms of the X-operators in the following way

$$S_i^+ = c_{i,\uparrow}^\dagger c_{i,\downarrow} = X_i^{+-}\,, \quad S_i^- = c_{i,\downarrow}^\dagger c_{i,\uparrow} = X_i^{-+}\,, \quad S_i^z = \frac{1}{2}(X_i^{++} - X_i^{--})\,. \tag{34}$$

4. Effective Hamiltonian in the half-filled band case

In what follows we focus on the case of the half-filled band. In this particular case the minimum of the interacting energy is reached in the subspace with one electron per each site. Therefore, no hops are possible without increasing the number of doubly occupied sites and therefore, for any state in this subspace $\mid \Psi_{LS}\rangle$

$$T_{-1}^{e} \mid \Psi_{LS}\rangle = 0 \qquad T_{-1}^{o} \mid \Psi_{LS}\rangle = 0 \qquad T_0 \mid \Psi_{LS}\rangle = 0\,. \tag{35}$$

Equation (35) may be generalized to higher orders

$$T^k[m] \mid \Psi_{LS}\rangle = 0\,, \tag{36}$$

if $M_n^k[m] \equiv \sum_{i=n}^k m_i < 0$ for at least one value of n. Equation (36) can be used to eliminate many terms from the expansion for $\mathcal{H}'$ in the minimum $\langle \hat{V}\rangle$ subspace. Thus, in the fourth order of $\hat{T}$, the perturbed Hamiltonian has the form:

$$\mathcal{H}^{(4)} = -\frac{T^o_{-1}T^o_1}{U-\Delta} - \frac{T^e_{-1}T^e_1}{U+\Delta} - \frac{T^o_{-1}T^{0o}_0T^{0e}_0T^o_1}{U(U-\Delta)^2} + \frac{T^o_{-1}T^{do}_0T^{de}_0T^o_1}{U(U-\Delta)^2}$$
$$-\frac{T^e_{-1}T^{de}_0T^{do}_0T^e_1}{U(U+\Delta)^2} - \frac{T^e_{-1}T^0_0T^0_0T^e_1}{U(U+\Delta)^2} + \frac{T^e_{-1}T^e_1T^e_{-1}T^e_1}{(U+\Delta)^3}$$
$$-\frac{T^o_{-1}T^o_{-1}T^o_1T^o_1}{2(U-\Delta)^3} - \frac{T^e_{-1}T^e_{-1}T^e_1T^e_1}{2(U+\Delta)^3} + \frac{T^o_{-1}T^o_1T^o_{-1}T^o_1}{(U-\Delta)^3}$$
$$-\frac{T^o_{-1}T^{0o}_0T^{do}_0T^e_1}{U(U^2-\Delta^2)} - \frac{T^e_{-1}T^{de}_0T^{0e}_0T^o_1}{U(U^2-\Delta^2)} - \frac{T^o_{-1}T^{do}_0T^{0o}_0T^e_1}{U(U^2-\Delta^2)}$$
$$-\frac{T^e_{-1}T^{0e}_0T^{de}_0T^o_1}{U(U^2-\Delta^2)} - \frac{T^o_{-1}T^e_{-1}T^o_1T^e_1}{2U(U^2-\Delta^2)} - \frac{T^e_{-1}T^o_{-1}T^e_1T^o_1}{2U(U^2-\Delta^2)}$$
$$-\frac{T^e_{-1}T^o_{-1}T^o_1T^e_1}{2U(U+\Delta)^2} - \frac{T^o_{-1}T^e_{-1}T^e_1T^o_1}{2U(U+\Delta)^2}$$
$$+\frac{U}{(U^2-\Delta^2)^2}\left[T^o_{-1}T^o_1T^e_{-1}T^e_1 + T^e_{-1}T^e_1T^o_{-1}T^o_1\right]. \quad (37)$$

Using Eqs. (30)-(34), one can easily rewrite the products of T-terms in (37) via the Hubbard X and therefore the spin $S = 1/2$ operators, in the following way:

$$T^o_{-1}T^o_1 = -2t_\uparrow t_\downarrow \sum_i (S^x_iS^x_{i+1} + S^y_iS^y_{i+1}) - (t^2_\uparrow + t^2_\downarrow)\sum_i (S^z_iS^z_{i+1} - \frac{1}{4})$$
$$+ (t^2_\uparrow - t^2_\downarrow)\sum_i (-1)^i S^z_i\,, \quad (38)$$

$$T^e_{-1}T^e_1 = -2t_\uparrow t_\downarrow \sum_i (S^x_iS^x_{i+1} + S^y_iS^y_{i+1}) - (t^2_\uparrow + t^2_\downarrow)\sum_i (S^z_iS^z_{i+1} - \frac{1}{4})$$
$$- (t^2_\uparrow - t^2_\downarrow)\sum_i (-1)^i S^z_i\,. \quad (39)$$

Thus, within the second order approximation, the effective Hamiltonian reads

$$\mathcal{H}^{(2)}_{eff} = -\frac{1}{U-\Delta}T^o_{-1}T^o_1 - \frac{1}{U+\Delta}T^e_{-1}T^e_1$$
$$= J\sum_i \left(S^x_iS^x_{i+1} + S^y_iS^y_{i+1} + \gamma S^z_iS^z_{i+1}\right) + h_0\sum_i (-1)^i S^z_i\,, \quad (40)$$

where

$$J = \frac{4t_\uparrow t_\downarrow}{U(1-\lambda^2)}; \quad \gamma = \frac{t^2_\uparrow + t^2_\downarrow}{2t_\uparrow t_\downarrow}; \quad h_0 = \frac{2\Delta(t^2_\uparrow - t^2_\downarrow)}{U^2(1-\lambda^2)}\,, \quad (41)$$

and $\lambda = \Delta/U$.

Similarly, using Eqs. (30)-(34),after straightforward but lengthly calculations one can express all 4-th order in T term in (37) the spin $S = 1/2$ operators, in the following way:

$$\mathcal{H}_{eff} = \sum_i \left[J_\perp \left(S_i^x S_{i+1}^x + S_i^y S_{i+1}^y \right) + J_\parallel \left(S_i^z S_{i+1}^z - 1/4 \right) \right] + h_0 \sum_i (-1)^i S_i^z$$
$$+ \sum_i \left[J'_\perp \left(S_i^x S_{i+2}^x + S_i^y S_{i+2}^y \right) + J'_\parallel \left(S_i^z S_{i+2}^z - 1/4 \right) \right]$$
$$+ \sum_i (-1)^i \left[W_\perp \left(S_i^x S_{i+1}^x + S_i^y S_{i+2}^y \right) S_{i+2}^z + W_\parallel S_i^z S_{i+1}^z S_{i+2}^z \right], \qquad (42)$$

where

$$J_\perp = \frac{4t_\uparrow t_\downarrow}{U(1-\lambda^2)} - \frac{8t_\uparrow t_\downarrow (t_\uparrow^2 + t_\downarrow^2)}{U^3(1-\lambda^2)^3}(1+3\lambda^2), \qquad (43)$$

$$J_\parallel = \frac{2(t_\uparrow^2 + t_\downarrow^2)}{U(1-\lambda^2)} - \frac{4t_\uparrow^2 t_\downarrow^2 + 6(t_\uparrow^4 + t_\downarrow^4)}{U^3(1-\lambda^2)^3}(1+3\lambda^2) \qquad (44)$$

$$J'_\perp = \frac{4t_\uparrow^2 t_\downarrow^2}{U^3(1-\lambda^2)^3}\left(1+4\lambda^2-\lambda^4\right), \qquad (45)$$

$$J'_\parallel = \frac{2(t_\uparrow^4 + t_\downarrow^4)^2}{U^3(1-\lambda^2)^3}\left(2+3\lambda^2-\lambda^4\right) - \frac{4t_\uparrow^2 t_\downarrow^2}{U^3(1-\lambda^2)^3}\left(1-\lambda^2\right), \qquad (46)$$

$$W_\perp = 32\,\frac{\Delta(t_\uparrow^2 - t_\downarrow^2)t_\uparrow t_\downarrow}{U^4(1-\lambda^2)^3}, \qquad (47)$$

$$W_\parallel = 32\,\frac{\Delta(t_\uparrow^4 - t_\downarrow^4)}{U^4(1-\lambda^2)^3}, \qquad (48)$$

$$h_0 = \frac{2\Delta(t_\uparrow^2 - t_\downarrow^2)}{U^2(1-\lambda^2)} + \frac{4\Delta(t_\uparrow^4 - t_\downarrow^4)}{U^4(1-\lambda^2)^3}(7+3\lambda^2). \qquad (49)$$

The obtained effective Hamiltonian is that of a frustrated Heisenberg chain with alternating three spin term and alternating magnetic field.

5. Conclusion

In this paper we have derived the effective spin Hamiltonian which describes the low-energy sector of the one-dimensional half-filled spin-asymmetric ionic-Hubbard model in the limit of strong on-site repulsion. We have shown that the effective Hamiltonian, which describes spin degrees of freedom of the initial lattice electron system is given by the Hamiltonian of the frustrated XXZ Heisenberg chain with an additional three spin coupling term in the presence of alternating magnetic field. The intensity of this unconventional three spin coupling and the amplitude of the magnetic field are proportional to the multiple of the ionic parameter Δ with the difference of "up" and "down" spin hopping amplitudes $t_{\uparrow}-t_{\downarrow}$. Therefore, in marked contrast with the spin isotropic case, infrared properties of the spin-asymmetric ionic Hubbard model are described by the spin-chain model with implicitly broken by the alternating magnetic field translation symmetry.

Acknowledgments

It is our pleasure to thank Dionys Baeriswyl and Alvaro Ferraz for many interesting discussions and comments. GIJ would like to acknowledge the hospitality of the International Institute of Physics UFRN, Natal, Brazil where part of this work was done and the Ministry of Science and Innovations of Brazil for support. This work was supported by the Georgian National Science Foundation through the Grant No. ST09/4-447 and by the SCOPES Grant IZ73Z0-128058.

References

1. N. F. Mott, *Metal-Insulator Transitions*, 2nd edition (London, Taylor and Francis 1990).
2. F. Gebhard, *The Mott Metal-Insulator Transitions* (Berlin Springer, 1997).
3. M. Imada, A. Fujimori, and Y. Tokura, *Rev. Mod. Phys.* **70**, 1039 (1998).
4. D. H. Lee and R. Shankar, *Phys. Rev. Lett.* **65**, 1490 (1990).
5. J. Hubbard, *Proc. R. Soc. London Ser. A* **276**, 238 (1963).
6. E. Lieb and F. W. Wu, *Phys. Rev. Lett.* **20**, 1445 (1968).
7. F. H. L. Essler and H. Frahm, *Phys. Rev. B* **60**, 8540 (1999).
8. H. Frahm and V. E. Korepin, *Phys. Rev. B* **42**, 10553 (1990).
9. P. W. Anderson, *Phys. Rev* **115**, 2 (1959).
10. M.Takahashi, *Jour. Phys. C: Solid St. Phys.* **10**, 1289 (1977).
11. A. H. MacDonald, S. M. Girvin and D. Yoshioka, *Phys. Rev. B* **37**, 9753 (1988).
12. N. Datta, R. Fernández and J. Frölich, *J. Stat. Phys.* **96**, 545 (1999).
13. N. Nagaosa and J. Takimoto, *J. Phys. Soc. Jpn.* **55**, 2735 (1986). N. Nagaosa, ibid. **55**, 2754 (1986).
14. T. Egami, S. Ishihara, and M. Tachiki, *Science* **261**, 1307 (1993).
15. S. Ishihara, T. Egami, and M. Tachiki, *Phys. Rev. B* **49**, 8944 (1994).
16. R. Resta and S. Sorella, *Phys. Rev. Lett.* **74**, 4738 (1995).
17. M. Fabrizio, A.O. Gogolin and A.A. Nersesyan, *Phys. Rev. Lett.* **83** 2014 (1999).
18. N. Gidopoulos, S. Sorella, and E. Tosatti, *Eur. Phys. J. B* **14**, 217 (2000).
19. T. Wilkens and R. M. Martin, *Phys. Rev. B* **63**, 235108 (2001).

20. M. E. Torio, A. A. Aligia, and H. A. Ceccatto, *Phys. Rev. B* **64**, 121105 (2001).
21. A. P. Kampf, M. Sekania, G. I. Japaridze and P. Brune, *J. of Phys.C: Cond. Mat.* **15**, 5895 (2003).
22. A. A. Aligia, *Phys. Rev. B* **69**, 041101 (2004).
23. S. R. Manmana, V. Meden, R. M. Noack and K. Schoenhammer, *Phys. Rev. B* **70**, 155115 (2004).
24. M. E. Torio, A. A. Aligia, G. I. Japaridze and B. Normand, *Phys. Rev. B* **73**, 115109 (2006).
25. L. Tincani, R. M. Noack and D. Baeriswyl, *Phys. Rev. B* **79**, 165109 (2009).
26. J. Hubbard and J. B. Torrance, *Phys. Rev. Lett.* **47**, 1750 (1981).
27. U. Brandt, *J. Low Temp. Phys.* **84**, 477 (1991).
28. R. Lyzwa and Z. Domanski, *Phys. Rev. B* **50**, 11381 (1994).
29. G. Fäth, Z. Domański and R. Lemański, *Phys. Rev. B* **52** 13910 (1995);
30. Z. Domanski, R. Lemanski, and G. Fath, *J. Phys.: Condens. Matter* **8**, L261 (1996).
31. L. M. Falicov and J. C. Kimball, *Phys. Rev. Lett.* **22**, 997 (1969).
32. W. V. Liu, F. Wilczek, and P. Zoller, *Phys. Rev. A* **70**, 033603 (2004).
33. D. Ueltschi, *J. Stat. Phys.* **116**, 681 (2004).
34. A. M. C. Souza and C. A. Macedo, *Physica B* **384**, 196 (2006).
35. Z. G. Wang, Y. G. Chen, and S. J. Gu, *Phys. Rev. B* **75**, 165111 (2007).
36. J. Silva-Valencia, R. Franco, and M. S. Figueira, *Physica B* **398**, 427 (2007).
37. S. J. Gu, R. Fan, and H. Q. Lin, *Phys. Rev. B* **76**, 125107 (2007).
38. A. F. Ho, M. A. Cazalilla, and T. Giamarchi, *Phys. Rev. B* **77**, 085110 (2008).
39. O. Mandel el al., *Phys. Rev. Lett.* **91**, 010407 (2003); O. Mandel et al., *Nature* (London) **425**, 937 (2003).
40. M. Takahashi, Thermodynamics of the one-dimensional solvable models,(Cambridge, University Press 1999).
41. A. Luther, *Phys. Rev. B* **14**, 2153 (1976).
42. J. Hubbard, *Proc. R. Soc. London Ser. A* **285** 542 (1965); *ibid Proc. R. Soc. London Ser. A* **296**, 82 (1965).

HYPERBOLICITY VS RANDOMNESS IN COSMOLOGICAL PROBLEMS

V.G.Gurzadyan[1,2] and A.A.Kocharyan[2,3]

[1,2] *Center for Cosmology and Astrophysics, Alikhanian National Laboratory, and Yerevan State University, Armenia*
[3] *School of Mathematical Sciences, Monash University, Clayton, Australia*

We discuss the hyperbolicity in perturbed Friedmann-Robertson-Walker Universe and the cummulative role of inhomogeneities and their link to observations. The degree of the randomness defined by the Kolmogorov stochasticity parameter, acts as an efficient descriptor for astrophysical datasets such as those of the cosmic microwave background radiation and X-ray clusters of galaxies.

1. Introduction

Gauge invariant expressions in the relativistic and Newtonian treatments have been used for the study of the formation of the large scale matter structure of the Universe, e.g.[1] We will turn to the large scale structure issue, recalling that, the behavior of the propagating photon beams in the spaces with Roberston-Walker metric, determined by the scalar curvature, is well known. The behavior of the propagating photon beams in the spaces with Roberston-Walker metric, determined by the scalar curvature, is well known. The inhomogeneous distribution of matter in the Universe adds new features to those properties, as we discuss below. Particularly, it appears that the underdense regions, the voids, known observationally from the galactic surveys,[2] can behave as regions inducing the hyperbolicity of the photon beams. Although the physical parameters of the voids are not well known yet, their cummulative effect may be observed imprinted in the properties of the Cosmic Microwave Background (CMB).[3]

We then discuss another feature, the degree of randomness of signals, which can also have observational conseqeunces. Kolmogorov stochasticity parameter[4,5] enables to compare the randomness in various sequences. Kolmogorov's parameter appears an informative descriptor when applied, for example, to the CMB temperature data sequences, namely, it allows to locate regions of different randomness in the sky maps[6] among those is a non-Gaussian anomaly, known as Cold Spot, with enhanced value of that parameter. Moreover the latter also varies along the radius as predicted for the voids in the large scale matter distribution in the Universe; the void nature of the Cold Spot is supported also by other studies.

2. Instability of null geodesics in perturbed Friedmann-Robertson-Walker Universe

The properties of null geodesics in perturbed Friedmann-Robertson-Walker Universe can be studied by the equation of deviation of close geodesics. Consider $V = M \times R^1$ (3+1)-D smooth manifold with a perturbed RW metric

$$\begin{aligned} ds^2 &= g_{\mu\nu} dx^\mu dx^\nu \\ &= -(1+2\phi)dt^2 + (1-2\phi)a^2(t)d\sigma^2, \end{aligned} \tag{1}$$

where

$$\gamma_{mn} = \left(1 + \tfrac{k}{4}\left[(x^1)^2 + (x^2)^2 + (x^3)^2\right]\right)^{-2} \delta_{mn}$$

is the metric of 3-sphere ($k = 1$), 3-hyperboloid ($k = -1$), flat 3-space ($k = 0$), respectively. We assume (cf.[8]) that

$$|\phi| \ll 1,$$

also,

$$\left(\frac{\partial \phi}{\partial t}\right)^2 \ll a^{-2} \, \|\nabla\phi\|^2,$$

and

$$\|\nabla\phi\|^2 \ll |\Delta\phi|,$$

where

$$\|\nabla\phi\|^2 = \gamma^{mn} \frac{\partial \phi}{\partial x^m} \frac{\partial \phi}{\partial x^n}, \quad \Delta\phi = -\nabla^2\phi.$$

Substituting $\psi = \phi$ and $f = 2\phi$ and in view of $e^{2\phi} \sim 1 + 2\phi$ and $e^{-2\phi} \sim 1 - 2\phi$, we see that the metric gets the same form as for $d = 3$, see.[9]

The Lyapunov exponents can be obtained from the Riemann scalar curvature for the reduced space metric $\tilde{\gamma} = e^{-4\phi}\gamma$. Then

$$R = 6 \, e^{4\phi} \left[k - \tfrac{4}{3}\left(\Delta\phi + \|\nabla\phi\|^2\right)\right] \tag{2}$$

where[8]

$$-\Delta\phi = 4\pi a^2 \delta\rho.$$

Therefore,

$$\frac{R}{6} \simeq k - \frac{4}{3}\,\Delta\phi = k + \frac{16\pi}{3}\,a^2\delta\rho.$$

Then, from

$$\begin{aligned}
\bar{\rho} &= \rho_0\,\frac{a_0^3}{a^3},\\
\left(\frac{\dot{a}}{a}\right)^2 &= -\frac{k}{a^2} + \frac{\Lambda}{3} + \frac{8\pi\rho_0}{3}\,\frac{a_0^3}{a^3},\\
\frac{a}{a_0} &= \frac{1}{1+z}; \quad H = \frac{\dot{a}}{a},
\end{aligned}$$

$$\eta(t_0) - \eta(t) = \int_t^{t_0} \frac{d\tau}{a(\tau)} = \frac{1}{a_0}\int_0^z \frac{d\xi}{H(\xi)}$$

we get

$$\begin{aligned}
1 &= \Omega_k + \Omega_\Lambda + \Omega_m,\\
H^2 &= H_0^2\left[\Omega_\Lambda + [1 - \Omega_\Lambda + \Omega_m z]\,(1+z)^2\right],
\end{aligned}$$

where

$$\begin{aligned}
H_0^2 &= -\frac{k}{a_0^2} + \frac{\Lambda}{3} + \frac{8\pi\rho_0}{3},\\
\Omega_k &= -\frac{k}{a_0^2 H_0^2}, \quad \Omega_\Lambda = \frac{\Lambda}{3H_0^2}, \quad \Omega_m = \frac{8\pi\rho_0}{3H_0^2}.
\end{aligned}$$

Thus,

$$\eta(t_0) - \eta(t) = (a_0 H_0)^{-1}\,\lambda(z, \Omega_\Lambda, \Omega_m),$$

where

$$\lambda(z, \Omega_\Lambda, \Omega_m) = \int_0^z \frac{d\xi}{\sqrt{\Omega_\Lambda + [1 - \Omega_\Lambda + \Omega_m \xi]\,(1+\xi)^2}}. \qquad (3)$$

And

$$\frac{R}{6} = (a_0 H_0)^2\,(-\Omega_k + 2\delta_0\Omega_m) = (a_0 H_0)^2\,r,$$

where

$$\delta_0 \equiv \frac{\delta\rho_0}{\rho_0}.$$

Finally, the equation of deviation of geodesics gets the form

$$\frac{d^2\ell}{d\lambda^2} + r\,\ell = 0, \tag{4}$$

where

$$r = -\Omega_k + 2\delta_0\Omega_m. \tag{5}$$

Since by definition $\delta_0 \geq -1$, then, $r \geq -\Omega_k - 2\Omega_m$.

These equations describe the mixing properties of null geodesics, including for the propagation of cosmic microwave background both due the global geometry and local perturbations of metric (lensing). When $\delta_0 = 0$, and $r = -\Omega_k$, while in the perturbed FRW Universe the resulting effect is different: namely, the underdense regions with $\delta_0 \leq 0$, *the voids, contribute to the hyperbolicity* even if the global curvature is zero.[9] Importantly, this conclusion does not require compactness of M.

So, if in a RW Universe the scalar curvature $k = 0, \pm 1$ completely defines the behavior of the propagation of photon beams, we see that even when $k = 0$, $\Omega_k = 0$ the density contrast affects the propagation.

The underdense regions, the voids with

$$\delta_{void} = (\rho_{void} - \rho_0)/\rho_0 < 0$$

act as hyperbolic regions deviating the null geodesics close to the walls of the voids.

Then, a criterion for the cumulative role of voids in an inhomogeneous Universe in terms of observable characteristic parameters of the filaments, will be of particular interest.

For $\Omega_k = 0$, and $\Omega_\Lambda + \Omega_m = 1$, we have

$$\frac{d^2\ell}{d\tau^2} + \delta_0\,\ell = 0, \tag{6}$$

where

$$\tau(z, \Omega_m) = \sqrt{2\Omega_m} \int_0^z \frac{d\xi}{\sqrt{1 + \Omega_m[(1+\xi)^3 - 1]}}. \tag{7}$$

For example, at periodicity in the line-of-sight distribution of voids,[10] i.e. δ_0 periodic, $\delta_0(\tau + \tau_\kappa + \tau_\omega) = \delta_0(\tau)$ and

$$\delta_0 = \begin{cases} +\kappa^2 & 0 < \tau < \tau_k\ , \\ -\omega^2 & \tau_\kappa < \tau < \tau_\kappa + \tau_\omega\ , \end{cases} \tag{8}$$

the solution to Eq.(6) is unstable if (see e.g.[7]) $\mu = \nu - 2 > 0$, where

$$\nu = \left| 2\cos(\omega\tau_\omega)\cosh(\kappa\tau_\kappa) + \left(\frac{\kappa}{\omega} - \frac{\omega}{\kappa}\right)\sin(\omega\tau_\omega)\sinh(\kappa\tau_\kappa) \right|.$$

At $0 < \omega\tau_\omega \ll 1$ and $0 < \kappa\tau_\kappa \ll 1$. It is easy to see that if

$$\bar{\delta_0} \equiv \int_0^{\tau_\kappa+\tau_\omega} \delta_0 d\tau = \kappa^2\tau_\kappa - \omega^2\tau_\omega \neq 0, \tag{9}$$

then $\mu \approx \bar{\delta_0}(\tau_\kappa + \tau_\omega)$. Thus, $\mu < 0$, if $\bar{\delta_0} < 0$, and $\mu > 0$, if $\bar{\delta_0} > 0$. On the other hand, if $\bar{\delta_0} = 0$ then one can show that

$$\mu \approx -\frac{[(\kappa\tau_\kappa)^2 + (\omega\tau_\omega)^2]^2}{12} < 0. \tag{10}$$

The solutions are stable, if $\bar{\delta_0} \leq 0$ and are unstable, if $\bar{\delta_0} > 0$, i.e.

$$\bar{\delta_0} \sim -\delta_{void}L_{void} - \delta_{wall}L_{wall}; \tag{11}$$

$L_{void}, L_{wall}, \delta_{void}, \delta_{wall}$ are the scales and the density contrasts for the voids and the walls, respectively.

3. The degree of randomness of cosmic microwave background datasets

First let us recall the definition of the Kolmogorov stochasticity parameter.[4,5] Let $\{X_1, X_2, \ldots, X_n\}$ be n independent values of the same real-valued random variable X ordered in an increasing manner $X_1 \leq X_2 \leq \cdots \leq X_n$ and let

$$F(x) = P\{X \leq x\}$$

be a **cumulative** distribution function (CDF) of X. Their *empirical distribution function* $F_n(x)$ is defined by the relations

$$F_n(x) = \begin{cases} 0\,, & x < X_1\,; \\ k/n\,, & X_k \leq x < X_{k+1},\quad k = 1, 2, \ldots, n-1\,; \\ 1\,, & X_n \leq x\,. \end{cases}$$

Kolmogorov's stochasticity parameter λ_n has the following form

$$\lambda_n = \sqrt{n}\,\sup_x |F_n(x) - F(x)|\,. \tag{12}$$

Kolmogorov proved that for any continuous **CDF** F

$$\lim_{n\to\infty} P\{\lambda_n \leq \lambda\} = \Phi(\lambda)\,,$$

where $\Phi(0) = 0$,

$$\Phi(\lambda) = \sum_{k=-\infty}^{+\infty} (-1)^k\, e^{-2k^2\lambda^2}\,,\quad \lambda > 0\,, \tag{13}$$

the convergence is uniform, and Φ (Kolmogorov's distribution) is independent on F. KSP is applied to measure the objective stochasticity degree of datasets.[5] It is easy to see that the inequalities $0.3 \leq \lambda_n \leq 2.4$ can be considered as practically certain.[4]

It is very important to realise that KSP is different from the K-S (Kolmogorov, Smirnov) test. K-S test is a test for the equality of continues 1D probability distributions. KSP is an objectively measurable degree of randomness of observable events. As can be easily seen they are different with completely different objectives.

For large enough n and random sequence x_n the Kolmogorov stochasticity parameter λ_n will have a distribution close to $\Phi(\lambda)$. If the sequence is not random, the distribution will be different. So, $\Phi(\lambda)$ denotes the degree of randomness of a sequence. For each studied sequence one can get only one value of λ_n. But that is not enough to conclude on the degree of randomness of that sequence. Therefore, considering sequences of large enough length n, we has to split them into subsequences, and then calculate λ_n for each subsequence.

A broad class of sequences can be represented as follows

$$z_n = \alpha x_n + (1 - \alpha) y_n, \tag{14}$$

where x_n are random sequences and

$$y_n = \frac{a\, n \,\mathrm{mod} b}{b}, \tag{15}$$

are regular sequences, a and b are mutually fixed prime numbers; both sequences are of mod(0,1) and have uniform distribution, α thus indicating the fraction of random and regular sequences.[12]

This technique has been applied to estimate the degree of randomness of the CMB datasets obtained by the Wilkinson Microwave Anisotropy Probe (WMAP).[11] CMB is a remarkable sequence for this purpose since its cumulative distribution function happens to be known, namely, it is close to Gaussian. Then for Gaussian distribution function F and via a properly developed strategy,[6] one can obtain the distribution of the function Φ for CMB over the sky; the resulting Kolmogorov 3D-map is shown in Figure 1. This not only separates readily the non-cosmological signal of the Galactic disk from CMB but also has other remarkable consequences, i.e. it enables to detect point sources (quasars, blazars) in the CMB maps, to study the properties of the voids, to reveal the random component in the CMB signal.[12] When applied to X-ray data of the clusters of galaxies obtained by XMM-Newton satellite, one gets the correlated signal peculiar to those objects.[13]

Acknowledgments

VG acknowledges the financial support by CREST and NASA Research Centers, NC.

Fig. 1. The 3D view of the Kolmogorov map for the cosmic microwave background data obtained by the Wilkinson Microwave Anisotropy Probe.

References

1. Flender S.F., Schwarz D.J., *arXiv:1207.2035*
2. Hoyle F., Vogeley M.S., Pan D., *arXiv:1205.1843*
3. Mennella A., for the Planck Collaboration, *arXiv:1110.2051*
4. Kolmogorov A.N., *Giorn.Ist.Ital.Attuari*, **4**, 83 (1933)
5. Arnold V.I., *Usp. Mat. Nauk*, **63** (2008) 5
6. Gurzadyan V.G. & Kocharyan A.A., *A & A*, **492** (2008) L33
7. Arnold V.I., *Mathematical Methods in Classical mechanics*, Springer, 1989.
8. Holz D.E., Wald R.M., *Phys. Rev.*, **D58** (1998) 063501
9. Gurzadyan V.G. & Kocharyan A.A., *Europhys.Lett.* **86** (2009) 29002
10. Gurzadyan V.G. & Kocharyan A.A., *A & A*, **493** (2009) L61
11. Jarosik N., Bennett C.L., et al, *ApJS*, **192** (2011) 14
12. Gurzadyan V.G., Allahverdyan A.E., Ghahramanyan T., Kashin A.L., Khachatryan H.G., Kocharyan A.A., Mirzoyan S., Poghosian E., Vetrugno D., Yegorian G., *A & A*, **525** (2011) L7
13. Gurzadyan V.G., Durret F., Ghahramanyan T., Kashin A.L., Khachatryan H.G., Poghosian E., *Europhys. Lett.* **95**, (2011) 69001

ELECTROMAGNETIC PROPERTIES OF NEUTRINOS AT AN INTERFACE

A. N. Ioannisian[a,b], D. A. Ioannisian[a,b] and N. A. Kazarian[b]

[a] *Yerevan Physics Institute, Alikhanian Br. 2, 375036 Yerevan, Armenia*
[b] *Institute for Theoretical Physics and Modeling, 375036 Yerevan, Armenia*

We have calculated the electromagnetic properties of neutrinos at an interface. Particulary, we have measured the transition radiation process by neutrinos $\nu \to \nu\gamma$ at an interface of two media. The medium fulfills the dual purpose of inducing an effective neutrino-photon vertex and of modifying the photon dispersion relation. The transition radiation occurs when at least one of those two quantities have different values in different media. The neutrinos are taken to be with only standard-model couplings and neutrino mass is ignored due to its negligible contribution. We present a result for the probability of the transition radiation which is both accurate and analytic. For example, $E_\nu = 1$MeV neutrino crossing polyethylene-vacuum interface the transition radiation probability is about 10^{-39} and the energy intensity is about 10^{-34}eV. At the surface of the neutron stars the transition radiation probability may be $\sim 10^{-20}$.

Introduction.- As it is well known, in many astrophysical environments the absorption, emission, or scattering of neutrinos occurs in medium, in the presence of magnetic fields[1] or at the interface of two media. Of particular conceptual interest are those reactions which have no counterpart in vacuum, notably the plasmon decay $\gamma \to \bar{\nu}\nu$, the Cherenkov and transition radiation processes $\nu \to \nu\gamma$. These reactions do not occur in vacuum because they are kinematically forbidden and because neutrinos do not couple to photons. In the presence of a medium (or a magnetic field), neutrinos acquire an effective coupling to photons by virtue of intermediate charged particles. Medium (or external field) modify the dispersion relations of all particles so that phase space is opened for neutrino-photon reactions of the type $1 \to 2 + 3$. The violation of the translational invariance at the direction from one medium into another leads to the non conservation of the momentum at the same direction so that transition radiation becomes kinematically allowed.

The theory of transition radiation by charged particle is build on classical theory of electrodynamics and has been developed in[2] and.[3] In[4] the quantum field theory was used for describing the phenomenon. But the neutrinos have very tiny masses, therefore one has to use the quantum field theory approach in order to study transition radiation by neutrinos.

The dominant source for neutrinos in many types of stars is the plasma process and thus is of great practical importance in astrophysics.[1] The plasma process $\gamma \to \bar{\nu}\nu$ was first studied in.[5] The ν-γ-coupling is enabled by the presence of the electrons of the background medium,[6] and the process is kinematically allowed because the photons acquire essentially an effective mass. In a plasma there are electromagnetic excitations, namely, the longitudinal plasmons, γ_L. The four-momentum of those excitations are space like for certain values of energy. The Cherenkov decay $\nu \to \nu\gamma_L$ was studied in.[7]

In the presence of magnetic field an effective ν-γ-coupling is induced. The Cherenkov decay in a magnetic field was calculated in.[8] The $\gamma \to \bar{\nu}\nu$ decay rate was calculated in,[9] under the assumption that phase space is opened by a suitable medium- or field-induced modification of the photon refractive index.

The medium causes an effective ν-γ-vertex by standard-model neutrino couplings to the background electrons. In addition, the medium changes the photon dispersion relation. We neglect neutrino masses and medium-induced modifications of their dispersion relation due to their negligible role. Therefore, in the preset paper the studies of neutrino properties are extended to the case of neutrinos with only standard-model couplings.

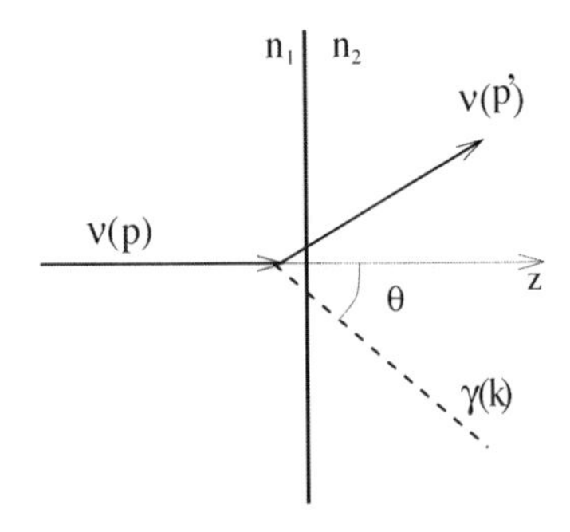

Fig. 1. Transition radiation by neutrino at the interface of two media with refractive indexes n_1 and n_2.

The transition radiation $\nu \to \nu\gamma$ at the interface of two media with different refractive indices was studied in[10] with the assumption of a neutrino magnetic dipole moment.

A detailed literature search reveals that neutrino transition radiation has been first studied earlier in.[11] They used vacuum induced ν-γ vertex ("neutrino toroid dipole moment") for the $\nu \to \nu\gamma$ matrix element. We disagree with their treatment of the process. The medium itself induces ν-γ vertex and the vacuum induced vertex can be treated as a radiation correction to the medium induced one. We found that the result of[11] for the transition radiation probability is by more than three orders of magnitude, $(\frac{8\alpha}{\pi})^2$, smaller than our result.

We proceed by deriving a general expression for the transition radiation probability (assuming a general ν-γ-vertex) in the quantum field theory. We derive the standard-model effective vertex in a background of electrons, then we calculate the transition radiation probability by performing semi-analytical integrations and summarize our findings.

Transition Radiation.- Let's consider a neutrino crossing of those two media with refraction indices n_1 and n_2 . (see Fig. 1). In terms of the matrix element $\mathcal{M}$ the transition radiation probability of the process $\nu \to \nu\gamma$ is

$$W = \frac{1}{(2\pi)^3}\frac{1}{2E\beta_z}\int \frac{d^3\mathbf{p}'}{2E'}\frac{d^3\mathbf{k}}{2\omega}\sum_{pols}\left|\int_{-\infty}^{\infty} dz e^{i(p_z-p'_z-k_z)z}\mathcal{M}\right|^2 \\ \times\ \delta(E-E'-\omega)\delta(p'_x+k_x)\delta(p'_y+k_y)\ . \tag{1}$$

The, $p = (E, \mathbf{p})$, $p' = (E', \mathbf{p}')$, and $k = (\omega, \mathbf{k})$ are the four momenta of the incoming neutrino, outgoing neutrino, and photon, respectively and $\beta_z = p_z/E$. The sum is over the photon polarizations.

We have to neglect the neutrino masses and the deformation of its dispersion relations due to the forward scattering. Thus we assume that the neutrino dispersion relation is precisely light-like so that $p^2 = 0$ and $E = |\mathbf{p}|$.

The formation zone length of the medium is

$$|p_z - p'_z - k_z|^{-1}. \tag{2}$$

The integral over z in eq. (1) oscillates beyond the length of the formation zone. Therefore, the contributions to the process from the depths over the formation zone length may be neglected. The neutrino transfers the z momentum $(p_z - p'_z - k_z)$ to the medium. Since photon propagation in the medium suffers from the attenuation(absorption) the formation zone length must be limited by the attenuation length of the photons in the medium when the latter is shorter than the formation zone length.

After integration of (1) over $\mathbf{p}'$ and z we find

$$W = \frac{1}{(2\pi)^3}\frac{1}{8E\beta_z}\int \frac{|\mathbf{k}|^2 d|\mathbf{k}|}{\omega E'\beta'_z}\sin\theta\, d\theta\, d\varphi$$

$$\times \sum_{pols}\left|\frac{\mathcal{M}^{(1)}}{p_z - p_z'^{(1)} - k_z^{(1)}} - \frac{\mathcal{M}^{(2)}}{p_z - p_z'^{(2)} - k_z^{(2)}}\right|^2, \tag{3}$$

where $\beta'_z = p'_z/E'$, θ is the angle between the emitted photon and incoming neutrino. $\mathcal{M}^{(1,2)}$ are matrix elements of the $\nu \to \nu\gamma$ in each medium. $k_z^{(i)}$ and $p_z'^{(i)}$ are z components of momenta of the photon and of the outgoing neutrino in each medium.

The main contribution to the process comes from the large formation zone lengths and, thus, small angle θ. Therefore, the rate of the process does not depend on the angle between the momenta of the incoming neutrino and the boundary surface of two media (if that angle is not close to zero). The integration over φ drops out and we may replace $d\varphi \to 2\pi$. $k_z^{(i)}$ and $p_z'^{(i)}$ have the forms

$$k_z^{(i)} = n^{(i)}\omega\cos\theta,\quad p_z'^{(i)} = \sqrt{(E-\omega)^2 - {n^{(i)}}^2\omega^2\sin^2\theta}\ , \tag{4}$$

where $n^{(1,2)} = |\mathbf{k}|^{(1,2)}/\omega$.

Let us consider if medium is isotropic and homogeneous. In this case, the polarization tensor, $\pi^{\mu\nu}$, of the photon is uniquely characterized by a pair of two polarization functions which are often chosen to be the longitudinal and transverse polarization functions. They can be projected from the full polarization matrix. In this paper we are interested in transverse photons, since they may propagate in vacuum as well. The transverse polarization function is

$$\pi_t = \frac{1}{2}T_{\mu\nu}\pi^{\mu\nu},\ T^{\mu\nu} = -g^{\mu i}(\delta_{ij} - \frac{\mathbf{k}_i\mathbf{k}_j}{\mathbf{k}^2})g^{j\nu}. \tag{5}$$

The dispersion relation for the photon in the medium is the location of its pole in the effective propagator (that is gauge independent)

$$\frac{1}{\omega^2 - \mathbf{k}^2 - \pi_t} \tag{6}$$

So, after summation over transverse polarizations the photons density matrix has a form

$$\sum_{trans}\epsilon_\mu\epsilon_\nu = -T_{\mu\nu} = g_{\mu i}(\delta_{ij} - \frac{\mathbf{k}_i\mathbf{k}_j}{\mathbf{k}^2})g_{j\nu}\ . \tag{7}$$

Neutrino-photon vertex- In the medium the photons couple to neutrinos via interactions with electrons by the amplitudes shown in Fig 2. One may take into account similar graphs with nuclei as well, but their contribution are usually negligible. When the photon energy is below the weak scale ($E \ll M_W$) one may use four-fermion interactions and the matrix element for the ν-γ vertex can be written in the form

$$M = -\frac{G_F}{\sqrt{2}e} Z\epsilon_\mu \bar{u}(p')\gamma_\nu(1-\gamma_5)u(p)\ (g_V \pi_t^{\mu\nu} - g_A \pi_5^{\mu\nu}) \tag{8}$$

$$= \frac{G_F}{\sqrt{2}e} Z\epsilon_\mu \bar{u}(p')\gamma_\nu(1-\gamma_5)u(p)$$
$$\times g^{\mu i}\left(g_V \pi_t(\delta_{ij} - \frac{\mathbf{k}_i\mathbf{k}_j}{\mathbf{k}^2}) - i g_A \pi_5 \epsilon_{ijl}\frac{\mathbf{k}_l}{|\mathbf{k}|}\right) g^{j\nu}, \tag{9}$$

here

$$g_V = \left\{ \begin{array}{l} 2\sin^2\theta_W + \frac{1}{2} \text{ for } \nu_e \\ 2\sin^2\theta_W - \frac{1}{2} \text{ for } \nu_\mu, \nu_\tau \end{array} \right. , \quad g_A = \left\{ \begin{array}{l} +\frac{1}{2} \text{ for } \nu_e \\ -\frac{1}{2} \text{ for } \nu_\mu, \nu_\tau \end{array} \right. . \tag{10}$$

$\pi_t^{\mu\nu}$ is the polarization tensor for transverse photons, while $\pi_5^{\mu\nu}$ is the axialvecor-vector tensor.[14] Further, Z is photon's wave-function renormalization factor. For the physical circumstances of interest to us, the photon

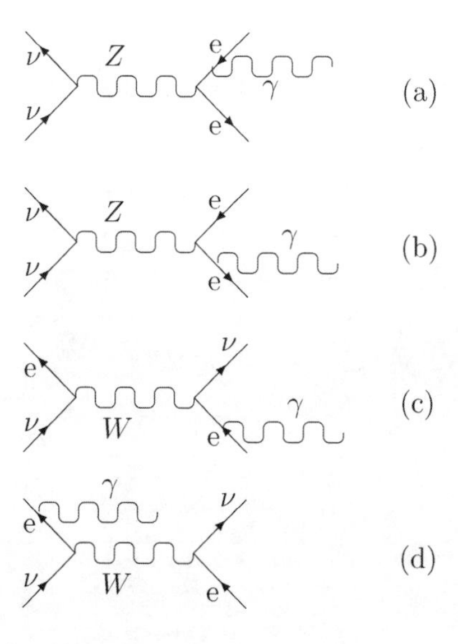

Fig. 2. Neutrino-photon coupling in electron background. (a,b) Z-γ-mixing. (c,d) "Penguin" diagrams (only for ν_e).

refractive index will be very close to unity so that we will be able to use the vacuum approximation $Z = 1$.

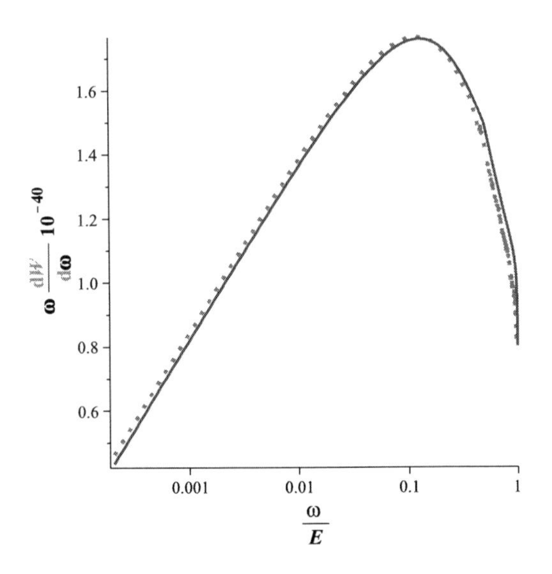

Fig. 3. The energy spectrum of the transition radiation by electron neutrinos at the interface of medium with plasma frequency $\omega_p = 20$eV and vacuum. The energy of the incoming neutrino is $E = 1$ MeV. The dot line is numerical and the solid line is semi-analytical integration over angle between photon and incoming neutrino momenta.

Transition radiation probability.- Armed with these results we may now turn to an evaluation of the $\nu \to \nu\gamma$ probability at the interface of medium and vacuum. We find that the transition probability is

$$W = \frac{G_F^2}{16\pi^3\alpha} \int \frac{\omega\ d\omega\ \sin\theta d\theta}{(E - p'_z - n\omega\cos\theta)^2}$$
$$\times \left[(g_V^2\pi_t^2 + g_A^2\pi_5^2) \left(1 - \frac{\cos\theta}{E-\omega}(p'_z\cos\theta - n\omega\sin^2\theta) \right) \right.$$
$$\left. -2g_V g_A \pi_t \pi_5 \left(\cos\theta - \frac{p'_z\cos\theta - n\omega\sin^2\theta}{E-\omega} \right) \right], \tag{11}$$

here

$$p'_z = \sqrt{(E-\omega)^2 - n^2\omega^2\sin^2\theta}\ . \tag{12}$$

The maximal allowed angle, θ_{max} , for the photon emission is $\pi/2$ when $\omega < \frac{E}{2}$ and $\sin\theta_{max} = \frac{E-\omega}{\omega}$ when $\omega > \frac{E}{2}$.

Now expand integrand into the series in small angle, since only in that case the denominator is small (and the formation zone length is large). Thus we write the transition probability in the form

$$W \simeq \frac{G_F^2}{16\pi^3\alpha}\int \frac{d\omega\ \theta^2 d\theta^2}{\omega\ [\ \theta^2 + (1-n^2)(1-\frac{\omega}{E})\ ]^2}\Big[(g_V^2\pi_t^2 + g_A^2\pi_5^2) \times (2 - 2\frac{\omega}{E} + \frac{\omega^2}{E^2}) - 2g_V g_A \pi_t\pi_5 \frac{\omega}{E}(2 - \frac{\omega}{E})\Big]\ . \tag{13}$$

Eq.(13) tells us that the radiation is forward peaked within an angle of order $\theta \sim \sqrt{1-n^2}$.

After integration over angle θ we get

$$W \simeq \frac{G_F^2}{16\pi^3\alpha}\int \frac{d\omega}{\omega}\Big[-\ln[(1-n^2)(1-\frac{\omega}{E})] + \ln[(1-n^2)(1-\frac{\omega}{E}) + \theta^2_{max}] - 1\Big] \times\Big[(g_V^2\pi_t^2 + g_A^2\pi_5^2)(2 - 2\frac{\omega}{E} + \frac{\omega^2}{E^2}) - 2g_V g_A \pi_t\pi_5 \frac{\omega}{E}(2 - \frac{\omega}{E})\Big]. \tag{14}$$

Numerically eq. (14) does not depend much on θ_{max}.

Usually the axialvector polarization function is much less than the vector one. For instance in nonrelativistic and nondegenerate plasma these functions are[14]

$$\pi_t = \omega_p^2 \ \text{ and } \ \pi_5 = \frac{|\mathbf{k}|}{2m_e}\frac{\omega_p^4}{\omega^2}, \tag{15}$$

where $\omega_p^2 = \frac{4\pi\alpha N_e}{m_e}$ is the plasma frequency. Therefore we may ignore the term proportional to π_5.

Since we are interested in the forward-peaked radiation in the gamma ray region, we assume that the index of refraction of the photon is

$$n^2 = 1 - \frac{\omega_p^2}{\omega^2} \tag{16}$$

and the photons from the medium to the vacuum propagate without any reflection or/and refraction.

In Fig.3 we plot the energy spectrum of the photons from the transition radiation by electron neutrinos with energy $E = 1\text{MeV}$.

After integration over the photon energy we find the neutrino transition radiation probability as

$$W = \int_{\omega_{min}}^{E} dW \simeq \frac{g_V^2 G_F^2 \omega_p^4}{16\pi^3\alpha}\left(2\ \ln^2\frac{E}{\omega_p} - 5\ln\frac{E}{\omega_p} + \delta\right) \tag{17}$$

here $\delta \simeq 5$ for $\omega_{min} = \omega_p$, $\delta \simeq -1$ for $\omega_{min} = 10\omega_p$.

The energy deposition by the neutrino in the medium due to the transition radiation is determined as

$$\int_{\omega_p}^{E} \omega \, dW_{\nu\to\nu\gamma} \simeq \frac{g_V^2 G_F^2 \omega_p^4}{16\pi^3\alpha} E\Big[\frac{8}{3}\ln\frac{E}{\omega_p} - 4.9 + 9\frac{\omega_p}{E} + O(\frac{\omega_p^2}{E^2})\Big] \qquad (18)$$

The eqs.(14),(17) and (18) represent the main results of our study.

For MeV electronic neutrinos the transition radiation probability is about $W \sim 10^{-39}$ and the energy deposition is about $1.4 \cdot 10^{-34}$ eV when they cross the interface of the medium with $\omega_p = 20$ eV to vacuum.

Unfortunately the transition radiation probability is extremely small and cannot be observed at the Earth.

On the other hand at the surface of the neutron stars thin layers of electrons with density $\sim m_e^3$ may form due to the fact that the electrons are not bound by the strong interactions and are displaced to the outside of the neutron star.[12] Therefore, MeV energy neutrinos emitted by the neutron star during its cooling processes would emit transition radiation with the probability of $\sim 10^{-20}$ and energy spectrum given in eq. (14).

In the previous calculation of neutrino transition radiation the vacuum indicted $\nu \to \nu\gamma$ matrix element was used ("neutrino toroid dipole moment").[11] The matrix element in[11] was $\sqrt{2} g_V G_F e\, \omega_p^2\, \bar{u}(p')\gamma_\nu(1-\gamma_5)u(p)\, \epsilon_\mu$. The vacuum induced matrix element can be treated as a radiation correction to the medium induced one. The ratio between squared matrix elements of (8) and vacuum inducted one is $|\frac{8\alpha}{\pi}|^{-2} \simeq 2850$. Therefore, our result is by the three orders of magnitude larger than those of previous calculations.[11]

Summary and Conclusion- We have calculated the neutrino transition radiation at the interface of two media. The charged particles of the medium provide an effective ν-γ vertex, and they modify the photon dispersion relation. Analytical expressions for the energy spectrum of the transition radiation, its probability and the energy deposition in the process have been obtained. The radiation is forward peaked within an angle of the order of $\frac{\omega_p}{\omega}$. The photon energy spectrum decreases almost linearly with the photon energy. Close to its maximum value ($\sim E_\nu$) it is further suppressed due to the smaller phase space.

Acknowledgments- It is pleasure to thanks the organizers for organizing a very interesting and enjoyable workshop in Yerevan and Tbilisi to celebrate the 80th birthday of professor Matinyan.

References

1. G. G. Raffelt, *Chicago, USA: Univ. Pr.* (1996) 664 p

2. V. L. Ginzburg and I. M. Frank, *J. Phys. (USSR)* **9** (1945) 353 [*Zh. Eksp. Teor. Fiz.* **16** (1946) 15].
3. G. M. Garibyan, *J. Exp. Theor. Phys.* **37** (1959) 527 [*Sov. Phys. JETP* **10** (1960) 372]
4. G. M. Garibyan, J. Exp. Theor. Phys. **39** (1960) 1630 [*Sov. Phys. JETP* **12** (1961) 1138]]; see also[13]
5. J. B. Adams, M. A. Ruderman and C. H. Woo, *Phys. Rev.* **129**, 1383 (1963); M. H. Zaidi, *Nuovo. Cimento* **40**, 502 (1965); see also[14]
6. see for instance J. C. D'Olivo, J. F. Nieves and P. B. Pal, *Phys. Rev. D* **40**, 3679 (1989).; J. F. Nieves and P. B. Pal, *Phys. Rev. D* **40**, 1693 (1989).
7. V. N. Oraevsky and V. B. Semikoz, *Physica* **142A** (1987) 135.; J. C. D'Olivo, J. F. Nieves and P. B. Pal, *Phys. Lett. B* **365** (1996) 178.
8. A. N. Ioannisian and G. G. Raffelt, *Phys. Rev. D* **55** (1997) 7038.
9. L. L. DeRaad, K. A. Milton and N. D. Hari Dass, *Phys. Rev. D* **14**, 3326 (1976); V. V. Skobelev, *Zh. Eksp. Teor. Fiz.* **71**, 1263 (1976); D. V. Galtsov and N. S. Nikitina, *Zh. Eksp. Teor. Fiz.* **62**, 2008 (1972).
10. M. Sakuda and Y. Kurihara, *Phys. Rev. Lett.* **74**, 1284 (1995); W. Grimus and H. Neufeld, *Phys. Lett. B* **344**, 252 (1995).
11. E. N. Bukina, V. M. Dubovik and V. E. Kuznetsov, *Phys. Lett. B* **435**, 134 (1998).
12. R. Picanco Negreiros, I. N. Mishustin, S. Schramm and F. Weber, *Phys. Rev. D* **82**, 103010 (2010); and references therein.
13. L. L. DeRaad, W. y. Tsai and T. Erber, *Phys. Rev. D* **18**, 2152 (1978).
14. E. Braaten and D. Segel, *Phys. Rev. D* **48**, 1478 (1993).

CORRELATIONS IN COSMOLOGICAL AND NON-COSMOLOGICAL SIGNALS

H.G. Khachatryan[1,2,3], G. Nurbaeva[1], D. Pfenniger[4], G. Meylan[1] and S. Sargsyan[2,3]

[1] *Laboratoire d'Astrophysique, Ecole Polytechnique Fédérale de Lausanne (EPFL), Observatoire de Sauverny, CH 1290 Versoix, Switzerland*
[2] *Center for Cosmology and Astrophysics, Alikhanyan National Science Laboratory, Alikhanyan brothers 2, Yerevan, Armenia*
[3] *Yerevan State University, A. Manoogian 1, Yerevan, Armenia*
[4] *University of Geneva, Geneva Observatory, CH-1290 Sauverny, Switzerland*

We analyze the correlations in the cosmological cosmic microwave background radiation and those of point sources, i.e. galaxies and quasars. The Wilkinson Microwave Anisotropy Probe 7-year data are used along with the method developed by Tegmark et al. for obtaining of foreground reduced maps, combined with the Kolmogorov parameter. We obtain the cross correlations in the temperature and K-maps, and report the detection of still unidentified point sources, along with the previously known ones, and hence, showing the power of the applied technique.

1. Introduction

The point sources are among the non-cosmological signals which are being separated in the cosmic microwave background (CMB) maps by various methods such as the wavelets and needlets.[8,38,39] The detected sources include radio sources, galaxies, quasars, and some sources remain unidentified;[25,43] the catalog is available in the.[44]

The correlations in the sequences can be studied via the Kolmogorov stochasticity parameter (KSP) also for this aim. We will use this technique[20] along with the method of reducing the foreground in the CMB maps developed by Tegmark et al.[42] We find non-cosmological signatures which partly coincide with known galaxies and quasars, however there are also sources which remain unidentified.

2. Correlations in cosmic microwave background maps

The parameter introduced by Kolmogorov[27] and developed by Arnold[3,5–7] has been already applied to cosmological datasets by Gurzadyan and collaborators.[18,20,22]

A remarkable property of the CMB, enabling to apply the Kolmogorov method, is its nearly Gaussian nature, so we can use the Gaussian distribution for the KSP method in order to obtain the KSP map. For the WMAP7 W band, the CMB map with $n_{\rm side} = 512$ resolution parameter (cf.[13]), we obtain a K-map with $n_{\rm side} = 32$ parameter. This means that the CMB map has $n_{\rm map} = 12n^2_{\rm side} = 3'145'728$ pixels and the K-map has $n_{\rm map} = 12'288$ pixels.

The spherical harmonic coefficients a_{lm} (Eq. (3) below), are studied also using the GLESP package.[9] For maps with low resolution (i.e. for K-map, $n_{\rm side} = 32$, $n_{\rm map} = 12'288$), the a_{lm} calculation and map reconstruction the use of HEALPIX package leads to calculation accuracy problems. To test whether the calculation error is small or not for a low resolution map, we construct a unity HEALPIX map with $n_{\rm side} = 32$, and another one with $n_{\rm side} = 512$, and run them through the a_{lm} calculation and map reconstruction procedures. For both cases this procedure adds some non-isotropic noise which has almost zero mean (e) and a very small standard deviation (σ). We obtain $\langle e_{512}\rangle = 10^{-6}$, $\langle e_{32}\rangle = 2\cdot 10^{-4}$, $\sigma_{512} = 2.3\cdot 10^{-4}$, $\sigma_{32} = 3.71\cdot 10^{-3}$. The fraction of pixels with larger errors near the poles for $n_{\rm side} = 32$ is less than 0.5%, i.e. for $n_{\rm side} = 32$, the error and standard deviation are small enough to enable us to calculate the a_{lm} and to construct the KSP map obtained by the Tegmark et al. method.

The Tegmark et al. method[18,34,35,42] is used to develop a cleaned CMB K-map for eight WMAP bands: Q1, Q2, V1, V2, W1, W2, W3, W4; it assigns a weight w^i_l to each map i and multipole l. In order not to distort the original map, w^i_l should obey the relation

$$\sum_i w^i_l = 1. \tag{1}$$

Since the w^i_l can be any real numbers, including negative ones, while the KSP must be within $0 \le \Phi \le 1$, the use of negative weights to construct the cleaned a_{lm} and then reconstruct the map can result in negative values of Φ for some pixels(see Fig. 2). To avoid this problem, the weights are normalized as

$$\acute{w}^i_l = \frac{w^i_l - w^{\rm min}_l}{1 - n\, w^{\rm min}_l}, \tag{2}$$

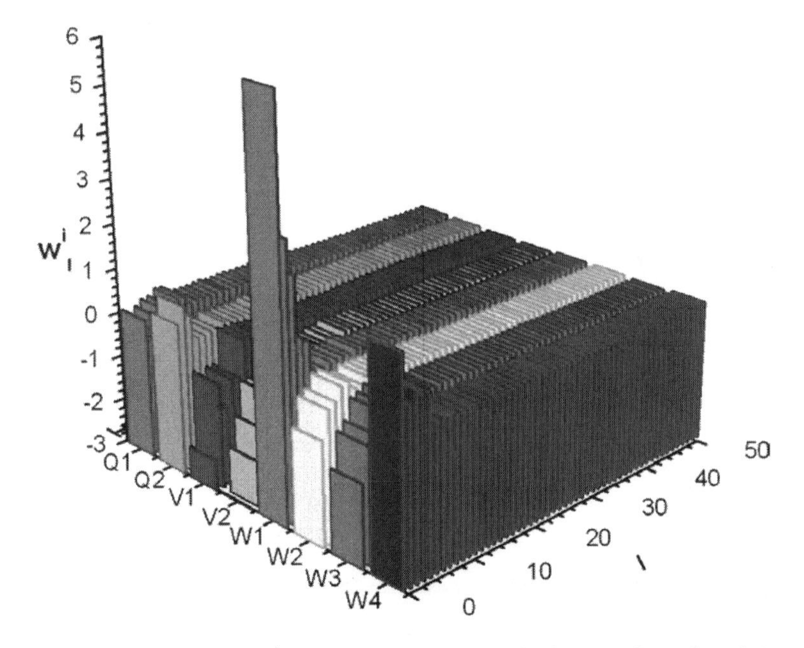

Fig. 1. Original weights w_l^i obtained using the standard Tegmark et al. technique.

where $w_l^{\min}$ is the minimal value of the original weights for fixed l multipole. For the new weights $\acute{w}_l^i$ the relation $\sum_i \acute{w}_l^i = 1$ is true and satisfies $\acute{w}_l^i \geq 0$ and minimally modifies the original w_l^i.

If a sequence of random numbers T_n obeys the distribution function $F(x)$ then N realizations of this sequence, in the Kolmogorov method, gives Φ_N, λ_N. The remarkable point is that Φ_N has a uniform distribution and the mean value $\langle\Phi\rangle = 0.5$. However, one can see from Fig. 2 that the frequency count for K-map is not uniform. Therefore for the CMB K-map the value 0.5 is a natural threshold for distinguishing non-Gaussian areas from Gaussian ones, since Gaussian temperature pixels cannot have $\Phi > 0.5$. The cleaned K-map mean value and sigma are respectively $\langle\Phi\rangle = 0.22$, $\sigma_\Phi = 0.14$, so pixels with $\Phi > 0.5$ mostly exceed the 3-σ region. There are in total 398 non-Gaussian pixels among 12'288 (about 3.2%). 27 among those are located outside the Galactic disk ($|b| < 20$) and the Large Magellanic Cloud ($l = 280.4136$, $b = 32.9310$) (see Fig. 2). So, these 27 regions are point source candidates. Indeed, most of them are found in previously published catalogs.[2,12,29] Seven of these point sources are listed in the Planck Early Release Compact Source Catalogue as high reliability point sources.[2] Finally, two possible point sources remain unidentified. The list of the point sources is given in Khachatryan et al.[26]

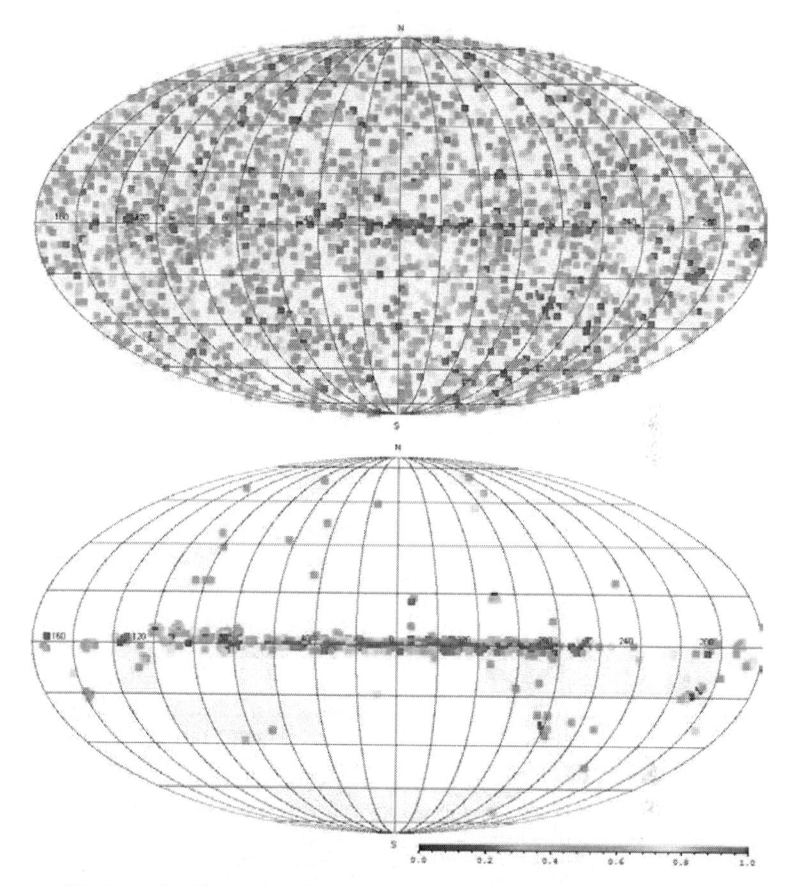

Fig. 2. Pixels with KSP value $\Phi > 0.5$ for the original WMAP7 W band (top) and cleaned (bottom) Kolmogorov maps.

3. CMB and K-map power spectra

Full sky map can be represented via a series of Legendre spherical functions $Y_{lm}(\theta, \varphi)$,[24] where

$$T(\theta, \varphi) = \sum_{l,m} a_{lm} Y_{lm}(\theta, \varphi),$$
$$a_{lm} = \int T(\theta, \varphi)\, Y^*_{lm}(\theta, \varphi) \sin\theta \, d\theta \, d\varphi. \tag{3}$$

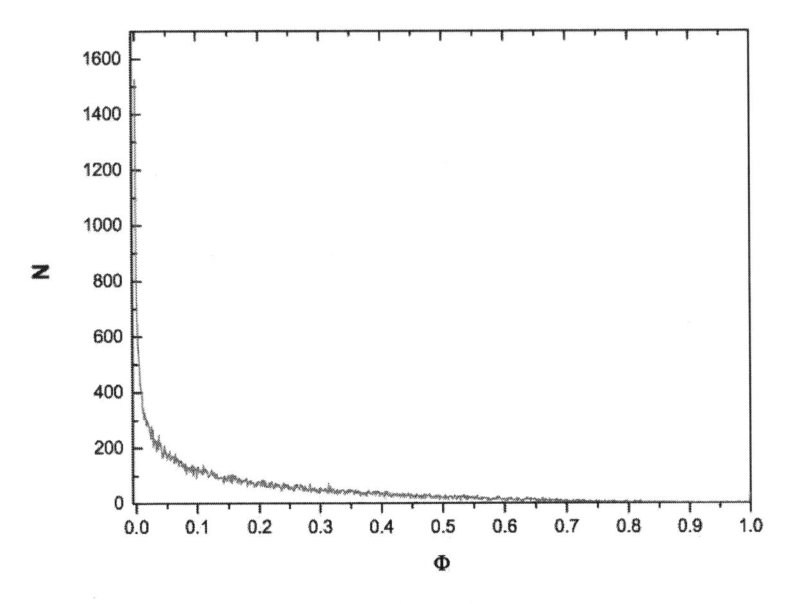

Fig. 3. Frequency count for the WMAP W K-map.

The coefficients C_l of the Legendre polynomials $P_l(\cos\theta)$ in the two point correlation function $C(\theta)$ of the power spectrum are related to the a_{lm} by

$$\begin{aligned} C_l &= \langle a^*_{lm} a_{lm} \rangle, \\ C(\theta) &= \frac{1}{4\pi} \sum_{l,m} (2l+1)\, C_l\, P_l(\cos\theta). \end{aligned} \tag{4}$$

To probe the correlation between CMB temperature and K-map, we degrade the resolution of the CMB temperature map to $n_{\text{side}} = 32$ and with a dimensionless parameter T we normalize to the same $0 \leq T \leq 1$ interval as Φ, to prevent any discrepancy between temperature and K-maps. For power spectrum and cross power spectrum calculations we use WMAP7[25] eight different bands T and the Kolmogorov maps (Q1, Q2, V1, V2, W1, W2, W3, W4). The foreground reduced map is obtained using Tegmark et al.[42] weighting technique for triplets of different bands for the calculation of the power spectra both for the temperature and the KSP correlation functions, as shown in Hinshaw et al.[23] and Gurzadyan et al.[17]

4. Cross correlations in cosmic microwave background temperature and K-maps

For cross power spectra calculation we use the common technique described in Hinshaw et al.[23] via taking the cross-correlation power spectrum coefficients for different type of a_{lm}, as

$$\tilde{C}_l = \langle a_{lm}^{*T} a_{lm}^{\Phi} \rangle. \tag{5}$$

Here we again use the foreground reduced maps[42] for the calculation of the cross power spectra between T and Φ, since the original ones are very noisy, while the aims is to get the smallest possible error bars in the final power spectrum. The power spectrum estimation is done without taking into account the Galactic disk plane region, i.e., we use the window function which is zero for the region $\theta = \pm 20$ for both T and Φ, and unity elsewhere. This cutting method influences mostly the even l inducing some unreasonable peaks (for example at $l = 2, 12, 22, \ldots$) around the window function power spectrum peaks. One could use Peebles[32] method to reduce this effect on even l values and adjust odd l values, but then one would have to calculate the power spectrum at least up to $l = 250$. The low resolution map of Φ allows the accurate estimation of the power spectrum up to $l = 97$ which makes impossible the use of this method. Therefore we have to use only odd l values. After tedious calculations no remarkable cross-correlations between temperature and Kolmogorov maps are found.[26]

5. Conclusions

We applied the Tegmark et al. cleaning technique not to the temperature but to the Kolmogorov parameter CMB maps. As a result, about 85% of the cleaned K-map is in the interval $0.0 < \Phi < 0.2$, which implies that the CMB maps are of Gaussian nature with high precision. The non-Gaussian pixels with $\Phi > 0.5$ are rather rare, only 398 of them. This is derived for the cleaned K-map but it implies that different types of noises add additional non-Gaussianities into the CMB maps, which may be analyzed by the methods given by de Oliveira Costa et al.[31] Outside the Galactic disk region, $|b| > 20$, this indicates the positions of point source candidates. While most of them have counterparts in existing catalogs, two of them are still unidentified.

KSP is a statistical parameter so for fixed numbers of elements, it must show statistical fluctuations. For the same reason, the full sky K-map should have uniform distribution which is not so.[22] In most cases these fluctuations have non-Gaussian nature. As it was shown in,[18] the power spectrum for CMB K-map reveals certain feature around $l = 25$; the simplest possible explanation is the existence of a characteristic size of voids in the Universe, i.e., $l = 25$ corresponds to angular size $\theta = 7.2^o$, as a result of the integrated Sachs-Wolfe effect. Certain models[10,28,36,37,41] predict primordial non-Gaussianities from inflation era, which can be discovered through CMB bi-spectrum. We studied the cross-correlation between CMB temperature and K-map. Numerical modeling of such a problem was done in Ghahramanyan et al.,[11] where the sensitivity of KSP to the small departures from the theoretical distribution was studied (Fig. 4). Certain non-Gaussian perturbations would appear in the K-map as KSP perturbations. If one uses a proper theoretical distribution function, no correlation between temperature and KSP should arise. Therefore, in regions outside the Galactic disk certain correlations could appear even in the presence of rather small non-Gaussianities.

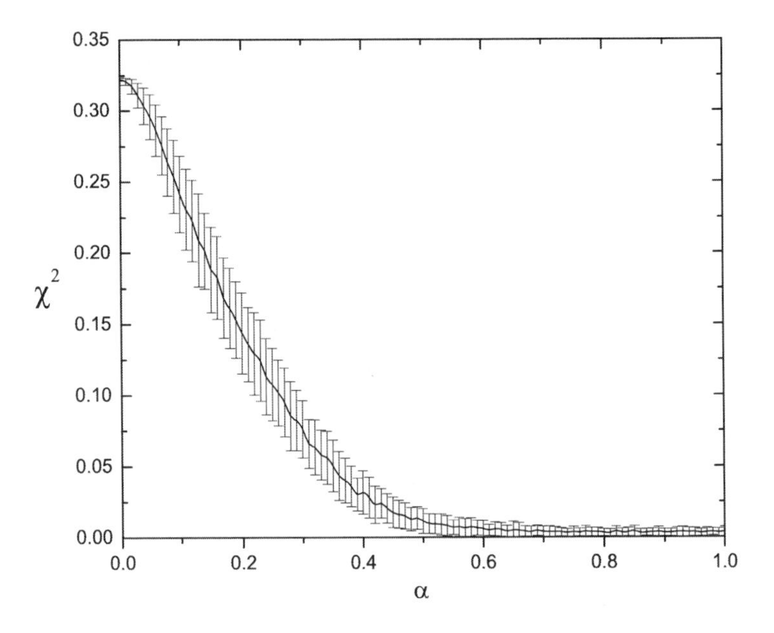

Fig. 4. The χ^2 for the Kolmogorov function vs. the ratio of the random to regular components of a signal[21]

Acknowledgments

We are grateful to V. Gurzadyan and colleagues in Center for Cosmology and Astrophysics for numerous comments and discussions. Many thanks also to O. Verkhodanov for information about GLESP.

References

1. Abramowitz M., Stegun I., (1970), Handbook of Mathematical Functions, New York: Dover Publications
2. Ade P.A.R., et al. 2011, A&A, 536, A7
3. Arnold V.I., 2008, Nonlinearity, 21, T109
4. Arnold V.I., 2008, ICTP/2008/001, Trieste
5. Arnold V.I., 2008, Uspekhi Mat. Nauk, 63, 5
6. Arnold V.I., 2009, Trans. Moscow Math. Soc., 70, 31
7. Arnold V.I., 2009, Funct. Anal. Other Math. 2, 139
8. Batista R.A., Kemp E., Daniel B. 2011, IJMPE, 20, 61,
9. Doroshkevich A.G., et al. 2005, Int. J. Mod. Phys. D 14, 275
10. Gangui A. 1994, Phys.Rev. D 50, 3684
11. Ghahramanyan T., et al. 2009, Mod.Phys.Lett. A24, 1187
12. Gold, B., et al. 2011, ApJS. 192, 15
13. Górski, K. M., Hivon, E., Banday, A. J., et al. 2005, ApJ, 622, 759
14. Gurzadyan V.G., de Bernardis P., et al. 2005, Mod.Phys.Lett. A20, 893
15. Gurzadyan V.G., Starobinsky A.A., et al. 2008, A&A, 490, 929
16. Gurzadyan V.G., Allahverdyan A.E., et al. 2009, A&A, 497, 343
17. Gurzadyan V.G., et al. 2009, A&A, 498, L1
18. Gurzadyan V.G., et al. 2009, A&A, 506, L37
19. Gurzadyan V.G. & Kocharyan A.A. 2009, A&A, 493, L61; Europhys.Lett.86, 29002
20. Gurzadyan V.G., Kashin A.L., et al. 2010, Europhys. Lett., 91, 19001
21. Gurzadyan V.G., Ghahramanyan T., Sargsyan S. Europhys.Lett. 95 (2011) 19001
22. Gurzadyan V.G., et al. 2011, A&A, 525 L7
23. Hinshaw G., et al. 2003, ApJS, 148, 135
24. Hivon E., et al. 2002, ApJ, 567, 2
25. Jarosik N., Bennett C.L., et al. 2010, arXiv:1001.4744
26. Khachatryan H. G., et al. 2012, submitted to A&A, arXiv:1206.7121
27. Kolmogorov A.N. 1933, Giorn.Ist.Ital.Attuari, 4, 83
28. Komatsu E. and Spergel D. N. 2001, Phys. Rev. D, 63, 063002
29. Lanz L.F., et al. 2011, submitted to MNRAS, preprint arXiv:1110.6877v2
30. Mather J.C., Cheng E.S., et al. 1994, ApJ, 420, 439
31. de Oliveira Costa A., et al. 2004, Phys. Rev. D, 69, 063516
32. Peebles P.J.E. 1973, ApJ, 185, 413
33. Rocha G., et al. 2005, MNRAS, 357, 1R
34. Saha R., Jain P., Souradeep T. 2006, ApJ, 645, L89
35. Saha R., Prunet S., Jain P., Souradeep T. 2008, Phys. Rev. D, 78, 023003

36. Salopek D.S. and Bond J. R. 1990, Phys. Rev. D, 42, 3936
37. Salopek D.S. and Bond J. R. 1991, Phys. Rev. D, 43, 1005
38. Scodeller S.S., Hansen F.K., Marinucci D. 2012, preprint arXiv:1201.5852
39. Scodeller S.S., Hansen F.K. 2012, preprint arXiv:1207.2315
40. Smoot G.F., Bennett C.L., et al. 1992, ApJ, 396, L1
41. Takeuchi Y., Ichiki K., Matsubara T. 2012, Phys. Rev. D, in press, arXiv:1111.6835v2
42. Tegmark M., de Oliveira Costa A., Hamilton A. 2003, Phys. Rev. D 68, 123523
43. Wright E., et al. 2009, ApJS, 180, 283
44. NASA WMAP team web page, http://lambda.gsfc.nasa.gov/product/map/current/

BIG BOUNCE AND INFLATION FROM GRAVITATIONAL FOUR-FERMION INTERACTION

I.B. Khriplovich

Budker Institute of Nuclear Physics
11 Lavrentjev pr., 630090 Novosibirsk, Russia
khriplovich@inp.nsk.su

The four-fermion gravitational interaction is induced by torsion. It gets dominating below the Planck scale. The regular, axial-axial part of this interaction by itself does not stop the gravitational compression. However, the anomalous, vector-vector interaction results in a natural way both in big bounce and in inflation.

Keywords: Planck scale; gravitational four-fermion interaction; big bounce; inflation.

1. The observation that, in the presence of torsion, the interaction of fermions with gravity results in the four-fermion interaction of axial currents, goes back at least to.[1]

We start our discussion of the four-fermion gravitational interaction with the analysis of its most general form.

As has been demonstrated in,[2] the common action for the gravitational field can be generalized as follows:

$$S_g = -\frac{1}{16\pi G}\int d^4x\,(-e)\,e^\mu_I e^\nu_J \left(R^{IJ}_{\mu\nu} - \frac{1}{\gamma}\tilde{R}^{IJ}_{\mu\nu}\right); \tag{1}$$

here and below G is the Newton gravitational constant, $I, J = 0, 1, 2, 3$ (and M, N below) are internal Lorentz indices, $\mu, \nu = 0, 1, 2, 3$ are space-time indices, e^I_μ is the tetrad field, e is its determinant, and e^μ_I is the object inverse to e^I_μ. The curvature tensor is

$$R^{IJ}_{\mu\nu} = -\partial_\mu \omega^{IJ}{}_\nu + \partial_\nu \omega^{IJ}{}_\mu + \omega^{IK}{}_\mu\, \omega_K{}^J{}_\nu - \omega^{IK}{}_\nu\, \omega_K{}^J{}_\mu,$$

here ω^{IJ}_μ is the connection. The first term in (1) is in fact the common action of the gravitational field written in tetrad components.

The second term in (1), that with the dual curvature tensor

$$\tilde{R}^{IJ}_{\mu\nu} = \frac{1}{2}\,\varepsilon^{IJ}_{\;\;KL} R^{KL}_{\mu\nu}\,,$$

does not vanish in the presence of spinning particles generating torsion.

As to the so-called Barbero-Immirzi parameter γ, its numerical value

$$\gamma = 0.274 \qquad (2)$$

was obtained for the first time in,[3] as the solution of the "secular" equation

$$\sum_{j=1/2}^{\infty} (2j+1) e^{-2\pi\gamma\sqrt{j(j+1)}} = 1. \qquad (3)$$

Interaction of fermions with gravity results, in the presence of torsion, in the four-fermion action which looks as follows:

$$S_{ff} = \frac{3}{2}\pi G \frac{\gamma^2}{\gamma^2+1} \int d^4x \sqrt{-g}$$

$$\left[\eta_{IJ} A^I A^J + \frac{\alpha}{\gamma}\eta_{IJ}(V^I A^J + A^I V^J) - \alpha^2\,\eta_{IJ} V^I V^J\right]; \qquad (4)$$

here and below g is the determinant of the metric tensor, A^I and V^I are the total axial and vector neutral currents, respectively:

$$A^I = \sum_a A^I_a = \sum_a \bar{\psi}_a \gamma^5 \gamma^I \psi_a\,; \qquad V^I = \sum_a V^I_a = \sum_a \bar{\psi}_a \gamma^I \psi_a\,; \quad (5)$$

the sums over a in (5) extend over all sorts of elementary fermions with spin 1/2.

The AA contribution to expression (4) corresponds (up to a factor) to the action derived long ago in.[1] Then, this contribution was obtained in the limit $\gamma \to \infty$ in[4] (when comparing the corresponding result from[4] with (4), one should note that our convention $\eta_{IJ} = \mathrm{diag}(1,-1,-1,-1)$ differs in sign from that used in[4]). The present form of the AA interaction, given in (4), was derived in.[5]

As to VA and VV terms in (4), they were derived in[6] as follows. The common action for fermions in gravitational field

$$S_f = \int d^4x \sqrt{-g}\,\frac{1}{2}\,[\bar{\psi}\gamma^I e^\mu_I\, i\boldsymbol{\nabla}_\mu\psi - i\,\overline{\boldsymbol{\nabla}_\mu\psi}\,\gamma^I e^\mu_I \psi] \qquad (6)$$

can be generalized to:

$$S_f = \int d^4x \sqrt{-g}\,\frac{1}{2}\,[(1-i\alpha)\,\bar{\psi}\gamma^I e^\mu_I\, i\boldsymbol{\nabla}_\mu\psi - (1+i\alpha)\,i\,\overline{\boldsymbol{\nabla}_\mu\psi}\,\gamma^I e^\mu_I\psi]; \quad (7)$$

here $\boldsymbol{\nabla}_\mu = \partial_\mu - \frac{1}{4}\,\boldsymbol{\omega}^{IJ}{}_\mu\,\gamma_I\gamma_J$; $\boldsymbol{\omega}^{IJ}{}_\mu$ is the connection. The real constant α introduced in (7) is of no consequence, generating only a total derivative, if the theory is torsion free. However, in the presence of torsion this constant gets operative. In particular, as demonstrated in,[6] it generates the VA and VV terms in the gravitational four-fermion interaction (4).

Simple dimensional arguments demonstrate that interaction (4), being proportional to the Newton constant G and to the particle number density squared, gets essential and dominates over the common interactions only at very high densities and temperatures, i.e. on the Planck scale and below it.

The list of papers where the gravitational four-fermion interaction is discussed in connection with cosmology, is too lengthy for this short note. Therefore, I refer here only to the most recent,[7] with a quite extensive list of references. However, in all those papers the discussion is confined to the analysis of the axial-axial interaction.

In particular, in my paper[8] it was claimed that VA and VV terms in formula (4) are small as compared to the AA one. The argument was as follows. Under these extreme conditions, the number densities of both fermions and antifermions increase, due to the pair creation, but the total vector current density V^I remains intact.

By itself, this is correct. However, the analogous line of reasoning applies to the axial current density A^I. It is in fact the difference of the left-handed and right-handed axial currents: $A^I = A^I_L - A^I_R$. There is no reason to expect that this difference changes with temperature and/ or pressure.

Moreover, the fermionic number (as distinct from the electric charge) is not a long-range charge. Therefore, even the conservation of fermionic number could be in principle violated, for instance, by the decay of a neutral particle (majoron) into two neutrinos. (I am grateful to A.D. Dolgov for attracting my attention to this possibility.)

So, we work below with both currents, A and V.

2. Let us consider the energy-momentum tensor (EMT) $T_{\mu\nu}$ generated by action (4). Therein, the expression in square brackets has no explicit dependence at all either on the metric tensor, or on its derivatives. The metric tensor enters action S_{ff} via $\sqrt{-g}$ only, so that the corresponding EMT is given by relation

$$\frac{1}{2}\,\sqrt{-g}\,T_{\mu\nu} = \frac{\delta}{\delta g_{\mu\nu}} S_{ff}\,. \qquad (8)$$

Thus, with identity

$$\frac{1}{\sqrt{-g}}\frac{\partial\sqrt{-g}}{\partial g^{\mu\nu}} = -\frac{1}{2}\,g_{\mu\nu}\,, \tag{9}$$

we arrive at the following expression for the EMT:

$$T_{\mu\nu} = -\frac{3\pi}{2}G\,\frac{\gamma^2}{\gamma^2+1}\,g_{\mu\nu}$$

$$\left[\eta_{IJ}A^I A^J + \frac{\alpha}{\gamma}\,\eta_{IJ}\,(V^I A^J + A^I V^J) - \alpha^2\,\eta_{IJ}V^I V^J\right],$$

or, in the tetrad components,

$$T_{MN} = -\frac{3\pi}{2}\,G\,\frac{\gamma^2}{\gamma^2+1}\,\eta_{MN}$$

$$\left[\eta_{IJ}A^I A^J + \frac{\alpha}{\gamma}\,\eta_{IJ}\,(V^I A^J + A^I V^J) - \alpha^2\,\eta_{IJ}V^I V^J\right]. \tag{10}$$

We note first of all that this EMT in the locally inertial frame corresponds to the equation of state

$$p = -\,\varepsilon; \tag{11}$$

here and below $\varepsilon = T_{00}$ is the energy density, and $p = T_{11} = T_{22} = T_{33}$ is the pressure.

Let us analyze the expressions for ε and p in our case of the interaction of two ultrarelativistic fermions (labeled a and b) in their locally inertial center-of-mass system.

The axial and vector currents of fermion a are, respectively,

$$A^I_a = \frac{1}{4E^2}\,\phi^\dagger_a\,\{E\,\boldsymbol{\sigma}_a\,(\mathbf{p}'+\mathbf{p}),$$
$$(E^2 - (\mathbf{p}'\mathbf{p}))\,\boldsymbol{\sigma}_a + \mathbf{p}'(\boldsymbol{\sigma}_a\mathbf{p}) + \mathbf{p}\,(\boldsymbol{\sigma}_a\mathbf{p}') - i\,[\mathbf{p}'\times\mathbf{p}]\}\,\phi_a$$

$$= \frac{1}{4}\,\phi^\dagger_a\,\{\boldsymbol{\sigma}_a\,(\mathbf{n}'+\mathbf{n}),\;(1-(\mathbf{n}'\mathbf{n}))\,\boldsymbol{\sigma}_a + \mathbf{n}'(\boldsymbol{\sigma}_a\mathbf{n}) + \mathbf{n}\,(\boldsymbol{\sigma}_a\mathbf{n}') - i\,[\mathbf{n}'\times\mathbf{n}]\}\,\phi_a; \tag{12}$$

$$V^I_a = \frac{1}{4E^2}\,\phi^\dagger_a\,\{E^2 + (\mathbf{p}'\mathbf{p}) + i\,\boldsymbol{\sigma}_a\,[\mathbf{p}'\times\mathbf{p}],\;E\,(\mathbf{p}'+\mathbf{p} - i\,\boldsymbol{\sigma}_a\times(\mathbf{p}'-\mathbf{p}))\}\,\phi_a$$

$$= \frac{1}{4}\,\phi^\dagger_a\,\{1 + (\mathbf{n}'\mathbf{n}) + i\,\boldsymbol{\sigma}_a\,[\mathbf{n}'\times\mathbf{n}],\;\mathbf{n}'+\mathbf{n}\,-i\,\boldsymbol{\sigma}_a\times(\mathbf{n}'-\mathbf{n})]\}\,\phi_a; \tag{13}$$

here E is the energy of fermion a, $\mathbf{n}$ and $\mathbf{n}'$ are the unit vectors of its initial and final momenta $\mathbf{p}$ and $\mathbf{p}'$, respectively; under the discussed extreme conditions all fermion masses can be neglected. In the center-of-mass system, the axial and vector currents of fermion b are obtained from these expressions by changing the signs: $\mathbf{n} \to -\mathbf{n}$, $\mathbf{n}' \to -\mathbf{n}'$. Then, after averaging over the directions of $\mathbf{n}$ and $\mathbf{n}'$, we arrive at the following semiclassical expressions for the nonvanishing components of the energy-momentum tensor, i.e. for the energy density ε and pressure p (for the correspondence between ε, p and EMT components see,[9] S 35):

$$\varepsilon = T_{00} = -\frac{\pi}{48}\, G\, \frac{\gamma^2}{\gamma^2+1}$$

$$\sum_{a,\,b} \rho_a\, \rho_b\, [(3 - 11 \, < \boldsymbol{\sigma}_a \boldsymbol{\sigma}_b >) - \alpha^2 (60 - 28 \, < \boldsymbol{\sigma}_a \boldsymbol{\sigma}_b >)]$$

$$= -\frac{\pi}{48}\, G\, \frac{\gamma^2}{\gamma^2+1}\, \rho^2\, [(3 - 11\,\zeta) - \alpha^2 (60 - 28\,\zeta)]\,; \tag{14}$$

$$p = T_{11} = T_{22} = T_{33} = \frac{\pi}{48}\, G \frac{\gamma^2}{\gamma^2+1} \sum_{a,\,b} \rho_a\, \rho_b\, [(3 - 11 \, < \boldsymbol{\sigma}_a \boldsymbol{\sigma}_b >)$$

$$- \alpha^2 (60 - 28 \, < \boldsymbol{\sigma}_a \boldsymbol{\sigma}_b >)]$$

$$= \frac{\pi}{48}\, G\, \frac{\gamma^2}{\gamma^2+1}\, \rho^2\, [(3 - 11\,\zeta) - \alpha^2 (60 - 28\,\zeta)]\,; \tag{15}$$

here and below ρ_a and ρ_b are the number densities of the corresponding sorts of fermions and antifermions, $\rho = \sum_a \rho_a$ is the total density of fermions and antifermions, the summation $\sum_{a,\,b}$ extends over all sorts of fermions and antifermions; $\zeta \; = < \boldsymbol{\sigma}_a \boldsymbol{\sigma}_b >$ is the average value of the product of corresponding $\boldsymbol{\sigma}$-matrices, presumably universal for any a and b. Since the number of sorts of fermions and antifermions is large, one can neglect here for numerical reasons the contributions of exchange and annihilation diagrams, as well as the fact that if $\boldsymbol{\sigma}_a$ and $\boldsymbol{\sigma}_b$ refer to the same particle, $< \boldsymbol{\sigma}_a \boldsymbol{\sigma}_b > = 3$. The parameter ζ, just by its physical meaning, in principle can vary in the interval from 0 (which corresponds to the complete thermal incoherence or to the antiferromagnetic ordering) to 1 (which corresponds to the complete ferromagnetic ordering).

It is only natural that after the performed averaging over $\mathbf{n}$ and $\mathbf{n}'$, the P-odd contributions of VA to ε and p vanish.

3. Though for $\alpha \sim 1$ the VV interaction dominates numerically the results (14) and (15), it is instructive to start the analysis with the discussion

of the case $\alpha = 0$, at least, for the comparison with the previous investigations. We note in particular that, according to (14), the contribution of the gravitational spin-spin interaction to energy density is positive, i.e. the discussed interaction is repulsive for fermions with aligned spins. This our conclusion agrees with that made long ago in.[4]

To simplify the discussion, we confine from now on to the region around the Planck scale, so that one can neglect effects due to the common fermionic EMT, originating from the Dirac Lagrangian and linear in the particle density ρ.

A reasonable dimensional estimate for the temperature τ of the discussed medium is

$$\tau \sim m_{\text{Pl}} \tag{16}$$

(here and below m_{Pl} is the Planck mass). This temperature is roughly on the same order of magnitude as the energy scale ω of the discussed interaction

$$\omega \sim G\rho \sim m_{\text{Pl}} \,. \tag{17}$$

Numerically, however, τ and ω can differ essentially. Both options, $\tau > \omega$ and $\tau < \omega$, are conceivable.

If the temperature is sufficiently high, $\tau \gg \omega$, it destroys the spin-spin correlations in formulas (14) and (15). In the opposite limit, when $\tau \ll \omega$, the energy density (14) is minimized by the antiferromagnetic spin ordering. Thus, in both these limiting cases the energy density and pressure simplify to

$$\varepsilon = -\frac{\pi}{16}\frac{\gamma^2}{\gamma^2+1}\,G\rho^2; \qquad p = \frac{\pi}{16}\frac{\gamma^2}{\gamma^2+1}\,G\rho^2 \,. \tag{18}$$

The energy density ε, being negative and proportional to ρ^2, decreases with the growth of ρ. On the other hand, the common positive pressure p grows together with ρ. Both these effects result in the compression of the fermionic matter, and thus make the discussed system unstable.

A curious phenomenon could be possible if initially the temperature is sufficiently small, $\tau < \omega$, so that equations (18) hold. Then the matter starts compressing, its temperature increases, and the correlator $\zeta = <\boldsymbol{\sigma}_a\boldsymbol{\sigma}_b>$ could arise. When (and if!) ζ exceeds its critical value $\zeta_{cr} = 3/11$, the compression changes to expansion. Thus, we would arrive in this case at the big bounce situation.

However, I am not aware of any physical mechanism which could result here in the transition from the initial antiferromagnetic ordering to the ferromagnetic one with positive $\zeta = <\boldsymbol{\sigma}_a\boldsymbol{\sigma}_b>$.

Here one should mention also quite popular idea according to which the gravitational collapse can be stopped by a positive spin-spin contribution to the energy. However, how such spin-spin correlation could survive under the discussed extremal conditions? The naïve classical arguments do not look appropriate in this case.

4. Let us come back now to equations (14), (15). In this general case, with nonvanishing anomalous VV interaction, the big bounce takes place if the energy density (14) is positive (and correspondingly, the pressure (15) is negative). In other words, the anomalous, VV interaction results in big bounce under the condition

$$\alpha^2 \geq \frac{3 - 11\zeta}{4\,(15 - 7\zeta)}\,. \tag{19}$$

For vanishing spin-spin correlation ζ, this condition simplifies to

$$\alpha^2 \geq \frac{1}{20}\,. \tag{20}$$

The next remark refers to the spin-spin contribution to energy density (14)

$$\varepsilon_\zeta = -\,\frac{\pi}{48}\,\frac{\gamma^2}{\gamma^2+1}\,G\,\rho^2\,(28\,\alpha^2 - 11)\,\zeta\,. \tag{21}$$

It could result in the ferromagnetic ordering of spins if $\alpha^2 > 11/28$. Whether or not this ordering takes place, depends on the exact relation between $G\rho$ and temperature, both of which are on the order of magnitude of m_{Pl}.

5. One more comment related to equations (14), (15). As mentioned already, according to them, the equation of state, corresponding to the discussed gravitational four-fermion interaction, is

$$p = -\varepsilon. \tag{22}$$

It is rather well-known that this equation of state results in the exponential expansion of the Universe. Let us consider in this connection our problem.

We assume that the Universe is homogeneous and isotropic, and thus is described by the well-known Friedmann-Robertson-Walker (FRW) metric

$$ds^2 = dt^2 - a(t)^2[dr^2 + f(r)(d\theta^2 + \sin^2\theta\, d\phi^2)]; \tag{23}$$

here $f(r)$ depends on the topology of the Universe as a whole:

$$f(r) = r^2, \quad \sin^2 r, \quad \sinh^2 r$$

for the spatial flat, closed, and open Universe, respectively. As to the function $a(t)$, it depends on the equation of state of the matter.

The Einstein equations for the FRW metric (23) reduce to

$$H^2 \equiv \left(\frac{\dot{a}}{a}\right)^2 = \frac{8\pi G\varepsilon}{3} - \frac{k}{a^2}\,, \tag{24}$$

$$\frac{\ddot{a}}{a} = -\frac{4\pi G}{3}\,(\varepsilon + 3p\,). \tag{25}$$

They are supplemented by the covariant continuity equation, which can be written as follows:

$$\dot{\varepsilon} + 3\,H(\varepsilon + p\,) = 0; \qquad H = \frac{\dot{a}}{a}\,. \tag{26}$$

For the energy-momentum tensor (14), (15), dominating below the Planck scale, and resulting in $\varepsilon = -p$, this last equation reduces to

$$\dot{\varepsilon} = 0, \qquad \text{or} \qquad \varepsilon = \text{const}. \tag{27}$$

In its turn, equation (25) simplifies to

$$\frac{\ddot{a}}{a} = \frac{8\pi G\varepsilon}{3} = \text{const}. \tag{28}$$

In this way, for $\varepsilon > 0$, we arrive at the following expansion law:

$$a \sim \exp(Ht), \qquad \text{where} \qquad H = \sqrt{\frac{8\pi G\varepsilon}{3}} = \text{const} \tag{29}$$

(as usual, the second, exponentially small, solution of eq. (28) is neglected here).

Thus, the discussed gravitational four-fermion interaction results in the inflation starting below the Planck scale.

Acknowledgments

I am grateful to D.I. Diakonov, A.D. Dolgov, A.A. Pomeransky, and A.S. Rudenko for useful discussions.

The investigation was supported in part by the Russian Ministry of Science, by the Foundation for Basic Research through Grant No. 11-02-00792-a, by the Federal Program "Personnel of Innovational Russia" through Grant No. 14.740.11.0082, and by the Grant of the Government of Russian Federation, No. 11.G34.31.0047.

References

1. T.W.B. Kibble, *J. Math. Phys.* **2** (1961) 212.
2. S. Holst, *Phys. Rev.* **D53** (1996) 5966; gr-qc/9511026.
3. I.B. Khriplovich and R.V. Korkin, *J. Exp. Theor. Phys.* **95** (2002) 1; gr-qc/0112074.
4. G.D. Kerlick, *Phys. Rev.* **D12**, 3004 (1975).
5. A. Perez and C. Rovelli, *Phys. Rev.* **D73**, 044013 (2006); gr-qc/0505081.
6. L. Freidel, D. Minic, and T. Takeuchi, *Phys. Rev.* **D72**, 104002 (2005); hep-th/0507253.
7. G. de Berredo-Peixoto, L. Freidel, I.L. Shapiro, and C.A. de Souza; gr-qc/12015423.
8. I.B. Khriplovich, *Phys. Lett.* **B709**, 111 (2012); gr-qc/12014226.
9. L.D. Landau and E.M. Lifshitz, *The Classical Theory of Fields* (Butterworth - Heinemann, 1975).

CAPTURE AND EJECTION OF DARK MATTER BY THE SOLAR SYSTEM

I.B. Khriplovich

Budker Institute of Nuclear Physics
11 Lavrentjev pr., 630090 Novosibirsk, Russia
khriplovich@inp.nsk.su

We consider the capture and ejection of dark matter by the Solar System. Both processes are due to the gravitational three-body interaction of the Sun, a planet, and a dark matter particle. Simple estimates are presented for the capture cross-section, as well as for density and velocity distribution of captured dark matter particles close to the Earth.

Keywords: dark matter; Solar system; restricted three-body problem.

1. Introduction

The local density of dark matter (dm) in our Galaxy can be estimated, according to[1] (see also references therein), as

$$\rho_g = (8.31 \pm 0.59(\text{stat}) \pm 1.37(\text{syst})) \cdot 10^{-25}\ \text{g/cm}^3\,. \qquad (1)$$

However, only upper limits on the level of 10^{-19} g/cm^3 (see, e.g.,[2,3]) are known for the density of dark matter particles (dmp) in the Solar System (SS). Besides, even these limits are derived under the quite strong assumption that the distribution of dm density in the SS is spherically-symmetric with respect to the Sun. Meanwhile, information on dm density in SS is very important, in particular for the experiments aimed at the detection of dark matter.

The capture of dark matter by the SS was addressed previously in.[4–7] In particular, in[7] the total mass of the captured dark matter was estimated analytically. In the present note the analytical estimates are given for the capture cross-section, as well as for the density and velocity distribution of captured dm close to the Earth.

Of course, a particle cannot be captured by the Sun alone. The interaction with a planet is necessary for it, i.e. this is essentially a three-body (the Sun, planet and dmp) problem. Obviously, the capture is dominated by the particles with orbits close to parabolic ones with respect to the Sun; besides, the distances between their perihelia and the Sun should be comparable with the radius of the planet orbit r_p. Just the trajectories of these particles are most sensitive to the attractive perturbation by the planet.

The capture can be effectively described by the so-called restricted three-body problem (see, e.g.,[8]). In this approach the interaction between two heavy bodies (the Sun and a planet in our case) is treated exactly. As exactly is treated the motion of the third, light body (a dmp in our case) in the gravitational field of the two heavy ones. One neglects however the back reaction of a light particle upon the motion of the two heavy bodies. Obviously, this approximation is fully legitimate for our purpose.

Still, the restricted three-body problem is rather complicated, its solution requires both subtle analytical treatment and serious numerical calculations (see, e.g.,[9]). Under certain conditions the dynamics of light particle becomes chaotic. The "chaotic" effects are extremely important for the problem. However their quantitative investigation is quite complicated and remains beyond the scope of the present note. We confine here instead to simple estimates which could be also of a methodological interest by themselves. On the other hand, thus derived results for the total mass and density of the captured dark matter constitute at least an upper limit for their true value. As to the velocity distribution of dmp's given here, together with the mentioned result for the dark matter density, it could be possibly of some practical interest for planning the experimental searches for dm.

2. Total mass of dark matter captured by the Earth

The Solar System is immersed in the halo of dark matter and moves together with it around the center of our Galaxy. To simplify the estimates, we assume that the Sun is at rest with respect to the halo. The dark matter particles in the halo are assumed also to have in the reference frame, comoving with the halo, the Maxwell distribution:

$$f(v)\,dv = \sqrt{\frac{54}{\pi}}\,\frac{v^2 dv}{u^3}\,\exp\left(-\frac{3}{2}\frac{v^2}{u^2}\right), \tag{2}$$

with the local rms velocity $u \simeq 220$ km/s. Let us note that the velocities v discussed in this section are the asymptotic ones, they refer to large distances from the Sun, so that their values start at $v = 0$ and formally extend to infinity.

The amount of dm captured by the SS can be found by means of simple estimates*. The total mass captured by the Sun (its mass is M) together with a planet with the mass m_p, during the lifetime

$$T \simeq 4.5 \cdot 10^9 \text{ years} \simeq 10^{17} \text{ s} \tag{3}$$

of the SS, can be written as follows:

$$\mu_p = \rho_g T < v\,\sigma >; \tag{4}$$

here σ is the capture cross-section. The product σv is averaged over distribution (2); with all typical velocities in the SS much smaller than u, this distribution simplifies to

$$f(v)\,dv = \sqrt{\frac{54}{\pi}}\,\frac{v^2 dv}{u^3}\,. \tag{5}$$

To estimate the average value $< v\,\sigma >$, we resort to dimensional arguments, supplemented by two rather obvious physical requirements: the masses m_p and M of the two heavy components of our restricted three-body problem should enter the result symmetrically, and the mass of the dmp should not enter the result at all in virtue of the equivalence principle. Thus, we arrive at

$$< v\,\sigma > \sim \sqrt{54\pi}\;\frac{k^2\, m_p\, M}{u^3}\,, \tag{6}$$

or

$$\int_0^\infty dv\, v^3\, \sigma \;\sim \pi\, k^2\, m_p\, M\,; \tag{7}$$

here k is the Newton gravitation constant; an extra power of π, inserted into these expressions, is perhaps inherent in σ. Since the capture would be impossible if the planet were not bound to the Sun, it is only natural that the result is proportional to the corresponding effective "coupling constant" $k\, m_p\, M$. One more power of k corresponds to the gravitational interaction of the dark matter particle. The final estimate for the captured mass is

$$\mu_p \sim \rho_g T \sqrt{54\pi}\;\; k^2\, m_p\, M/u^3\,. \tag{8}$$

For the Earth it constitutes

$$\mu_E \sim 8 \cdot 10^{18}\,\text{g}\,. \tag{9}$$

*These estimates were given previously in.[7] Here we repeat them, as well as results (8), (9), (20) (see below), since they are essential for the present discussions.

3. Capture cross-section

By the same dimensional reasons (and in complete correspondence with formula (7)), the total capture cross-section for the Earth should look as follows:

$$\sigma \sim \pi k^2 m_E M / \tilde{v}^4 , \tag{10}$$

where m_E is the mass of the Earth, and $\tilde{v}$ is some velocity which can be estimated as follows. It is natural to assume that the capture of dm particles occurs when they are close to the Earth, i.e. at the distances $\sim r_E$ from the Sun. As natural are the following assumptions: 1) the initial velocities of the captured dmp's exceed only slightly the parabolic one $v_{\rm par}$ ($v_{\rm par}^2 = 2kM/r_E$); 2) their final velocities are only slightly less than v_{par}. To our accuracy, here we omit the factor of 2 in the definition of $v_{\rm par}^2$, and thus put $\tilde{v}^2 \sim v_E^2 = kM/r_E$ ($v_E = 30$ km/s is the velocity of the Earth). Thus, the capture cross-section is

$$\sigma \sim \pi k^2 m_E M / v_E^4 . \tag{11}$$

This formula can be also conveniently rewritten as

$$\sigma \sim \pi r_E^2 (m_E/M) . \tag{12}$$

Let us note here that the impact parameter corresponding to formula (12), i.e. the typical distance between a dmp and the Earth crucial for the capture, is

$$r_{\rm imp} \sim r_E (m_E/M)^{1/2} \ll r_E . \tag{13}$$

In fact, this impact parameter corresponds to the distance at which the attraction to the Earth equals the attraction to the Sun, i.e. where

$$km/r^2 > kM/r_E^2 \quad (r \ll r_E) . \tag{14}$$

Up to now, in all relevant formulae, (7), (10), (11), we dealt with the capture cross-section averaged over the directions of the dmp velocity $\mathbf{v}$. However, this cross-section depends essentially on the mutual orientation of $\mathbf{v}$ and $\mathbf{v}_E$. Certainly, it is maximum when these velocities are parallel and as close as possible by modulus. Besides, the impact parameter $r_{\rm imp}$ of the collision is much less than the radius r_E of the Earth orbit (see (13)), and thus within the distances $\simeq r_{\rm imp}$ both the Earth and dmp trajectories can be treated as rectilinear. Therefore, it looks quite natural to identify $\tilde{v}$

in (10) with the relative velocity $|\mathbf{v}-\mathbf{v}_E|$ of the dmp and the Earth, i.e. to generalize formula (11) as follows:

$$d\sigma \sim \frac{k^2\, m_p\, M}{(\mathbf{v}-\mathbf{v}_E)^4}\, \frac{1}{4}\, d\Omega \tag{15}$$

(factor 1/4 is introduced here for the correspondence with factor π in (11): $(1/4)\int d\Omega = \pi$).

Thus derived total cross-section is

$$\sigma \sim \frac{1}{4}\int d\Omega\, \frac{k^2\, m_p\, M}{(\mathbf{v}-\mathbf{v}_E)^4} = \frac{\pi\, k^2\, m_p\, M}{(v^2 - v_E^2)^2}\,. \tag{16}$$

Clearly, just the particles, moving initially with the velocities only slightly above the parabolic one $\sqrt{2}\, v_E = 42$ km/s, are captured predominantly, and thus, with $v = \sqrt{2}\, v_E$, cross-sections (11) and (16) practically coincide.

On the other hand, it follows from (16) that in the vicinity of the Earth the captured particles move with respect to it with the velocities close to $(\sqrt{2}-1)v_E \simeq 12$ km/s.

4. Space distribution of captured dark matter

The captured dmp's had initial trajectories predominantly close to parabolas focussed at the Sun, and the velocities of these dmp's change only slightly as a result of scattering. Therefore, their trajectories become elongated ellipses with large semimajor axes, still focussed at the Sun. The ratio of their maximum $r_{\max}$ and minimum $r_{\min}$ distances from the Sun is (see[10])

$$\frac{r_{\max}}{r_{\min}} = \frac{1+e}{1-e}\,, \tag{17}$$

where e is the eccentricity of the trajectory. In the numerator of this ratio, we can safely put with our accuracy $1+e \simeq 2$ both for the initial parabolas and final ellipses. In our case, as a result of the capture, the eccentricity changes from $1+\varepsilon_1$ to $1-\varepsilon_2$, where $\varepsilon_{1,2} \ll 1$. This loss of eccentricity is due to the gravitational perturbation by the Earth, and therefore both ε_1 and ε_2 are proportional to m_E. In particular, for the final ellipse

$$r_{\max} \sim 1/m_E\,. \tag{18}$$

On the other hand, $r_{\min}$ is close to the radius r_E of the Earth orbit. Thus, for dimensional reasons, we arrive (see[7]) at

$$r_{\max} \sim r_E\, (M/m_E)\,. \tag{19}$$

Let us note here that the analogous estimate for the case of Jupiter complies qualitatively with the results of corresponding numerical calculations presented in.[9]

Obviously, the semimajor axis a_{dmp} of the trajectory of a captured dmp is on the same order of magnitude as r_{max}. Then, the time spent by a dmp, with the characteristic velocity close to v_E and at the distance from the Sun close to r_E, is comparable to the orbital period of the Earth $T_E = 1$ year. Besides, the orbital period T is related to the semimajor axis a as follows (see[10]): $T \sim a^{3/2}$. Thus, we arrive at the following estimate for the orbital period of the captured dmp †:

$$T_{\text{dmp}} \sim T_E \, (M/m_E)^{3/2} \,. \tag{20}$$

In other words, the relative time spent by a dmp at the distances $\sim r_E$ from the Earth can be estimated as $(m_E/M)^{3/2}$. Moreover, the typical distances from the Earth at which a dmp can be captured, should be less than the impact parameter $r_{\text{imp}} \sim r_E \, (m_E/M)^{1/2}$ (see (13)). Thus, the relative time spent by a dmp sufficiently close to the Earth to be captured, can be estimated as $(m_E/M)^2$.

With the impact parameter (13), the corresponding volume V, centered at the Earth and crucial for the capture, can be estimated as

$$V \sim \frac{4\pi}{3} r_{\text{imp}}^3 \sim \frac{4\pi}{3} r_E^3 \, (m_E/M)^{3/2} \ll \frac{4\pi}{3} r_E^3 . \tag{21}$$

Let us combine formula (9) for the total captured mass with the effective volume (21) occupied by this mass and with the estimate $(m_E/M)^2$ for the relative time spent by a dmp within the impact parameter (13) with respect to the Earth. In this way we arrive at the following estimate for the density of dark matter, captured by the SS, in the vicinity of the Earth:

$$\rho_E \sim 10^{-24} \ \text{g/cm}^3 \,. \tag{22}$$

This estimate practically coincides with the value (1) for the galactic dm density.

In fact, the result (22) for the density of the captured dm, as well as the estimates (8) and (9) for its total mass, should be considered as upper limits only, since we have neglected therein the inverse process, that of the ejection of the captured dm from the SS. The characteristic time of the inverse process is not exactly clear now. Therefore, it cannot be excluded

---

†In the case of the Earth, this orbital period is huge, $\sim 10^8$ years. Still, it is much less than the lifetime of the SS, $\sim 5 \cdot 10^9$ years.

that it is comparable to, or even larger than, the lifetime T of the SS (see[7]). In this case our estimates for the total mass and the density of the captured dm are valid.

If this is the case indeed, then the dm around the Earth consists essentially of two components with comparable densities. In line with the common component with the typical velocity around $u \sim 220$ km/s, there is one more, with the velocity relative to the Earth $\gtrsim 12$ km/s.

5. Back to the capture process

So, the initial trajectory of a captured dmp is hyperbola close to parabola, its final trajectory is elongated ellipse, and both trajectories are focussed at the Sun.

The transitions between these stationary states (i.e. states of fixed energies) are due to the time-dependent perturbation caused by the planet rotating around the Sun. Obviously, the intensities of both these transitions, up and down, are equal.

Therefore, if densities of dmp's on the hyperbolic and elliptic trajectories were initially equal, these densities remain equal. If density of dmp's on the hyperbolic trajectories was initially higher or lower than that on the elliptic ones, these densities get equal finally.

Thus, the following conclusion looks quite natural: if time is sufficiently long, the density of the dmp's captured by the Solar System is the same as the density of dmp's in the Galaxy.

Acknowledgments

I am grateful to A.S. Rudenko, V.V. Sokolov, and V.G. Serbo for useful discussions. The investigation was supported in part by the Russian Ministry of Science, by the Foundation for Basic Research through Grant No. 11-02-00792-a, by the Federal Program "Personnel of Innovational Russia" through Grant No. 14.740.11.0082, and by the Grant of the Government of Russian Federation, No. 11.G34.31.0047.

References

1. M. Pato, O. Agertz, G. Bertone, B. Moore, and R. Teyssier, *Phys. Rev.* **82**, 023531 (2010).
2. I. B. Khriplovich, E. V. Pitjeva, *Int. J. Mod. Phys. D* **15**, 615 (2006).
3. I. B. Khriplovich,*Int. J. Mod. Phys. D* **16**, 1475 (2007).
4. A. Gould, S. M. K. Alam, *Astrophys. J.* **549**, 72 (2001).

5. J. Lundberg, J. Edsjo, *Phys. Rev. D.* **69**, 123505 (2004).
6. A. H. G. Peter, *Phys. Rev. D* **79**, p. 1003531, 1003532, 1003533 (2009).
7. I. B. Khriplovich, D. L. Shepelyansky, *Int. J. Mod. Phys. D* **18**, 1903 (2009).
8. V. Szebehely, *Theory of Orbits*, Academic Press, N.Y. (1964).
9. T. Y. Petrosky, *Phys. Lett. A* **117**, 328 (1986).
10. L. D. Landau, E. M. Lifshitz, *Mechanics*, Nauka, Moscow (1988), S15.

TOPOLOGICAL THEORY OF QHE

M. Kohmoto*

Institute for Solid State Physics, University of Tokyo,
Kashiwa, Chiba 277-8581, Japan
**E-mail: kohmoto@issp.u-tokyo.ac.jp*

Electrons in a two-dimensional periodic potential with a magnetic field is considered. The Hall conductances are shown to be integers, i.e Chen numbers. Thus these electronic bands give one of the first examples of topologically ordered state of matter.

Keywords: Quantum Hall effect; Chern Number; Topological Order.

1. Introduction

Much of condensed matter is about how different kinds of order emerge from interactions between many simple constituents. Until 1980 all ordered phase could be understood as "symmetry breaking". An ordered state appears at low temperature when the system spontaneously loses one of the symmetries present at high temperature.

Quantum Hall effect takes place in two-dimensional electrons in a magnetic field at low temperature. Hall conductance is precisely quantized to be an integer (multiple of e^2/h) or a fractional.[1] These new states of matter do not involve breaking of any symmetry. Instead, the precise quantizations indicates topological origin, namely topological invariants do not depend on details of parameters of the systems unless a gap of the ground state closes(quantum phase transition). In this article we consider the Hall conductance of a two-dimensional electron gas in a uniform magnetic field and a periodic substrate potential.[2] The Kubo formula is written in a form that makes apparent the quantization when the Fermi energy lies in a gap. In fact the quantization is due to the topological nature of the system. From the wavefunctions, a fiber bundle is constructed on the magnetic Brillouin zone which is a torus T^2. The topological Chern number gives a contribution of the Hall conductance from a single magnetic subband.[3]

2. Hofstadter Butterfly

Let us consider non-interacting electrons hopping nearest neighbors on the square lattice in a magnetic field. See Fig. 1.

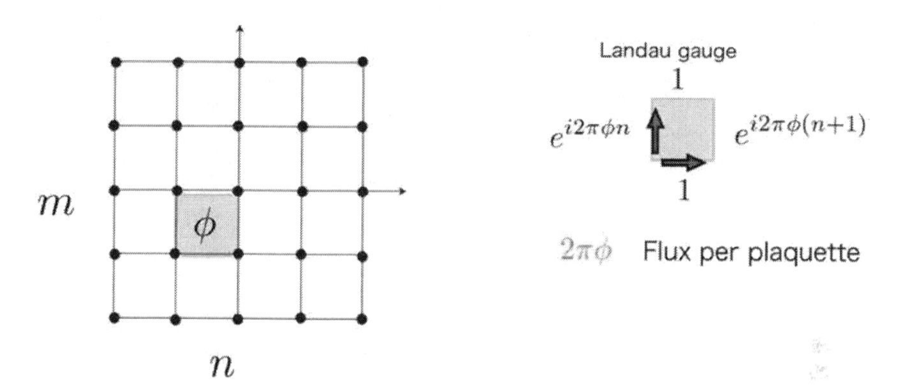

Fig. 1. Square lattice and flux

The Schrödinger equation is

$$\psi_{n+1,m} + \psi_{n-1,m} + e^{i2\pi\phi n}\psi_{n,m+1} + e^{-i2\pi\phi n}\psi_{n-1,m} = E\psi_{n,m}, \quad (1)$$

where hoppings are set to be unity, and the exponential factors represent the magnetic flux; $2\pi\phi$ is magnetic flux per plaquette. Since this is translationally invariant along y-direction one can set $\psi_{nm} = e^{ik_y m}\psi_n$, so we have

$$\psi_{n+1} + \psi_{n-1} + 2\cos(2\pi\phi n + k_y) = E\psi_n. \quad (2)$$

This is the famous Harper equation. Suppose ϕ is rational, then it is written $\phi = p/q$ where p and q are mutually prime integers and Eq. (2) has period q. Since the number of bands is q, the energy spectrum of Eq. (2) as a function of ϕ gives highly intricate structures, see Fig. 2.

This shows self-similar behaviors and Cantor-set structure. For irrational ϕ, $p \to \infty$, and $q \to \infty$ and numbers of bands are infinite.

We are now consider the cases when ϕ is rational (p and q are finite).

3. Bloch Electrons in a Uniform Magnetic Field

Consider two-dimensional periodic system in a uniform magnetic field perpendicular to the plain.

$$H\Psi = [\frac{1}{2m}(p + eA)^2 + U(x, y)]\Psi = E\Psi, \quad (3)$$

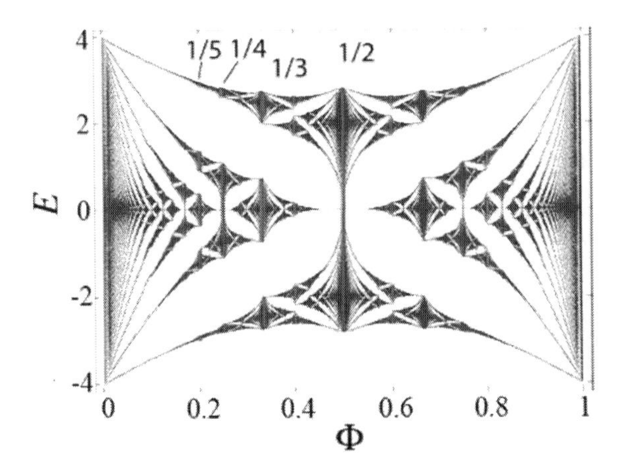

Fig. 2. Hofstdater butterfly

where $p = -i\hbar\nabla$ is a momentum operator, A is a gauge potential in $x-y$ plain, and $U(x+a,y) = U(x,y+b) = U(x,y)$ is the 2d periodic potential. Note that the system is translationally invariant in x-direction by a, and y-direction by b, but the Hamiltonian(3) is not translationally invariant due to the gauge potential $A(x,y)$. The Bravais lattice is given by

$$R = na + mb, \tag{4}$$

where n and m are integers. For each Bravais lattice vector we define a translational operator T_R which. when operating on any smooth function $f(r)$, shift the argument by R:

$$T_R f(r) = f(r+R). \tag{5}$$

This operator is explicitly written as

$$T_R = \exp[(i/\hbar)R\cdot p]. \tag{6}$$

If T_R is applied to the Hamiltonian (3), the potential is left invariant. However, the gauge potential is transformed to $A(r+R)$ which is not equal to $A(r)$, but they differ by a gradient of a scalar function since the agnetic field is uniform:

$$A(r+R) = A(r) + \nabla g(r). \tag{7}$$

Let us consider the magnetic translation operators[4,5]

$$\hat{T}_R = \exp\{(i/\hbar)R\cdot[p + e(r\times B)/2]\} \tag{8}$$

$$= T_R \exp\{(ie/\hbar)(B\times R)\cdot r/2\} \tag{9}$$

If the symmetric gauge $A = (B \times r)/2$ is taken, $\hat{T}_R$ leaves the Hamiltonian invariant,

$$[\hat{T}_R, H] = 0. \tag{10}$$

Now, we look for eigenstates which simultaneously diagonalize $\hat{T}_R$ and H. However note that the magnetic translations do not commute with each other in general

$$\hat{T}_a \hat{T}_b = \exp(2\pi i \phi) \hat{T}_b \hat{T}_a, \tag{11}$$

where $\phi = (eB/h)ab$ is a number of magnetic flux in the unit cell. When ϕ is a rational number, $\phi = p/q$ (p and q are mutually prime), we have a subset of translations which commute with each other. We take an enlarged unit cell (magnetic unit cell) which an integral multiple of magnetic flux goes through. For example, if the Bravais lattice vectors of a form

$$R' = nqa + mb \tag{12}$$

are taken, then p magnetic flux quanta are in the magnetic unit cell which is formed by the vectors qa and b. The magnetic translation operator $\hat{T}_{R'}$ which correspond to these new Bravais lattice vectors commute with each other.

Let ψ be an eigenfunction which diagonalizes H and $\hat{T}_{R'}$ simultaneously, then it is easy to see that the eigenvalues of $\hat{T}_{qa}$ and $\hat{T}_b$ are given by

$$\begin{aligned} \hat{T}_{qa}\psi &= e^{ik_1 qa}\psi, \\ \hat{T}_b\psi &= e^{ik_2 b}\psi, \end{aligned} \tag{13}$$

where k_1 and k_2 are generalized crystal momenta, and can be restricted in the magnetic Brillouin zone: $0 \le k_1 \le 2\pi/qa, 0 \le k_2 \le 2\pi/b$. The eigenfunctions are labeled by k_1 and k_2 in addition to a band index α , and written in a Bloch form as

$$\psi^{(\alpha)}_{k_1 k_2}(x, y) = e^{ik_1 x + ik_2 y} u^{(\alpha)}_{k_1 k_2}(x, y). \tag{14}$$

Eq.(13) with Eq.(9) give the properties of $u^{(\alpha)}_{k_1 k_2}(x, y)$:

$$u^{(\alpha)}_{k_1 k_2}(x + qa, y) = e^{-i\pi py/b} u^{(\alpha)}_{k_1 k_2}(x, y) \tag{15}$$

$$u^{(\alpha)}_{k_1 k_2}(x, y + b) = e^{i\pi px/qa} u^{(\alpha)}_{k_1 k_2}(x, y). \tag{16}$$

These are the generalized Bloch conditions.

4. Linear Response Formula for the Hall Conductance

It is useful to write the Schrödinger equation (3) in a form

$$\hat{H}(k_1, k_2)u^{\alpha}_{k_1k_2} = E^{\alpha}u^{\alpha}_{k_1k_2}, \tag{17}$$

with

$$\hat{H}(k_1, k_2) = \frac{1}{2m}(-i\hbar\nabla + \hbar k + eA)^2 + U(x, y), \tag{18}$$

where k is a vector whose $x-$ and $y-$ components are k_1 and k_2, respectively. Note that the eigenvalue E^{α} depends on k continuously. For a fixed band index α, a set of possible value of E^{α} with k varying in the magnetic Brillouin zone forms a band (magnetic subband).

When a small electric field is applied, a resulting current may be given by the Kubo linear response formula. A linear response of current in the perpendicular direction to the applied electric field is represented by the Hall conductance

$$\sigma_{xy} = \frac{e^2\hbar}{i} \sum_{E^{\alpha}<E_F<E^{\beta}} \frac{(v_y)_{\alpha\beta}(v_x)_{\beta\alpha} - (v_x)_{\alpha\beta}(v_y)_{\beta\alpha}}{(E^{\alpha} - E^{\beta})^2}, \tag{19}$$

where E_F is a Fermi energy and the summation implies the sum over all the states below and above the Fermi energy. The indices α and β label bands. One needs k to specify a state in addition to the band index. The existence of k must implicitly be understood wherever a band index appears. To obtain the matrix elements of the velocity operator $v = (-i\hbar\nabla + eA)/m$, it is sufficient to integrate over one magnetic unit cell

$$(v)_{\alpha\beta} = \delta_{k_1k'_1}\delta_{k_2k'_2} \int_0^{qa} dx \int_0^b dy u^{\alpha *}_{k_1k_2} v u^{\beta}_{k'_1k'_2}, \tag{20}$$

here the states are normalized as $\int_0^{qa} dx \int_0^b dy |u|^2 = 1$. In Eq.(19) the velocity operators can be replaced by partial derivative of the k-dependent Hamiltonian (18), since only off-diagonal matrix elements are considered

$$\begin{aligned} (v_x)_{\alpha\beta} &= \frac{1}{\hbar}\langle\alpha|\frac{\partial\hat{H}}{\partial k_1}|\beta\rangle, \\ (v_y)_{\alpha\beta} &= \frac{1}{\hbar}\langle\alpha|\frac{\partial\hat{H}}{\partial k_2}|\beta\rangle. \end{aligned} \tag{21}$$

Furthermore the matrix elements of the partial derivative of $\hat{H}$ are written as

$$\langle\alpha|\frac{\partial\hat{H}}{\partial k_j}|\beta\rangle = (E^\beta - E^\alpha)\langle\alpha|\frac{\partial u^\beta}{\partial k_j}\rangle$$
$$= -(E^\beta - E^\alpha)\langle\frac{\partial u^\alpha}{\partial k_j}|\beta\rangle, j = 1\text{or}2 \tag{22}$$

From Eqs. (21) and (22) Eq.(19) is written as

$$\sigma_{xy} = \frac{e^2}{i\hbar}\sum_{E^\alpha<E_F<E^\beta}(\langle\frac{\partial u^\alpha}{\partial k_2}|\beta\rangle\langle\beta|\frac{\partial u^\alpha}{\partial k_1}\rangle - \langle\frac{\partial u^\alpha}{\partial k_1}|\beta\rangle\langle\beta|\frac{\partial u^\alpha}{\partial k_2}\rangle). \tag{23}$$

Using the identity $\sum_{E^\alpha<E_F<E^\beta}(|\alpha\rangle\langle\alpha| + \beta\rangle\langle\beta| = 1$, we have

$$\sigma_{xy}^{(\alpha)} = \frac{e^2}{h}\frac{1}{2\pi i}\int d^2k\int d^2r(\frac{\partial u^{\alpha*}_{k_1k_2}}{\partial k_2}\frac{\partial u^\alpha_{k_1k_2}}{\partial k_1} - \frac{\partial u^{\alpha*}_{k_1k_2}}{\partial k_1}\frac{\partial u^\alpha_{k_1k_2}}{\partial k_2}), \tag{24}$$

where $\sigma_{xy}^{(\alpha)}$ is a contribution of the Hall conductance from a completely filled band α. Let us define a vector field in the magnetic Brillouin zone by

$$\hat{A}(k_1,k_2) = \int d^2r u^*_{k_1,k_2}\nabla_k u_{k_1,k_2} = \langle u_{k_1,k_2}|\nabla_k|u_{k_1,k_2}\rangle, \tag{25}$$

where ∇_k is a vector operatorwhose components are $\partial/\partial k_1$ and $\partial/\partial k_2$. The band index α is omitted from the wavefunctions, since we will consider only a contribution from a single band from now on. The contribution is written from Eqs.(24) and (25) as

$$\sigma_{xy}^{(\alpha)} = \frac{e^2}{h}\frac{1}{2\pi i}\int d^2k[\nabla_k \times \hat{A}(k_1,k_2)]_3, \tag{26}$$

where $[\ ]_3$ represents the third component of the vector.

The integration is over the magnetic Brillouin zone: $0 \leq k_1 \leq 2\pi/qa, 0 \leq k_2 \leq 2\pi/b$. The important observation here is that the magnetic Brillouin zone is topologically a torus T^2 rather than a rectangular in k-space. Two points in k-space $k_1 = 0$ and $2\pi/qa$ (or, $k_2 = 0$ and $2\pi/b$) must be identified as the same point. Since a torus does not have a boundary, the application of Stokes' theorem to Eq.(26) would give $\sigma_{xy}^{(\alpha)} = 0$ if $\hat{A}(k_1,k_2)$ is uniquely defined on the entire torus T^2. A possible non-zero value of $\sigma_{xy}^{(\alpha)}$ is a consequence of a non-trivial topology of $\hat{A}(k_1,k_2)$. Note that the identification of the magnetic Brillouin zone as a torus T^2 is essential here. Non-trivial $\hat{A}(k_1,k_2)$ can only be constructed when the global topology of the base space in non-contractible.

In order to understand the topology of $\hat{A}(k_1,k_2)$ let us first discuss a "gauge transformation" of a special kind. Suppose $u_{k_1k_2}(x,y)$ satisfies the

Schrödinger equation (17), so does $u_{k_1k_2}(x,y)e^{if(k_1,k_2)}$, where $f(k_1,k_2)$ is an arbitrary smooth function of k_1 and k_2, and is independent of x and y. This introduces a transformation

$$u'_{k_1k_2}(x,y) = u_{k_1k_2}(x,y)\exp[if(k_1,k_2)]. \tag{27}$$

Since this is a change of the overall phase of the wavefunction, any physical quantity remains invariant. From Eq. (25) the corresponding transformation of $\hat{A}(k_1,k_2)$ is given by

$$\hat{A}'(k_1,k_2) = \hat{A}(k_1,k_2) + i\nabla_k f(k_1,k_2). \tag{28}$$

From this it is easy to see σ_{xy} given in Eq. (26) remains invariant.

The non-trivial topology arises when the phase of the wavefunction cannot be determined uniquely and smoothly in the entire magnetic Brillouin zone. The transformation (27) implies that the overall phase factor for each state vector $|u_{k_1k_2}\rangle$ can be chosen arbitrary. This phase can be determined, for example, by demanding that a component of the state vector $u_{k_1k_2}(x^{(0)},y^{(0)}) = \langle x^{(0)},y^{(0)}|u_{k_1k_2}\rangle$ is real. However this convention is not enough to fix the phase on the entire magnetic Brillouin zone, since $u_{k_1k_2}(x^{(0)},y^{(0)})$ vanishes for some (k_1,k_2). For the sake of simplicitly, consider the case where $u_{k_1k_2}(x^{(0)},y^{(0)})$ vanishes only at one point $(k_1^{(0)},k_2^{(0)})$. in the magnetic Brillouin zone. See Fig.3

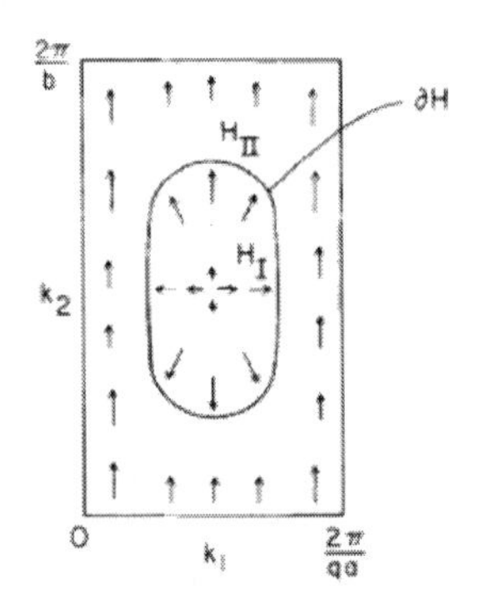

Fig. 3. Schematic diagram of a phase of a wavefunction in the magnetic Brillouin zone

Devide T^2 into two pieces H_I and H_{II} such that H_I contains $(k_1^{(0)},k_2^{(0)})$. We adopt a different convention in H_I so that another component of the state vector $u_{k_1k_2}(x^{(1)},y^{(1)}) = \langle x^{(1)},y^{(1)}|u_{k_1k_2}\rangle$ is real, where $(x^{(1)},y^{(1)})$

and H_I are chosen that $u_{k_1k_2}(x^{(1)}, y^{(1)}$ does not vanish. Thus the overall phase is uniquely determined on the entire T^2. In Fig. 3 a phase of one component of the state vector $u_{k_1k_2}(x^{(0)}, y^{(0)}) = \langle x^{(0)}, y^{(0)}|u_{k_1k_2}\rangle$ is schematically drawn.

Note that the overall phase of the state vector is well defined at $(k_1^{(0)}, k_2^{(0)})$ even through a phase of a single component $u_{k_1k_2}(x^{(0)}, y^{(0)}) = \langle x^{(0)}, y^{(0)}|u_{k_1k_2}\rangle$ cannot be defined there. At the boundary ∂H of H_I and H_{II} we have a phase mismatch

$$|u^{II}_{k_1k_2}\rangle = \exp[i\chi(k_1, k_2)]|u^{I}_{k_1k_2}\rangle, \tag{29}$$

where $\chi(k_1, k_2)$ is a smooth function of (k_1, k_2) on ∂H.

This non-trivial topology of $|u_{k_1k_2}\rangle$ is simply carried over to that of $\hat{A}(k_1, k_2)$. Smooth vector fields $\hat{A}_I(k_1, k_2)$ and $\hat{A}_{II}(k_1, k_2)$ are defined on H_I and H_{II}, respectively, by Eq. (25). The phase mismatch of the state vector given by Eq. (29) induces the following relation between $\hat{A}_I(k_1, k_2)$ and $\hat{A}_{II}(k_1, k_2)$ on ∂H:

$$\hat{A}_{II}(k_1, k_2) = \hat{A}_I(k_1, k_2) + i\nabla_k\chi(k_1, k_2). \tag{30}$$

Now, in Eq.(26) we can apply Stokes' theorem to H_I and H_{II} separately

$$\begin{aligned}\sigma_{xy}^{(\alpha)} &= \frac{e^2}{h}\frac{1}{2\pi i}\{\int_{H_I} d^2k[\nabla_k \times \hat{A}_I(k_1, k_2)]_3 + \int_{H_{II}} d^2k[\nabla_k \times \hat{A}_{II}(k_1, k_2)]_3\} \\ &= \frac{e^2}{h}\frac{1}{2\pi i}\int_{\partial H} dk \cdot [\hat{A}_I(k_1, k_2) - \hat{A}_{II}(k_1, k_2)],\end{aligned} \tag{31}$$

where $\int_{\partial H} dk$ represents a line integral on ∂H and the sign change occurs because ∂H has oposite orientation for H_I and H_{II}. Using the relation between $\hat{A}_I$ and $\hat{A}_{II}$, Eq. (30), we find

$$\sigma_{xy}^{(\alpha)} = \frac{e^2}{\hbar}n, \tag{32}$$

with

$$n = \frac{1}{2\pi}\int_{\partial H} dk \cdot \nabla_k\chi(k_1, k_2). \tag{33}$$

Here n must be an integer for each of the state vectors must fit together exactly when we complete a full revolution around ∂H. Thus the Hall conductance is quantized.

A generalization of the above which allow $u_{k_1,k_2}(x^{(0)}, y^{(0)})$ to have more than one zero can be done using the theory of fiber bundle in the next Section.

5. Chern Number, Hall Conductance, and Fiber Bundle

We have been considering the normalized eigenstate $|u_{k_1,k_2}\rangle$. Since $\exp[if(k_1,k_2)]|u_{k_1k_2}\rangle$ is also a normalized eigenstate, it is natural to consider a principal $U(1)$ bundle over T^2. A torus is covered by four neighborhoods H_i, $i = 1, ..., 4$ where each H_i is a subspace of R^2. For example, if we define four regions by

$$H_1' = \{(k_1,k_2)|0 < k_1 < \pi/qa, 0 < k_2 < \pi/b\}, \tag{34}$$
$$H_2' = \{(k_1,k_2)|\pi/qa < k_1 < 2\pi/qa, 0 < k_2 < \pi/b\}, \tag{35}$$
$$H_3' = \{(k_1,k_2)|\pi/qa < k_1 < 2\pi/qa, \pi/b < k_2 < 2\pi/b\}, \tag{36}$$
$$H_4' = \{(k_1,k_2)|0 < k_1 < 2\pi/qa, \pi/b < k_2 < 2\pi/b\}, \tag{37}$$

then we can choose H_i, $i = 1, ..., 4$ to be slightly larger regions each of which completely includes H_i', $i = 1, ..., 4$. A principal $U(1)$ bundle is a topological space which is locally isomorphic to $H_i \times U(1)$ in each neighborhood of H_i. We consider a specific fiber whose global topology is determined by the eigen state $|u_{k_1k_2}\rangle$. A construction of a fiber bundle is as follows: Take a component of the state vector $u_{k_1k_2}(x^{(0)}, y^{(0)}) = \langle x^{(0)}y^{(0)}|u_{k_1k_2}\rangle$ which does not vanish in the overlaps of H_i. Sonce H_i is contractible, it is possible to choose a phase convention such that the phase factor $\exp[i\theta^{(i)}(k_1k_2)] = u^{(i)}_{k_1k_2}(x^{(0)}, y^{(0)})/|u^{(i)}_{k_1k_2}(x^{(0)}, y^{(0)})|$ is smooth in each neighborhood H_i except at zeroes of $u_{k_1k_2}(x^{(0)}, y^{(0)})$. As exemplified in the last Section it is not possible in general to have a global phase convention which is good to all the neighborhoods. As a result we have a transition function Φ_{ij} in the overlap between two neighborhoods, $H_i \cap H_j$:

$$\Phi_{ij} = \exp i[\theta^{(i)}(k_1,k_2) - \theta^{(j)}(k_1,k_2)] = \exp[if^{(ij)}(k_1,k_2)]. \tag{38}$$

If we regard Φ_{ij} to be a map $\Phi_{ij} : U(1) \to U(1)$, a principal $U(1)$ bundle over T^2 is completely specified by this transition function.

Thus we have constructed a nontrivial fiber bundle. Fiber bundles are classified by certain integers characterizing the transition functions. These integers also correspond to integrals involving a bundle curvature when we pit connections on the bundles. We may write a connection 1-form as

$$\omega = g^{-1}Ag + g^{-1}dg \tag{39}$$
$$- A + id\chi, \tag{40}$$

where $g = e^{i\chi} \in U(1)$ is a fiber. We choose a 1-form by

$$A(k_1,k_2) = \hat{A}_\mu(k_1,k_2)dk_\mu = \langle u_{k_1k_2}|\frac{\partial}{\partial k_\mu}|u_{k_1k_2}\rangle dk_\mu. \tag{41}$$

The transition function of the form (38) act on fibers by left multiplication. IN an overlap of two neighborhoods H_i and H_j, a transition function $\Phi = \Phi_{ij}$ relates the local fiber coordinates g and g' in H_i and H_j as

$$g' = \Phi g. \tag{42}$$

This is equivalent to the "gauge transformation" Eq. (27). From Eqs. (28), (38), and (42), a transformation of $A(k_1, k_2)$ is given by

$$A' = \Phi A \Phi^{-1} + \Phi d\Phi^{-1} = A - i\frac{\partial f}{\partial k_\mu} dk_\mu. \tag{43}$$

It can be shown that ω is invaruant under the transformations (42) and (43). So, ω is indeed a legitimate connection 1-form with a choice of A in Eq. (41).

Since a connection is given, we have a differential geometry on the topological space. The curvature is given by

$$F = dA = \frac{\partial \hat{A}_\mu}{\partial k_\nu} dk_\nu \wedge dk_\mu. \tag{44}$$

Since $c_1 = (i/2\pi)F$ is the first Chern form, an integral of c_1 over the entire manifold T^2,

$$C_1 = \frac{i}{2\pi}\int F = \frac{i}{2\pi}\int dA = \frac{i}{2\pi}\int \frac{\partial \hat{A}_\mu}{\partial k_\nu} dk_\nu \wedge dk_\mu, \tag{45}$$

is the first Chern number. This number is an integer which is independent of a particular connection chosen. It is only given by the topology of the principal $U(1)$ bundle which is constructed from the state vector $|u_{k_1 k_2}\rangle$.

A comparison of Eqs. (26) and (45) gives

$$\sigma_{xy}^{(\alpha)} = -\frac{e^2}{h} C_1 \tag{46}$$

i.e., a contribution to the Hall conductance from a single filled band is, in unit of e^2/h is given by minus the first Chern number.

References

1. K. von Klitzing, G. Dorda, and M. Pepper, *Phys. Rev. Lett.* **45**, 494 (1980).
2. D.J. Thouless, M. Kohmoto, P. Nightingale, M. den Nijs, *Phys. Rev. Lett.* **49**, 405 (1982).
3. M. Kohmoto, **Ann. Phys. 160**, 343 (1985).
4. E. Brown, *Phys. Rev.* **133**, A1038 (1964).
5. J. Zak, *Phys. Rev.* **134**, A1602 (1964).

6. T. Eguchi, P.B. Gilkey, and A.J. Hansson, Phys. Rep. 66, 213 (1980); C. Nash and S. Sen, "*Topology and Geometry for Physicists*" Academic Press, New Yotk,1983; M. Nakahara, "*Geometry, Topology and Physics*," IOP Publishing, Bristol and Philadelphia, 1990.

QCD STRING AS AN EFFECTIVE STRING

Y. Makeenko
Institute of Theoretical and Experimental Physics, B. Cheremushkinskaya 25, 117218 Moscow,
E-mail: makeenko@itep.ru

There are two cases where QCD string is described by an effective theory of long strings: the static potential and meson scattering amplitudes in the Regge regime. I show how the former can be solved in the mean-field approximation, justified by the large number of space-time dimensions, and argue that it turns out to be exact for the Nambu–Goto string. By adding extrinsic curvature I demonstrate how the tachyonic instability of the ground-state energy can be cured by operators less relevant in the infrared.

Based on the talks given at the NBI Summer Institute "Strings, gauge theory and the LHC", Copenhagen August, 22 – September, 02 2011 and the Workshop "Low dimensional physics and gauge principles", Yerevan September 21 – 26, 2011.

Keywords: QCD string, Lüscher term, mean-field approximation, extrinsic curvature, tachyonic instability

1. Introduction

As is well known since 1970's – early 1980's, QCD string is not fundamental but is rather formed by fluxes of the gluon field at distances larger than the confinement scale. It has the meaning of an *effective* string which makes sense as a string in the limit when it is long. It can be then consistently quantized in $d = 4$ dimensions order by order in the inverse length.

In the present talk I shall briefly review this approach and describe how the series in the inverse length can be summed up for a pure bosonic Nambu–Goto string by the mean-field method. From the viewpoint of an effective string it is the most relevant operator in the infrared. I compare the results with recent Monte-Carlo data for the spectrum of QCD string and argue why they are well described by the mean-field approximation. I also incorporate the next relevant operator – the extrinsic curvature – and demonstrate how the tachyonic instability of the ground-state energy can be cured by operators less relevant in the infrared.

2. QCD string as such

QCD string is formed by fluxes of the gluon field at the distances larger than the confinement scale $1/\Lambda_{\rm QCD}$, where the lines of force between static quarks are collimated into a tube. This picture is supported by numerous Monte-Carlo simulations and agrees with linear hadron Regge trajectories seen in experiment. Perturbative QCD works at small distances (thanks to asymptotic freedom), while an effective string theory works at large distances.

QCD string is not pure bosonic Nambu–Goto string, as was first shown by Migdal and Y.M. (1979)[1] from the loop equation. Extra (fermionic) degrees of freedom are required at the string worldsheet, as advocated by Migdal (1981),[2] to satisfy the loop equation for self-intersecting Wilson loops. But the asymptote of large loops is universal and described by a classical string

$$W(C) \stackrel{\text{large } C}{\propto} \mathrm{e}^{-KS_{\min}(C)} \qquad \Longrightarrow \quad \text{the area law} = \text{confinement}. \qquad (1)$$

Here K is the string tension.

Semiclassical fluctuations of a long string were elegantly calculated by Lüscher, Symanzik, Weisz (1980).[3] For a plane loop the result is given by a conformal anomaly:

$$W(C) \stackrel{\text{plane } C}{\propto} \mathrm{e}^{-KS_{\min}(C)+\frac{\#}{24\pi}\int \mathrm{d}^2 w \left(\partial_a \ln\left|\frac{\mathrm{d}z}{\mathrm{d}w}\right|\right)^2} \qquad w(z): \text{UHP} \Rightarrow D, \quad (2)$$

where the function $w(z)$ conformally maps the upper half-plane (UHP) onto the interior of the loop. Equation (2) takes a simple form for a $\mathcal{T}\times R$ ($\mathcal{T}\gg R$) rectangle

$$W(C) \stackrel{\text{rectangle}}{\propto} \mathrm{e}^{-KR\mathcal{T}+\frac{\#}{24}\frac{\pi \mathcal{T}}{R}} \qquad \Longrightarrow \quad \text{the Lüscher term}. \qquad (3)$$

The only unknown constant here is the number # of fluctuating degrees of freedom which equal $d-2$ for the bosonic string. It agrees with the results of numerical simulations pioneered by Ambjorn, Olesen, Peterson (1984)[4] and De Forcrand, Schierholz, Schneider, Teper (1985).[5]

The $1/R$ term is *universal* owing to Lüscher's roughening[6] which states that the typical transversal size of the string grows with R as

$$\left\langle x_\perp^2 \right\rangle \propto \alpha' \ln(R^2/\alpha') \gg \alpha' \qquad \alpha' = 1/2\pi K. \qquad (4)$$

Next orders in $1/R$ are not universal.

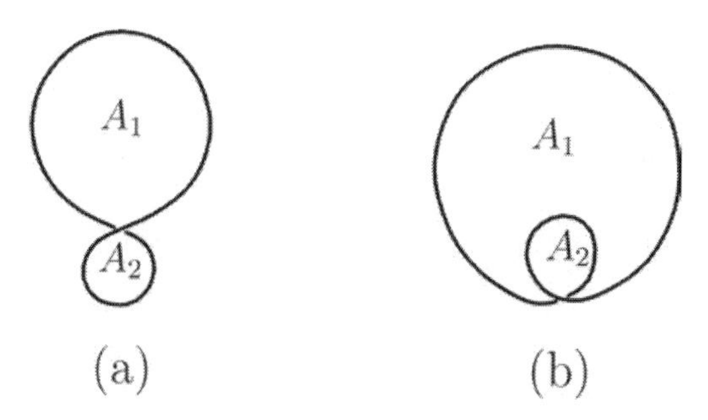

Fig. 1. Loops with one self-intersection: A_1 and A_2 denote the areas of the proper windows.

In the Polyakov (1981)[7] string formulation the Lüscher term (3) can be reproduced *á la* Durhuus, Olesen, Petersen (1984)[8] by conformally mapping UHP onto a $\mathcal{T} \times R$ rectangle.

3. What to expect: QCD$_2$ string

QCD$_2$ is solvable at large N as demonstrated by 't Hooft (1974).[9] The interaction vanishes in the axial gauge that immediately results in an exact area law for Wilson loops without self-intersections

$$W(C) = \mathrm{e}^{-A(C)} \qquad A(C) = \frac{g^2 N}{2} \text{Area}. \tag{5}$$

This looks like bosonic string in $d = 2$, but this is not the case for self-intersecting loops as observed by Kazakov, Kostov (1980),[10] Bralić (1980).[11]

The simplest loops with one self-intersection are depicted in Fig. 1. There is nothing special about the loop in Fig. 1a. Equation (5) still holds in this case with $A(C) = A_1 + A_2$ being the total area:

$$W(\mathrm{a}) = \mathrm{e}^{-A_1 - A_2} \qquad W(\mathrm{b}) = (1 - 2A_2)\,\mathrm{e}^{-A_1 - 2A_2}. \tag{6}$$

But for the loop in Fig. 1b the exponential of the total (folded) area $A(C) = A_1 + 2A_2$ is multiplied in Eq. (6) by a nontrivial polynomial which may have negative sign.

The appearance of such preexponential polynomials for the Wilson loops in QCD$_2$ makes their stringy interpretation more difficult but possible as shown by Gross, Taylor (1993).[12] These preexponential polynomials (and therefore self-intersections of the loop) become inessential for large C, so the asymptote (1) is always recovered.

4. Effective string theory

Closed string winding along a compact direction of large radius R is described by the Polchinski, Strominger (1991)[13] nonpolynomial action

$$S_{\text{eff}} = 2K \int \mathrm{d}^2 z \, \partial X \cdot \bar{\partial} X - \frac{\beta}{2\pi} \int \mathrm{d}^2 z \, \frac{\partial^2 X \cdot \bar{\partial}^2 X}{\partial X \cdot \bar{\partial} X} + \dots \qquad \beta = \frac{26-d}{12}. \quad (7)$$

It can be analyzed order by order in $1/R$ by expanding about the classical solution

$$X^\mu_{\text{cl}} = (e^\mu z + \bar{e}^\mu \bar{z})\, R \qquad e \cdot e = \bar{e} \cdot \bar{e} = 0 \quad e \cdot \bar{e} = -1/2. \qquad (8)$$

The action (7) emerges from the Polyakov formulation if we integrate over fast fluctuations and express the resulting effective action for slow fields (modulo total derivatives and the constraints) via an induced metric

$$\mathrm{e}^{\varphi_{\text{ind}}} = 2\, \partial X \cdot \bar{\partial} X \qquad (9)$$

(in the conformal gauge), which is not treated independently. This effective string theory has been analyzed[13–15] using the conformal field theory technique order by order in $1/R$, revealing the Arvis (1983)[16] spectrum

$$E_n = \sqrt{(KR)^2 + \left(n - \frac{d-2}{24}\right) 8\pi K} \qquad 8 \Longrightarrow 2 \quad \text{for open string} \qquad (10)$$

of the Nambu–Goto string in d-dimensions.

The conformal symmetry is maintained in $d \neq 26$ order by order in $1/R$:

$$\delta X^\mu = \epsilon(z) \partial X^\mu - \frac{\beta \alpha'}{4} \partial^2 \epsilon(z) \frac{\bar{\partial} X^\mu}{\partial X \cdot \bar{\partial} X} + \dots + \text{c.c.}. \qquad (11)$$

It transforms X^μ nonlinearly and the corresponding conserved energy-momentum tensor is

$$T_{zz} = -\frac{1}{\alpha'} \partial X \cdot \partial X + \frac{\beta}{2} \frac{\partial^3 X \cdot \bar{\partial} X}{\partial X \cdot \bar{\partial} X} + \mathcal{O}(R^{-2}). \qquad (12)$$

Expanding about the classical solution (8): $X^\mu = X^\mu_{\text{cl}} + Y^\mu_{\text{q}}$, we obtain

$$T_{zz} = -\frac{2R}{\alpha'} e \cdot \partial Y_{\text{q}} - \frac{1}{\alpha'} \partial Y_{\text{q}} \cdot \partial Y_{\text{q}} - \frac{\beta}{R} \bar{e} \cdot \partial^3 Y_{\text{q}} + \mathcal{O}(R^{-2}). \qquad (13)$$

The central charge is determined by

$$\langle T_{zz}(z_1) T_{zz}(z_2) \rangle = \frac{d + 12\beta}{2(z_1 - z_2)^4} + \mathcal{O}\left((z_1 - z_2)^{-2}\right) \qquad (14)$$

to be $d + 12\beta = 26$ and is cancelled by ghosts at any d.

5. Mean-field approximation for bosonic string

The ground state energy of the string determines the static potential between heavy quarks. It was first computed to all orders in $1/R$ by Alvarez (1981)[17] in the large-d limit, using the saddle point technique in the Nambu–Goto formulation. In the Polyakov formulation (= conformal gauge) these results were extended to any d by Y.M. (2011)[18] as follows.

Let us consider the (variational) mean-field ansatz with fluctuations included

$$X^1_{\rm mf}(\omega) = \frac{\omega_1}{\omega_R}R + \delta X^1(\omega) \quad X^2_{\rm mf}(\omega) = \frac{\omega_2}{\omega_T}T + \delta X^2(\omega) \quad X^\perp(\omega) = \delta X^\perp(\omega) \tag{15}$$

in the world-sheet parametrization, when $\omega_1, \omega_2 \in \omega_R \times \omega_T$ rectangle. These ω_R, ω_T change under reparametrizations of the loop. The ratio ω_R/ω_T is a variational parameter. It is a reminder of the reparametrization invariance of the boundary for the given parametrization.

The mean-field action with accounting for the Lüscher term reads

$$S_{\rm mf} = \frac{1}{4\pi\alpha'}\left(R^2\frac{\omega_T}{\omega_R} + T^2\frac{\omega_R}{\omega_T}\right) - \frac{\pi(d-2)}{24}\frac{\omega_T}{\omega_R}. \tag{16}$$

The minimization with respect to ω_T/ω_R reproduces the square root

$$\left(\frac{\omega_T}{\omega_R}\right)_* = \frac{T}{\sqrt{R^2-R_c^2}} \qquad S_{\rm mf*} = \frac{T}{2\pi\alpha'}\sqrt{R^2-R_c^2} \qquad R_c^2 = \pi^2\frac{(d-2)}{6}\alpha'. \tag{17}$$

For the upper half-plane parametrization, where vertices of the rectangle are at the values $s_1 < s_2 < s_3 < s_4 \in (-\infty, +\infty)$, we have

$$\frac{\omega_T}{\omega_R} = \frac{K\left(\sqrt{r}\right)}{K\left(\sqrt{1-r}\right)} \qquad r = \frac{s_{43}s_{21}}{s_{42}s_{31}} \qquad s_{ij} \equiv s_i - s_j, \tag{18}$$

as is obtained from the Schwarz–Christoffel mapping. Here K is the complete elliptic integral of the first kind and the ratio on the right-hand side of Eq. (18) is known as the Grötzsch modulus which is monotonic in r. Therefore, the minimization with respect to r gives the same result.

The mean-field approximation is applicable if fluctuations about the minimal surface are small. The ratio of the area of typical surfaces to the minimal area is

$$\frac{\langle {\rm Area}\rangle}{A_{\rm min}} = \frac{1}{RT}\frac{{\rm d}}{{\rm d}K}S_{\rm mf} = \frac{1-R_c^2/2R^2}{\sqrt{1-R_c^2/R^2}}, \tag{19}$$

which is plotted in Fig. 2. It approaches 1 for large R/R_c, implying small fluctuations, but diverges when $R \to R_c$ from above, so typical surfaces become very large and the mean-field approximation ceases to be applicable.

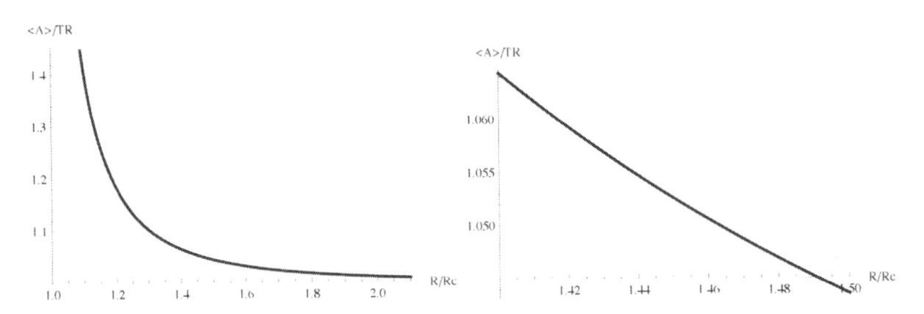

Fig. 2. Plot of the ratio (19). The region near $R/R_c \approx 1.4$ is magnified in the right figure.

The mean-field works generically at large d but is expected to be exact for the bosonic string at any d. The arguments are:
(1) it is true in the semiclassical approximation;
(2) it reproduces an exact result in $d = 26$;
(3) it agrees with the existence of a massless bound state in $N = \infty$ QCD_2 for massless quarks as shown by 't Hooft (1974).[9]

For QCD string we expect that the mean-field approximation works with an exponential accuracy $\exp(-C\, R/R_c)$, as is explained below.

The coefficient of $d-2$ in the above formulas is simply the number of fluctuating (transverse) degrees of freedom in the static gauge. In conformal gauge the path integral over reparametrizations of the boundary contributes 24 as is shown by Olesen, Y.M. (2010),[19] ghosts contribute 26, the fluctuations of X^μ contribute d. All together we get $d+24-26 = d-2$ again.

6. Comparison with Monte-Carlo data

The stringy spectrum (10) has been recently compared with the results of the very interesting Monte-Carlo computations by Athenodorou, Bringoltz, Teper (2010)[20] of the spectrum of closed winding string (flux tube) of circumstance R in 3+1 dimensional $SU(N)$ lattice gauge theory. The agreement is absolutely beautiful down to the distances

$$R/R_c \geq 1.4 \qquad \sqrt{K} R_c = \sqrt{\frac{d-2}{3}}\pi \approx 1.44\,. \tag{20}$$

The question immediately arises as to whether or not these results indicate that QCD string is indeed the Nambu–Goto one? To answer this question, let us look at Fig. 2, where it is seen that the ratio (19) is a rather sharp function, approaching infinity as $R/R_c \to 1$ from above. Large values of the ratio imply that typical surfaces have large area (in units of $\mathcal{T}R$) so that the fluctuations are large.* This is of course the case where the mean-field approximation does not work. But for $R/R_c \gtrsim 1.4$, where the Monte-Carlo data are available, the ratio is smaller than 1.1 which means that the mean field nicely works and we can restrict ourselves in the effective action of QCD string only by the quadratic operator which is most relevant in the infrared. For the values of R/R_c closer to 1, other operators will be apparently no longer negligible. We shall explicitly consider in the next section the operator of next relevance in the infrared – the extrinsic curvature.

7. QCD string as rigid string

String with extrinsic curvature in the action was introduced in the given context by Polyakov (1986)[22] and Kleinert (1986).[23] The original idea was that it may provide rigidity of the string that makes stringy fluctuation smoother. We shall momentarily see this is indeed the case after some subtleties will be resolved, in spite of some contradictory statements in the literature[24–26] (for a review see Ref.[27]).

The action of the bosonic string with the extrinsic curvature term reads

$$S_{\text{rigid string}} = \frac{K}{2}\int d^2\omega\, \partial_a X \cdot \partial_a X + \frac{1}{2\alpha}\int d^2\omega\, \frac{1}{\sqrt{g}}\Delta X \cdot \Delta X, \tag{21}$$

where α is dimensionless constant. It is to be distinguished from intrinsic (or scalar) curvature

$$R = D^2 X \cdot D^2 X - D^a D^b X \cdot D_a D_b X \tag{22}$$

that leads to the Gauss–Bonnet term in 2d $\Longrightarrow$ the Euler character. The original motivation was that rigidity smoothen crumpling of the surfaces.

Introducing $\rho = \sqrt{g}$ and the Lagrange multipliers λ^{ab}, we rewrite the action (21) as

$$S_{\text{r.s.}} = K\int d^2\omega\, \rho + \frac{1}{2\alpha}\int d^2\omega\, \frac{1}{\rho}\Delta X \cdot \Delta X + \frac{1}{2}\int d^2\omega\, \lambda^{ab}\left(\partial_a X \cdot \partial_b X - \rho\delta_{ab}\right). \tag{23}$$

*This crumpling of the surfaces is related to the tachyonic instability which is not expected to happen for QCD string as pointed out by Olesen (1985).[21]

We consider a mean-field (variational) ansatz, when only $X^\perp$ fluctuates. It is exact at large d but approximate at finite d (like summing bubble graphs for an $O(d)$-vector field). We write

$$X^1_{\rm mf}(\omega) = \frac{\omega_1}{\omega_R} R \quad X^2_{\rm mf}(\omega) = \omega_2 \; (\omega_T = T) \quad X^\perp(\omega) = \delta X^\perp(\omega)$$

$$\rho_{\rm mf}(\omega) = \rho \quad \lambda^{11}_{\rm mf}(\omega) = \lambda^{11} \quad \lambda^{22}_{\rm mf}(\omega) = \lambda^{22} \quad \lambda^{12}_{\rm mf}(\omega) = \lambda^{21}_{\rm mf}(\omega) = 0$$

$$\begin{aligned}\frac{1}{T} S_{\rm mf} &= \frac{1}{2}\left(\lambda_{11}\omega_R + \lambda_{22}\frac{R^2}{\omega_R}\right) + \rho\left(K - \frac{\lambda^{11}}{2} - \frac{\lambda^{22}}{2}\right)\omega_R \\ &\quad + \frac{d}{2T}\,\mathrm{tr}\,\ln\left(-\lambda^{11}\partial_1^2 - \lambda^{22}\partial_2^2 + \frac{1}{\alpha\rho}(\partial_1^2+\partial_2^2)^2\right).\end{aligned} \tag{24}$$

The determinant in the last line equals

$$\frac{d}{2T}\,\mathrm{tr}\,\ln(\ldots) \longrightarrow \begin{cases} 1) \; -\frac{\pi d}{6\omega_R}\sqrt{\frac{\lambda^{22}}{\lambda^{11}}} & \alpha \to \infty \\ 2) \; -\frac{\pi d}{3\omega_R} + \frac{d}{2}\sqrt{\alpha\rho\lambda^{11}} & \alpha \to 0 \end{cases} \qquad \text{(closed string)} \tag{25}$$

as $\alpha \to \infty$ or $\alpha \to 0$. Both limiting cases can be analyzed analytically:
1) the same mean field as above (large α) Alvarez (1981)[17]
2) solvable in square roots (small α) Polchinski, Yang (1992)[28]

For small α we have[28]

$$E_0 = \lambda^{11}\omega_R \quad \sqrt{\lambda^{11}} = \frac{3}{8}\frac{d\sqrt{\alpha}}{R} + \sqrt{\frac{9}{16}\frac{d^2\alpha}{R^2} + K - \frac{\pi d}{3R^2}}$$

$$\omega_R = \sqrt{R^2 - \frac{dR}{2}\sqrt{\frac{\alpha}{\lambda^{11}}}}. \tag{26}$$

The ground state energy E_0 is plotted versus R/R_c for various α in Fig. 3. The tachyonic singularity moves left to smaller values of R/R_c with decreasing α, and then returns back to $R/R_c = 1$ as $\alpha \to 0$.

The lines in Fig. 3 are drown using exact formulas by Olesen, Yang (1987),[24] Braaten, Pisarski, Tse (1987),[25] Germán, Kleinert (1988).[26] Integrating in Eq. (24) over dk_2 (as $T \to \infty$), regularizing via the the zeta function and introducing

$$\Lambda = \frac{\sqrt{\alpha\rho\lambda^{11}}\,\omega_R}{2\pi} \tag{27}$$

instead of ρ, we find

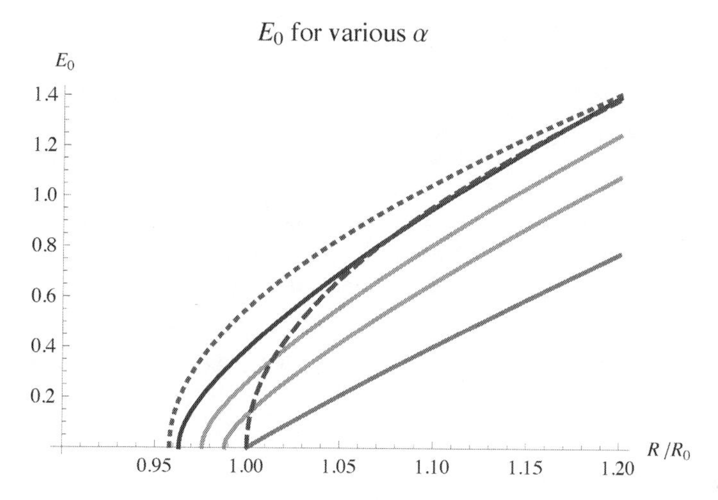

Fig. 3. Ground state energy versus R/R_c for various α. The (blue) dashed line emanated from $R/R_c = 1$ corresponds to $\alpha = \infty$ (*i.e.* no rigidity = pure Nambu–Goto). The other line corresponds to $\alpha \sim 10, 0.3, 0.2, 0.1, 0$ from left to right.

$$
\begin{aligned}
\frac{1}{T}S_{\mathrm{mf}} &= \frac{1}{2}\left(\lambda_{11}\omega_R + \lambda_{22}\frac{R^2}{\omega_R}\right) + \left(\frac{2K}{\lambda^{11}} - 1 - \frac{\lambda^{22}}{\lambda^{11}}\right)\frac{2\pi^2\Lambda^2}{\alpha\omega_R} \\
&\quad + \frac{2\pi d}{\omega_R}\left[-\frac{1}{6} + \frac{\Lambda}{2} + \frac{\Lambda^2}{4}\left(1 + \frac{\lambda^{22}}{\lambda^{11}}\right)\ln\frac{1}{\mu a_{\mathrm{UV}}}\right] \\
&\quad + \frac{2\pi d}{\omega_R}\sum_{n\geq 1}\left[\sqrt{\frac{\Lambda^2}{2} + n^2 + \Lambda\sqrt{\frac{\Lambda^2}{4} + \left(1 - \frac{\lambda^{22}}{\lambda^{11}}\right)n^2}}\right. \\
&\quad \left. + \sqrt{\frac{\Lambda^2}{2} + n^2 - \Lambda\sqrt{\frac{\Lambda^2}{4} + \left(1 - \frac{\lambda^{22}}{\lambda^{11}}\right)n^2}} - 2n - \frac{\Lambda^2}{4n}\left(1 + \frac{\lambda^{22}}{\lambda^{11}}\right)\right], \qquad (28)
\end{aligned}
$$

where a_{UV} is an ultraviolet cutoff, introduced by

$$\sum_{n\geq 1}\frac{1}{n} = \ln\frac{1}{\mu a_{\mathrm{UV}}}. \qquad (29)$$

The renormalization of the parameters K and α of the bare action (21) is to be performed in Eq. (28) by introducing (renormalized)

$$\alpha(\mu) = \frac{\alpha}{1 - \dfrac{\alpha d}{4\pi}\ln\dfrac{1}{\mu a_{\mathrm{UV}}}}, \qquad K(\mu) = \frac{K}{1 - \dfrac{\alpha d}{4\pi}\ln\dfrac{1}{\mu a_{\mathrm{UV}}}} \qquad (30)$$

as is prescribed by asymptotic freedom of the model.[22,23] Then UV divergences disappear in Eq. (28), so the result is finite.

8. Induced extrinsic curvature

A conclusion from the previous section is that adding extrinsic curvature improves the situation but does not cure the problem of tachyonic instability which is not present for QCD string. Thus, more operators of lower dimensions (not relevant in the infrared) have to be added within the effective string theory description of QCD string. They can be systematically induced by internal degrees of freedom of QCD string, *e.g.* massive fermions or higher dimensions, *á la* Sakharov's induced gravity.

The determinant of massless 2d Laplacian (or the Dirac operator squared) in the conformal gauge shows how it works:

$$\operatorname{tr}\ln\Delta = \frac{1}{12\pi}\int \mathrm{d}^2 z \left(\mu_0\,\mathrm{e}^{\varphi} \mp \partial\varphi\bar\partial\varphi\right). \tag{31}$$

For the induced metric $\mathrm{e}^{\varphi} = 2\partial X\cdot\bar\partial X$, we have

$$\int \mathrm{d}^2 z\,\partial\varphi\bar\partial\varphi = \frac{1}{4}\int \mathrm{d}^2 z\;\mathrm{e}^{-\varphi}\Delta X\cdot\Delta X \tag{32}$$

and there are no other operators of the same dimension as the extrinsic curvature.

A logarithmically divergent coefficient appears, when the extrinsic curvature is induced by 4d fermions pulled back to the string worldsheet, as calculated by Sedrakian, Stora (1987),[29] Wiegmann (1989),[30] Parthasarathy, Viswanathan (1999).[31] Likewise, the extrinsic curvature is induced by higher dimensions in the AdS/CFT correspondence with confining background if # of massless modes = # of modes that acquired mass, as demonstrated by Greensite, Olesen (1999).[32]

9. Conclusion

- QCD string can be viewed as an effective long string and analyzed in $d = 4$ by the mean-field method.
- Two important applications of this technique:
  – ground state energy of QCD string,
  – meson scattering amplitudes in the Regge regime[18] (not described in this talk).
- Monte Carlo data for the spectrum of QCD string can be well described by only the most relevant operator (Nambu–Goto).

- Extrinsic curvature softens the tachyonic problem and may be induced by additional degrees of freedom of QCD string.

Acknowledgement

I am indebted to A. Gonzalez-Arroyo, P. Olesen and M. Teper for useful discussions.

References

1. Y. M. Makeenko and A. A. Migdal, *Exact equation for the loop average in multicolor QCD,* Phys. Lett. B **88** (1979) 135.
2. A. A. Migdal, *QCD = Fermi string theory*, Nucl. Phys. B **189**, 253 (1981).
3. M. Lüscher, K. Symanzik, and P. Weisz, *Anomalies of the free loop wave equation in the WKB approximation,* Nucl. Phys. B **173**, 365 (1980).
4. J. Ambjorn, P. Olesen and C. Peterson, *Three-dimensional lattice gauge theory and strings,* Nucl. Phys. B **244**, 262 (1984).
5. P. de Forcrand, G. Schierholz, H. Schneider and M. Teper, *The string and its tension in SU(3) lattice gauge theory: towards definitive results,* Phys. Lett. B **160**, 137 (1985).
6. M. Lüscher, *Symmetry breaking aspects of the roughening transition in gauge theories,* Nucl. Phys. B **180**, 317 (1981).
7. A. M. Polyakov, *Quantum geometry of bosonic strings,* Phys. Lett. B **103**, 207 (1981).
8. B. Durhuus, P. Olesen, and J. L. Petersen, *On the static potential in Polyakov's theory of the quantized string*, Nucl. Phys. B **232**, 291 (1984).
9. G. 't Hooft, *A two-dimensional model for mesons*, Nucl. Phys. B **75**, 461 (1974).
10. V. A. Kazakov and I. K. Kostov, *Nonlinear strings in two-dimensional* $U(\infty)$ *theory*, Nucl. Phys. B **176**, 199 (1980).
11. N. Bralić, *Exact computation of loop averages in two-dimensional Yang–Mills theory*. Phys. Rev. D **22**, 3090 (1980).
12. D. G. Gross and W.I. Taylor *Two-dimensional QCD is a string theory, Nucl. Phys.* B **400**, 181 (1993).
13. J. Polchinski and A. Strominger, *Effective string theory*, Phys. Rev. Lett. **67**, 1681 (1991).
14. J. M. Drummond, *Universal subleading spectrum of effective string theory*, arXiv: hep-th/0411017.
15. O. Aharony and E. Karzbrun, *On the effective action of confining strings*, J. High Energy Phys. **0906**, 012 (2009) [arXiv:0903.1927 [hep-th]].
16. J. F. Arvis, *The exact $\bar{q}q$ potential in Nambu string theory*, Phys. Lett. B **127**, 106 (1983).
17. O. Alvarez, *Static potential in string theory*, Phys. Rev. D **24**, 440 (1981).
18. Y. Makeenko, *An interplay between static potential and Reggeon trajectory for QCD string*, Phys. Lett. B **699**, 199 (2011) [arXiv:1103.2269 [hep-th]].

19. Y. Makeenko and P. Olesen, *Quantum corrections from a path integral over reparametrizations,* Phys. Rev. D **82**, 045025 (2010) [arXiv:1002.0055 [hep-th]].
20. A. Athenodorou, B. Bringoltz, and M. Teper, *Closed flux tubes and their string description in D=3+1 SU(N) gauge theories,* J. High Energy Phys. **1102**, 030 (2011) [arXiv:1007.4720 [hep-lat]].
21. P. Olesen, *Strings and QCD,* Phys. Lett. B **160**, 144 (1985).
22. A. M. Polyakov, *Fine structure of strings,* Nucl. Phys. B **268**, 406 (1986).
23. H. Kleinert, *The membrane properties of condensing strings,* Phys. Lett. B **174**, 335 (1986).
24. P. Olesen and S.-K. Yang, *Static potential in a string model with extrinsic curvatures,* Nucl. Phys. B **283**, 73 (1987).
25. E. Braaten, R. D. Pisarski and S.-M. Tse, *The static potential for smooth strings,* Phys. Rev. Lett. **58**, 93 (1987)
26. G. Germán and H. Kleinert, *Perturbative two loop quark potential of stiff strings in any dimension,* Phys. Rev. D **40**, 1108 (1989).
27. G. Germán, *Some developments in Polyakov-Kleinert string with extrinsic curvature stiffness,* Mod. Phys. Lett. A **6**, 1815 (1991).
28. J. Polchinski and Z. Yang, *High temperature partition function of the rigid string,* Phys. Rev. D **46**, 3667 (1992) [arXiv:hep-th/9205043].
29. A. G. Sedrakian and R. Stora, *Dirac and Weyl fermions coupled to two-dimensional surfaces: determinants,* Phys. Lett. B **188**, 442 (1987).
30. P. B. Wiegmann, *Extrinsic geometry of superstrings,* Nucl. Phys. B **323**, 330 (1989).
31. R. Parthasarathy and K.S. Viswanathan, *Induced rigid string action from fermions,* Lett. Math. Phys. **48**, 243 (1999) [arXiv:hep-th/9811144].
32. J. Greensite and P. Olesen, *World sheet fluctuations and the heavy quark potential in the AdS/CFT approach,* J. High Energy Phys. **9904**, 001 (1999) [arXiv:hep-th/9901057].

THE MUG: On $E = mc^2$ AND RELATIVITY THEORY IN THE MASS CULTURE

L. B. Okun
ITEP, Moscow, Russia

This note is an attempt to explain in simple words why the famous relation $E = mc^2$ misrepresents the essence of Einstein's relativity theory. The first part of the note is addressed to high-school teachers, the rest – to those university professors of Physics who allow themselves to say that the mass of a body increases with its velocity.

1. Introduction

For more than twenty years I collect artifacts of mass culture (from post-cards and T- shirts to popular articles and books) sporting "the famous Einstein's formula". Recently my friends gave me a Relativity Floxy Noxy mug. You can type these words in the search line of your computer and look at it:

In a certain sense it contains the quintessence of my collection presenting the main popular science cliches and misconceptions. As they are quite often repeated in newspapers and textbooks, I decided to reproduce the text of the mug and to explain briefly what is wrong with it. Believe that it might

Fig. 1. The Relativity Mug

be useful to many.

2. The text on the mug

There are three columns of text on the mug: to the right of the handle (**1**), to the left of the handle (**2**), opposite the handle (**3**).

1. In 1905 at the age of 26, Einstein proposed the Special Theory of Relativity, using the equation: $E = mc^2$ where E=energy, m=mass, c=the speed of light. Special relativity expresses the concept that matter and energy are really different forms of the same thing. Any mass has an associated energy and vice versa.

2. Albert Einstein's SPECIAL THEORY OF RELATIVITY In the 1850-s it was calculated that light travelled at a fixed speed of 670 million mph. However, whatever speed we travelled at, we would never catch up with the speed of light. Einstein proposed that if the speed of light is always fixed, something else must give way, i.e. mass must change. An object must get heavier as it approaches the speed of light. He concluded that energy and mass must be interrelated.

3. His formula suggests that tiny amounts of mass can be converted into huge amounts of energy... ... which revealed the secret of how stars shine and unlocked the key to atomic energy.

3. My clarifications and comments

1. After Maxwell established in 1865 his equations for electromagnetic field it became clear that they are incompatible with Newton's Mechanics. When at the junction of XIX and XX centuries physicists started to study the motion of very fast particles, they saw that equations of Newton's mechanics should be modified. First they thought that Newton's equations could be preserved by assuming that mass increases with velocity. Than it turned out that the equations themselves should be changed in such a way that mass does not depend on velocity, but there exists an important connection between mass and energy. This had serious impact on the language and philosophy of physics.

In 1905 Einstein published a detailed article in which he presented his theory which later was named Einstein's theory of relativity. In the same year he published a short note in which he stated that in the framework of this theory the mass of a body depends on its energy content and that to any mass m there corresponds energy E such that $E = mc^2$. In his subsequent clarifications he used the more precise equation $E_0 = mc^2$, where E_0 is the rest energy of the body. (The total energy E is the sum of rest energy E_0 and kinetic energy E_K of the body.) He used it to stress that mass does not depend on the velocity of the body. Though sometimes, especially in his popular writings, he did not care about index zero, nevertheless in 1948 he warned L. Barnett – the author of the book "Universe of Dr. Einstein" – not to use the concept of mass depending on velocity. This semantic swirling is caused by the clash of two languages – the old non-relativistic and the new – consistently relativistic.

2. The statement that the speed of light is fixed is correct, but the date (1850's) is not. That the speed of light is finite, was established in 1676 by Römer. That it does not depend on the speed of the source and the observer, was discovered in 1887 by Michelson and Morley. The statement that energy and mass are interrelated is correct: $E_0 = mc^2$, while that the mass changes with velocity is definitely wrong. In the theory of relativity (unlike the mechanics of Newton) the measure of inertia is not mass m but the total energy E of the body. The momentum $\mathbf{p}$ of a body is connected with its velocity $\mathbf{v}$ not by the relation $p = m\mathbf{v}$, but by the relation $\mathbf{p} = (E/c^2)\mathbf{v}$. And only at zero velocity the total energy equals to rest energy: $E = E_0$.

3. Here everything is correct if one uses the equation $E_0 = mc^2$ and takes into account that in nuclear reactions in the stars, on the Sun and on the Earth a part of the rest energy of the particles which are burned transforms into kinetic energy of the products of burning. The same is valid for any process of burning.

4. Four dimensions of the world

Finally I would like to say a few words to those who are familiar with the concept of four-dimensional world introduced in the relativity theory in 1908 by Minkowski. The energy E of a body and three components of its momentum p form four components of the pseudoeucledean 4-vector. The scalar length of this 4-vector is given by the mass of the body m according to the equation $E^2c^{-4} - \mathbf{p}^2c^{-2} = m^2$. (The word pseudoeucledean indicates that the square of the hypothenuse is equal to the difference (not sum!) of the squares of two other legs of the triangle.) This main equation of relativity theory has been tested in thousands of experiments with accuracy up to ten digits. It implies that for a massive body whose momentum is zero $E_0 = mc^2$, while for a massless particle of light – the photon – the speed is always equal to c. The theory of relativity is impeccable. One cannot say the same about its image in the mass culture.

Unfortunately the sudden illness and death of Minkowski did not allow him to persuade his contemporaries to switch to the language of four-dimensional world, and they continued futile attempts to explain the meaning of relativity theory in terms of three-dimensional mechanics of Newton. Though Einstein used the four-dimensional apparatus in deriving the equations of his general theory of relativity for gravitational interaction, I failed to find the equation $E^2c^{-4} - \mathbf{p}^2c^{-2} = m^2$ on the pages of his writings. It appeared first in the works of Dirac in 1930 and in the book "Teoria Polya" by Landau and Lifshitz in 1941. Four-dimensional description is equally good for massive and massless particles of matter. It shows that mass and matter are not the same thing, that momentum is the measure of all motions in nature, while energy is the measure of all processes. As for the mass of the particles it becomes to be non-essential at high energies $E \gg mc^2$.

I am grateful to Erica Gulyaeva, Marek Karliner, Elya and Vitaly Kisin and Boris Okun who helped me, each in her/his own way, to write this note.

NEW CHERN-SIMONS DENSITIES IN BOTH ODD AND EVEN DIMENSIONS

E. Radu[†⋆] and T. Tchrakian[‡⋆⋄]

[‡] *School of Theoretical Physics – DIAS, 10 BurlingtonRoad, Dublin 4, Ireland*
[⋆] *Department of Computer Science, National University of Ireland Maynooth, Maynooth, Ireland*
[⋄] *Theory Division, Yerevan Physics Institute (YerPhI), AM-375 036 Yerevan 36, Armenia*

After reviewing briefly the dimensional reduction of Chern–Pontryagin densitie, we define new Chern–Simons densities expressed in terms of Yang-Mills and Higgs fields. These are defined in all dimensions, including in even dimensional spacetimes. They are constructed by subjecting the dimensionally reduced Chern–Pontryagin densites to further descent by two steps.

1. Introduction

The central task of these notes is to explain how to subject $n-$th Chern–Pontryagin (CP) density $\mathcal{C}^{(n)}$,

$$\mathcal{C}^{(n)} = \frac{1}{\omega(\pi)} \varepsilon_{M_1 M_2 M_3 M_4 \dots M_{2n-1} M_{2n}} \mathrm{Tr}\, F_{M_1 M_2} F_{M_3 M_4} \dots F_{M_{2n-1} M_{2n}} \tag{1}$$

defined on on the $2n-$dimensional space, to dimensional descent to R^D by considering (1) on the direct product space $R^D \times S^{2n-D}$. The resulting residual density on R^D will be denoted as $\Omega_D^{(n)}$.

The density $\mathcal{C}^{(n)}$ is by construction, a total divergence

$$\mathcal{C}^{(n)} = \boldsymbol{\nabla} \cdot \boldsymbol{\Omega}^{(n)} , \tag{2}$$

and it turns out that under certain retrictions, the dimensional descendant of (2) is also a total divergence. This result will be applied to the construction of Chern-Simons solitons in all dimensions.

Some special choices, or restrictions, are made for practical reasons. Firstly, we have restricted to the codimension S^{2n-D}, the $(2n-D)-$sphere, as this is the most symmetric compact coset space that is defined both in even and in odd dimensions. It can of course be replaced by any other symmetric and compact coset space.

Secondly, the gauge field of the bulk gauge theory is chosen to be a $2^{n-1} \times 2^{n-1}$ array with complex valued entries. Given our choice of spheres for the codimension, this leads to residual gauge fields on which take their values in the Dirac matrix representation of the residual gauge group $SO(D)$.

As a result of the above two choices, it is possible to make the symmetry imposition (namely the dimensional reduction) such that the residual Higgs field is described by a $D-$component isovector multiplet. With this choice, the asymptotic gauge fields describe a Dirac-Yang[1–3] monopoles.

The central result exploited here is that the CP density $\mathcal{C}_D^{(n)}$ on R^D descended from the $2n-$dimensional bulk CP density $\mathcal{C}^{(n)}$ is **also** a *total divergence*

$$\mathcal{C}_D^{(n)} = \boldsymbol{\nabla} \cdot \boldsymbol{\Omega}^{(n,D)} \tag{3}$$

like $\mathcal{C}^{(n)}$ formally is on the bulk. From the reduced density $\boldsymbol{\Omega}^{(n,D)} \equiv \Omega_i^{(n,D)}$, where the index i labels the coordinate of the residual space x_i with $i = 1, 2, \ldots, D$, one can identify a Chern–Simons (CS) density as the $D-$th component of $\boldsymbol{\Omega}^{(n,D)}$. This quantity can then be interpreted as a CS term on $(D-1)-$ dimensional Monkowski space, *i.e.*, on the spacetime (t, R^{D-2}). The solitons of the corresponding CS–Higgs (CSH) theory can be constructed systematically. Note, that this is not the usual CS term defined in terms of a pure Yang–Mills field on *odd* dimensional spacetime, but rather these new CS terms are defined by both the YM and, the Higgs fields. Most importantly, the definition of these new CS terms is not restricted to *odd* dimensional spacetimes, but covers also *even* dimensional spacetimes. Such CSH solutions have not been studied to date.

2. Dimensional reduction of gauge fields

The calculus of the dimensional reduction of Yang–Mills fields employed here is based on the formalism of A. S. Schwarz,[4] which is specially transparent due to the choice of displaying the results only at a fixed point of the compact symmetric codimensional space K^N (the North or South pole for S^N). Our formalism is a straightforward extension of.[4–6]

2.1. *Descent over S^N: N odd*

For the descent from the bulk dimension $2n = D+N$ down to **odd** D (over odd N), the components of the residual connection evaluated at the North pole of S^N are given by

$$\mathcal{A}_i = A_i(\vec{x}) \otimes \mathbb{I} \tag{4}$$

$$\mathcal{A}_I = \mathcal{F}(\vec{x}) \otimes \frac{1}{2}\Gamma_I \,. \tag{5}$$

The unit matrix in (4), like the $N-$dimensional gamma matrix in (5), are $2^{\frac{1}{2}(N-1)} \times 2^{\frac{1}{2}(N-1)}$ arrays. Choosing the $2^{n-1} \times 2^{n-1}$ bulk gauge group to be, say, $SU(n-1)$, allows the choice of $SO(D)$ as the gauge group of the residual connection $A_i(x)$. This choice is made such that the asymptotic connections describe a Dirac–Yang monopole.

For the same reason, the choice for the multiplet structure of the Higgs field is made to be less restrictive. The (anti-Hermitian) field $\mathcal{F}$, which is not necessarily traceless [*], can be and *is* taken to be in the algebra of $SO(D+1)$, in particular, in one or other of the chiral reprentations of $SO(D+1)$, $D+1$ here being even.

$$\mathcal{F} = \phi^{ab}\,\mathbf{\Sigma}_{ab} \,, \quad a = i,\, D+1 \,, \quad i = 1, 2, \ldots, D \,. \tag{6}$$

(Only in the $D = 3$ case does the Higgs field take its values in the algebra of $SO(3)$, since the representations $SO(3)$ coincide with those of chiral $SO(4)$.)

In anticipation of the corresponding situation of even D to be given next, one can specialise (6) to the a $D-$component *isovector* expression of the Higgs field

$$\mathcal{F} = \phi^i\,\mathbf{\Sigma}_{i,D+1} \,, \tag{7}$$

with the purpose of having a unified notation for both even and odd D, where the Higgs field takes its values in the components $\mathbf{\Sigma}_{i,D+1}$ orthogonal to elements $\mathbf{\Sigma}_{ij}$ of the algebra of $SO(D+1)$. This specialisation is not necessary, and is in fact inappropriate should one consider, *e.g.*, axially symmetric fields. It is however adequate for the presentation here and is sufficiently general to describe spherically symmetric monopoles [†].

---

[*] In practice, when constructing soliton solutions, $\mathcal{F}$ is taken to be traceless without loss of generality.

[†] While all concrete considerations in the following are restricted to spherically symmetric fields, it should be emphasised that relaxing spherical symmetry results in the Higgs multiplet getting out of the orthogonal complement $\mathbf{\Sigma}_{i,D+1}$ to $\mathbf{\Sigma}_{i,j}$. Indeed, subject to axial symmetry one has

$$\mathcal{F} = f_1(\rho,z)\mathbf{\Sigma}_{\alpha\beta}\hat{x}_\beta + f_2(\rho,z)\mathbf{\Sigma}_{\beta,D+1}\hat{x}_\beta + f_3(\rho,z)\mathbf{\Sigma}_{D,D+1} \,, \tag{8}$$

where $x_i = (x_\alpha, z)$, $|x_\alpha|^2 = \rho^2$ and with $\hat{x}_\alpha = x_\alpha/\rho$. Clearly, the term in (8) multiplying the basis $\mathbf{\Sigma}_{\alpha\beta}$ does not occur in (7).

In (4) and (5), and everywhere henceforth, we have denoted the components of the residual coordinates as $x_i = \vec{x}$. The dependence on the codimension coordinate x_I is suppressed since all fields are evaluated at a fixed point (North or South pole) of the codimension space.

The resulting components of the curvature are

$$\mathcal{F}_{ij} = F_{ij}(\vec{x}) \otimes 1\!\!1 \tag{9}$$

$$\mathcal{F}_{iI} = D_i\mathcal{F}(\vec{x}) \otimes \frac{1}{2}\Gamma_I \tag{10}$$

$$\mathcal{F}_{IJ} = S(\vec{x}) \otimes \Gamma_{IJ}\,, \tag{11}$$

where $\Gamma_{IJ} = -\frac{1}{4}[\Gamma_I, \Gamma_J]$ are the Dirac representation matrices of $SO(N)$, the stability group of the symmetry group of the $N-$sphere. In (10), $D_i\mathcal{F}$ is the covariant derivative of the Higgs field $\mathcal{F}$

$$D_i\mathcal{F} = \partial_i\mathcal{F} + [A_i, \mathcal{F}] \tag{12}$$

and S is the quantity

$$S = -(\eta^2\, 1\!\!1 + \mathcal{F}^2)\,, \tag{13}$$

where η is the inverse of the radius of the $N-$sphere.

2.2. *Descent over S^N: N even*

The formulae corresponding to (4)-(11) for the case of **even** D are somewhat more complex. The reason is the existence of a chiral matrix Γ_{N+1}, in addition to the Dirac matrices Γ_I, $I = 1, 2, \ldots, N$. Instead of (4)-(5) we now have

$$\mathcal{A}_i = A_i(\vec{x}) \otimes 1\!\!1 + B_i(\vec{x}) \otimes \Gamma_{N+1}$$

$$\mathcal{A}_I = \phi(\vec{x}) \otimes \frac{1}{2}\Gamma_I + \psi(\vec{x}) \otimes \frac{1}{2}\Gamma_{N+1}\Gamma_I\,,$$

where A_i, B_i, ϕ, and ψ are again antihermitian matrices, but with only A_i being traceless. The fact that B_i is not traceless here results in an Abelian gauge field in the reduced system.

Anticipating what follows, it is much more transparent to re-express these formulas in the form

$$\mathcal{A}_i = A_i^{(+)}(\vec{x}) \otimes P_+ + A_i^{(-)}(\vec{x}) \otimes P_- + \frac{i}{2}\,a_i(\vec{x})\,\Gamma_{N+1} \tag{14}$$

$$\mathcal{A}_I = \varphi(\vec{x}) \otimes \frac{1}{2}\,P_+\Gamma_I - \varphi(\vec{x})^\dagger \otimes \frac{1}{2}\,P_-\Gamma_I\,, \tag{15}$$

where now $P_\pm$ are the $2^{\frac{N}{2}} \times 2^{\frac{N}{2}}$ projection operators

$$P_\pm = \frac{1}{2} (\mathbb{I} \pm \Gamma_{N+1}) \,. \tag{16}$$

In (14), the residual gauge connections $A_i^{(\pm)}$ are anti-Hermitian and traceless $2^{\frac{D}{2}} \times 2^{\frac{D}{2}}$ arrays, and the Abelian connection a_i results directly from the trace of the field B_i. The $2^{\frac{D}{2}} \times 2^{\frac{D}{2}}$ "Higgs" field φ in (15) is neither Hermitian nor anti-Hermitian. Again, to achieve the desired breaking of the gauge group, to lead eventually to the requisite Higgs *isomultiplet*, we choose the gauge group in the bulk to be $SU(n-1)$, where $2n = D + N$.

The components of the curvaturs are readily calculated to give

$$\mathcal{F}_{ij} = F_{ij}^{(+)}(\vec{x}) \otimes P_+ + F_{ij}^{(-)}(\vec{x}) \otimes P_- + \frac{i}{2}\, f_{ij}(\vec{x})\, \Gamma_{N+1} \tag{17}$$

$$\mathcal{F}_{iI} = D_i\varphi(\vec{x}) \otimes \frac{1}{2} P_+\Gamma_I - D_i\varphi^\dagger(\vec{x}) \otimes \frac{1}{2} P_-\Gamma_I \tag{18}$$

$$\mathcal{F}_{IJ} = S^{(+)}(\vec{x}) \otimes P_+\Gamma_{IJ} + S^{(-)}(\vec{x}) \otimes P_-\Gamma_{IJ} \,, \tag{19}$$

the curvatures in (17) being defined by

$$F_{ij}^{(\pm)} = \partial_i A_j^{(\pm)} - \partial_j A_i^{(\pm)} + [A_i^{(\pm)}, A_j^{(\pm)}] \tag{20}$$

$$f_{ij} = \partial_i a_j - \partial_j a_i \,. \tag{21}$$

The covariant derivative in (18) now is defined as

$$D_i\varphi = \partial_i\varphi + A_i^{(+)}\,\varphi - \varphi\, A_i^{(-)} + i\, a_i\, \varphi \tag{22}$$

$$D_i\varphi^\dagger = \partial_i\varphi^\dagger + A_i^{(-)}\,\varphi^\dagger - \varphi^\dagger\, A_i^{(+)} - i\, a_i\, \varphi^\dagger \,, \tag{23}$$

and the quantities $S^{(\pm)}$ in (19) are

$$S^{(+)} = \varphi\,\varphi^\dagger - \eta^2 \quad , \quad S^{(-)} = \varphi^\dagger\,\varphi - \eta^2 \,. \tag{24}$$

In what follows, we will suppress the Abelian field a_i, since only when less stringent symmetry than spherical is imposed is it that it would contribute. In any case, using the formal replacement

$$A_i^{(\pm)} \leftrightarrow A_i^{(\pm)} \pm \frac{i}{2}\, a_i\, \mathbb{I}$$

yields the algebraic results to be derived below, in the general case.

We now refine our calculus of descent over even codimensions further. We see from (14) that $A_i^{(\pm)}$ being $2^{\frac{D}{2}} \times 2^{\frac{D}{2}}$ arrays, that they can take their values in the two chiral representations, repectively, of the algebra of $SO(D)$. It is therefore natural to introduce the full $SO(D)$ connection

$$A_i = \begin{bmatrix} A_i^{(+)} & 0 \\ 0 & A_\mu^{(-)} \end{bmatrix} \,. \tag{25}$$

Next, we define the $D-$component *isovector* Higgs field

$$\Phi = \begin{bmatrix} 0 & \varphi \\ -\varphi^\dagger & 0 \end{bmatrix} = \phi^i \,\Gamma_{i,D+1} \tag{26}$$

in terms of the Dirac matrix representation of the algebra of $SO(D+1)$, with $\Gamma_{i,D+1} = -\frac{1}{2}\Gamma_{D+1}\Gamma_i$.

Note here the formal equivalence between the Higgs multiplet (26) in even D, to the corresponding one (7) in odd D. This formal equivalence turns out to be very useful in the calulus employed in following Sections. In contrast with the former case of odd D however, the form (26) for even D is much more restrictive. This is because in this case the Higgs multiplet is restricted to take its values in the components $\Gamma_{i,D+1}$ orthogonal to the elements Γ_{ij} of $SO(D)$ by definition, irrespective of what symmetry is imposed. It is clear that relaxing the spherical symmetry here, does not result in $\mathcal{F}$ getting out of the orthogonal complement of Γ_{ij}, when D is even.

From (25), follows the $SO(D)$ curvature

$$F_{ij} = \partial_i A_j - \partial_j A_i + [A_i, A_j] = \begin{bmatrix} F_{ij}^{(+)} & 0 \\ 0 & F_{ij}^{(-)} \end{bmatrix} \tag{27}$$

and from (25) and (26) follows the covariant derivative

$$D_\mu \Phi = \partial_i \mathcal{F} + [A_i, \mathcal{F}] = \begin{bmatrix} 0 & D_i\varphi \\ -D_i\varphi^\dagger & 0 \end{bmatrix} . \tag{28}$$

From (26) there simply follows the definition of S for even D

$$S = -(\eta^2\,\mathbb{1} + \mathcal{F}^2) = \begin{bmatrix} S^{(+)} & 0 \\ 0 & S^{(-)} \end{bmatrix} . \tag{29}$$

3. New Chern–Simons terms

First, we recall the usual dynamical Chern–Simons in odd dimensions defined in terms of the non-Abelian gauge connection. Topologically massive gauge field theories in $2+1$ dimensional spacetimes were first introduced in.[7,8] The salient feature of these theories is the presence of a Chern-Simons (CS) dynamical term. To define a CS density one needs to have a gauge connection, and hence also a curvature. Thus, CS densities can be defined both for Abelian (Maxwell) and non-Abelian (Yang–Mills) fields. They can also be defined for the gravitational[9] field since in that system too one has a (Levi-Civita or otherwise) connection, akin to the Yang-Mills connection in that it carries frame indices analogous to the isotopic indices of ther YM connection. Here we are interested exclusively in the (non-Abelian) YM case, in the presence of an *isovector* valued Higgs field.

The definition of a Chern-Simons (CS) density follows from the definition of the corresponding Chern-Pontryagin (CP) density (1). As stated by (2), this quantity is a total divergence and the density $\mathbf{\Omega}^{(n)} = \Omega_M^{(n)}$ $(M = 1, 2, \ldots, 2n)$ in that case has $(2n)-$components. The Chern-Simons density is then defined as one fixed component of $\mathbf{\Omega}^{(n)}$, say the $2n-$th component,

$$\Omega_{\mathrm{CS}}^{(n)} = \Omega_{2n}^{(n)} \tag{30}$$

which now is given in one dimension less, where $M = \mu, 2n$ and $\mu = 1, 2, \ldots (2n-1)$.

This definition of a (dynamical) CS term holds in all odd dimensional spacetimes (t, R^D), with $x_\mu = (x_0, x_i)$, $i = 1, 2, \ldots, D$, with D being an even integer. That D must be even is clear since $D + 2 = 2n$, the $2n$ dimensions in which the CP density (1) is defined, is itself even.

The properties of CS densities are reviewed in.[10] Most remarkably, CS densities are defined in odd (space or spacetime) dimensions and are *gauge variant*. The context here is that of a $(2n-1)-$dimensional Minkowskian space. It is important to realise that dynamical Chern-Simons theories are defined on spacetimes with Minkowskian signature. The reason is that the usual CS densities appearing in the Lagrangian are by construction *gauge variant*, but in the definition of the energy densities the CS term itself does not feature, resulting in a Hamiltonian (and hence energy) being *gauge invariant* as it should be [‡].

Of course, the CP densities and the resulting CS densities, can be defined in terms of both Abelian and non-Abelian gauge connections and curvatures. The context of the present notes is the construction of soliton solutions [§], unlike in.[7,8] Thus in any given dimension, our choice of gauge group must be made with due regard to regularity, and the models chosen must be consistent with the Derrick scaling requirement for the finiteness

---

[‡]Should one employ a CS density on a space with Euclidean signature, with the CS density appearing in the static Hamiltonian itself, then the energy would not be *gauge invariant*. Hamiltonians of this type have been considered in the literature, *e.g.*, in.[11] Chern-Simons densities on Euclidean spaces, defined in terms of the composite connection of a sigma model, find application as the topological charge densities of Hopf solitons.

[§]The term soliton solutions here is used rather loosely, implying only the construction of regular and finite energy solutions, without insisting on topological stability in general.

of energy. Accordingly, in all but $2+1$ dimensions, our considerations are restricted to non-Abelian gauge fields.

Clearly, such constructions can be extended to all odd dimensional spacetimes systematically. We list $\Omega_{\rm CS}$, defind by (30), for $D = 2, 4, 6$, familiar densities

$$\Omega_{\rm CS}^{(2)} = \varepsilon_{\lambda\mu\nu}\mathrm{Tr}\,A_{\lambda}\left[F_{\mu\nu} - \frac{2}{3}A_{\mu}A_{\nu}\right] \tag{31}$$

$$\Omega_{\rm CS}^{(3)} = \varepsilon_{\lambda\mu\nu\rho\sigma}\mathrm{Tr}\,A_{\lambda}\left[F_{\mu\nu}F_{\rho\sigma} - F_{\mu\nu}A_{\rho}A_{\sigma} + \frac{2}{5}A_{\mu}A_{\nu}A_{\rho}A_{\sigma}\right] \tag{32}$$

$$\begin{aligned}\Omega_{\rm CS}^{(4)} = \varepsilon_{\lambda\mu\nu\rho\sigma\tau\kappa}\mathrm{Tr}\,A_{\lambda}\Big[&F_{\mu\nu}F_{\rho\sigma}F_{\tau\kappa} - \frac{4}{5}F_{\mu\nu}F_{\rho\sigma}A_{\tau}A_{\kappa} - \frac{2}{5}F_{\mu\nu}A_{\rho}F_{\sigma\tau}A_{\kappa} \\ &+\frac{4}{5}F_{\mu\nu}A_{\rho}A_{\sigma}A_{\tau}A_{\kappa} - \frac{8}{35}A_{\mu}A_{\nu}A_{\rho}A_{\sigma}A_{\tau}A_{\kappa}\Big]\,.\end{aligned} \tag{33}$$

Concerning the choice of gauge groups, one notes that the CS term in $D+1$ dimensions features the product of D powers of the (algebra valued) gauge field/connection infront of the Trace, which would vanish if the gauge group *is not larger than* $SO(D)$. In that case, the YM connection would describe only a 'magnetic' component, with the 'electric' component necessary for the the nonvanishing of the CS density would be absent. As in,[12] the most convenient choice is $SO(D+2)$. Since D is always even, the representation of $SO(D+2)$ are the *chiral* representation in terms of (Dirac) spin matrices. This completes the definition of the usual non-Abelian Chern-Simons densities in $D+1$ spacetimes.

From (31)-(33), it is clear that the CS density is *gauge variant*. The Euler"=Lagrange equations of the CS density is nonetheless *gauge invariant*, such that for the examples (31)-(33) the corresponding arbitarry variations are

$$\delta_{A_{\lambda}}\Omega_{\rm CS}^{(2)} = \varepsilon_{\lambda\mu\nu}F_{\mu\nu} \tag{34}$$

$$\delta_{A_{\lambda}}\Omega_{\rm CS}^{(3)} = \varepsilon_{\lambda\mu\nu\rho\sigma}F_{\mu\nu}F_{\rho\sigma} \tag{35}$$

$$\delta_{A_{\lambda}}\Omega_{\rm CS}^{(4)} = \varepsilon_{\lambda\mu\nu\rho\sigma\kappa\eta}F_{\mu\nu}F_{\rho\sigma}F_{\kappa\eta}\,. \tag{36}$$

This, and other interesting properties of CS densities are given in.[10] A remarkable property of a CS density is its transformation under the action of an element, g, of the (non-Abelian) gauge group. We list these for the two examples (31)-(32),

$$\Omega_{\rm CS}^{(2)} \to \tilde{\Omega}_{\rm CS}^{(2)} = \Omega_{\rm CS}^{(2)} - \frac{2}{3}\varepsilon_{\underset{\sim}{<}\mu\nu}\mathrm{Tr}\,\alpha_{\underset{\sim}{<}}\alpha_\mu\alpha_\nu - 2\varepsilon_{\underset{\sim}{<}\mu\nu}\,\partial_{\underset{\sim}{<}}\mathrm{Tr}\,\alpha_\mu\,A_\nu \quad (37)$$

$$\Omega_{\rm CS}^{(3)} \to \tilde{\Omega}_{\rm CS}^{(3)} = \Omega_{\rm CS}^{(3)} - \frac{2}{5}\,\varepsilon_{\underset{\sim}{<}\mu\nu\rho\boldsymbol{\sigma}}\mathrm{Tr}\,\alpha_{\underset{\sim}{<}}\alpha_\mu\alpha_\nu\alpha_\rho\alpha_{\boldsymbol{\sigma}}$$

$$+2\varepsilon_{\underset{\sim}{<}\mu\nu\rho\boldsymbol{\sigma}}\,\partial_{\underset{\sim}{<}}\mathrm{Tr}\,\alpha_\mu\left[A_\nu\left(F_{\rho\boldsymbol{\sigma}} - \frac{1}{2}A_\rho A_{\boldsymbol{\sigma}}\right) + \left(F_{\rho\boldsymbol{\sigma}} - \frac{1}{2}A_\rho A_{\boldsymbol{\sigma}}\right)A_\nu\right.$$

$$\left. -\frac{1}{2}\,A_\nu\,\alpha_\rho\,A_{\boldsymbol{\sigma}} - \alpha_\nu\,\alpha_\rho\,A_{\boldsymbol{\sigma}}\right], \quad (38)$$

where $\alpha_\mu = \partial_\mu g\, g^{-1}$, as distinct from the algebra valued quantity $\beta_\mu = g^{-1}\,\partial_\mu g$ that appears as the inhomogeneous term in the gauge transformation of the non-Abelian curvature (in our convention).

As seen from (37)-(38), the gauge variation of $\Omega_{\rm CS}$ consists of a term which is explicitly a total divergence, and, another term

$$\omega^{(n)} \simeq \varepsilon_{\mu_1\mu_2\dots\mu_{2n-1}}\mathrm{Tr}\,\alpha_{\mu_1}\alpha_{\mu_2}\dots\alpha_{\mu_{2n-1}}\,, \quad (39)$$

which is *effectively total divergence*, and in a concrete group representation parametrisation becomes *explicitly total divergence*. This can be seen by subjecting (39) to variations with respect to the function g, and taking into account the Lagrange multiplier term resulting from the (unitarity) constraint $g^\dagger\, g = g\, g^\dagger = \mathbb{I}$.

The volume integaral of the CS density then transforms under a gauge transformation as follows. Given the appropriate asymptotic decay of the connection (and hence also the curvature), the surface integrals in (37)-(38) vanish. The only contribution to the gauge variation of the CS action/energy then comes from the integral of the density (39), which (in the case of Euclidean signature) for the appropriate choice of gauge group yields an integer, up to the angular volume as a multiplicative factor.

All above stated properties of the Chern-Simons (CS) density hold irrespective of the signature of the space. Here, the signature is taken to be Minkowskian, such that the CS density in the Lagrangian does not contribute to the energy density directly. As a consequence the energy of the soliton is gauge invariant and does not suffer the gauge transformation (37)-(38). Should a CS density be part of a static Hamiltonian (on a space of Euclidean signature), then the energy of the soliton would change by a multiple of an integer.

3.1. *New Chern–Simons terms in all dimensions*

The plan to introduce a completely new type of Chern-Simons term. The usual CS densities $\mathbf{\Omega}_{\rm CS}^{(n)}$, (30), are defined with reference to the total divergence expression (2) of the $n-$th Chern-Pontryagin density (1), as the $2n-$th component $\Omega_{2n}^{(n)}$ of the density $\mathbf{\Omega}^{(n)}$. Likewise, the new CS terms are defined with reference to the total divergence expression (3) of the dimensionally reduced $n-$th CP density, with the dimension D of the residual space replaced formally by $\bar{D}$

$$\mathcal{C}_{\bar{D}}^{(n)} = \nabla \cdot \mathbf{\Omega}^{(n,\bar{D})} \,. \tag{40}$$

The densities $\mathbf{\Omega}^{(n,\bar{D})}$ can be read off from $\mathbf{\Omega}^{(n,D)}$ given in Section **5**, with the formal replacement $D \to \bar{D}$. The new CS term is now identified as the $\bar{D}-$th component of $\mathbf{\Omega}^{(n,\bar{D})}$. The final step in this identification is to assign the value $\bar{D} = D+2$, where D is the spacelike dimension of the $D+1$ dimensional Minkowski space, with the new Chern-Simons term defined as

$$\tilde{\Omega}_{\rm CS}^{(n,D+1)} \stackrel{def}{=} \Omega_{D+2}^{(n,D+2)} \tag{41}$$

The departure of the new CS densitiess from the usual CS densities is stark, and these differ in several essential respects from the usual ones described in the previous subsection. The most important new features in question are

- The field content of the new CS systems includes Higgs fields in addition to the Yang-Mills fields, as a consequence of the dimensional reduction of gauge fields described in Section **4**. It should be emphasised that the appearance of the Higgs field here is due to the imposition of symmetries in the descent mechanism, in contrast with its presence in the models[13–15] supporting $2+1$ dimensional CS vortices, where the Higgs field was introduced by hand with the expedient of satisfying the Derrick scaling requirement.
- The usual dynamical CS densities defined with reference to the $n-$th CP density live in $2n-1$ dimensional Minkowski space, *i.e.*, only in odd dimensional spacetime. By contrast, the new CS densities defined with reference to the $n-$th CP densities live in $D+1$ dimensional Minkowski space, for all D subject to

$$2n-2 \ge D \ge 2\,, \tag{42}$$

i.e., in both odd, as well as even dimensions. Indeed, in any given D there is an infinite tower of new CS densities characterised by the integer n subject to (42). This is perhaps the most important feature of the new CS densities.

- The smallest simple group consistent with the nonvanishing of the *usual* CS density in $2n-1$ dimensional spacetime is $SO(2n)$, with the gauge connection taking its values in the *chiral* Dirac representation. By contrast, the gauge groups of the new CS densities in $D+1$ dimensional spacetime are fixed by the prescription of the dimensional descent from which they result. As *per* the prescription of descent described in Section **4**, the gauge group now will be $SO(D+2)$, independently of the integer n, while the Higgs field takes its values in the orthogonal complement of $SO(D+2)$ in $SO(D+3)$. As such, it forms an iso-$(D+2)-$vector multiplet.
- Certain properties of the new CS densities are remarkably different for D even and D odd.

  - Odd D: Unlike in the usual case (31)-(32), the new CS terms are *gauge invariant*. The gauge fields are $SO(D+2)$ and the Higgs are in $SO(D+3)$. D being odd, $D+3$ is even and hence the fields can be parametrised with respect to the *chiral* (Dirac) representations of $SO(D+3)$. An important consequence of this is the fact that now, both (electric) A_0 and (magnetic) A_i fields lie in the same isotopic multiplets, in contrast to the *pseudo*$-$dyons described in the previous section.
  - Even D: The new CS terms now consist of a *gauge variant* part expressed only in terms of the gauge field, and a *gauge invariant* part expressed in terms of both gauge and Higgs fields. The leading, *gauge variant*, term differs from the corresponding usual CS terms (31)-(32) only due to the presence of a (chiral) Γ_{D+3} matrix infront of the Trace. The gauge and Higgs fields are again in $SO(D+2)$ and in $SO(D+3)$ respectively, but now, D being even $D+3$ is odd and hence the fields are parametrised with respect to the (chirally doubled up) full Dirac representations of $SO(D+3)$. Hence the appearance of the chiral matrix infront of the Trace.

As in the usual CS models, the regular finite energy solutions of the new CS models are not topologically stable. These solutions can be constructed numerically.

Before proceeding to display some typcal examples in the Subsection following, it is in order to make a small diversion at this point to make a clarification. The new CS densities proposed are functionals of both the

Yang–Mills, and, the "isovector" Higgs field. Thus, the systems to be described below are Chern-Simons–Yang-Mills-Higgs models in a very specific sense, namely that the Higgs field is an intrinsic part of the new CS density. This is in contrast with Yang-Mills–Higgs-Chern-Simons or Maxwell–Higgs-Chern-Simons models in $2+1$ dimensional spacetimes that have appeared ubiquitously in the literature. It is important to emphasise that the latter are entirely different from the systems introduced here, simply because the CS densities they employ are the *usual* ones, namely (31) or more often its Abelian ¶ version

$$\Omega^{(2)}_{\mathrm{U}(1)} = \varepsilon_{\lambda\mu\nu} A_{\lambda} F_{\mu\nu} ,$$

while the CS densities employed here are *not* simply functionals of the gauge field, but also of the (specific) Higgs field. To put this in perspective, let us comment on the well known *Abelian* CS-Higgs solitons in $2+1$ dimensions constructed in[13,14] support self-dual vortices, which happen to be unique inasfar as they are also topologically stable. (Their non-Abelian counterparts[15] are not endowed with topological stability.) The presence of the Higgs field in[13–15] enables the Derrick scaling requirement to be satisfied by virtue of the presence of the Higgs self-interaction potential. In the Abelian case in addition, it results in the topological stability of the vortices. If it were not for the topological stability, it would not be necessary to have a Higgs field merely to satisfy the Derrick scaling requirement. That can be achieved instead, *e.g.*, by introducing a negative cosmological constant and/or gravity, as was done in the $4+1$ dimensional case studied in.[12] Thus, the involvement of the Higgs field in conventional (*usual*) Chern-Simons theories is not the only option. The reason for emphasising the optional status of the Higgs field in the usual $2+1$ dimensional Chern-Simons–Higgs models is, that in the new models proposed here the Higgs field is intrinsic to the definition of the (new) Chern-Simons density itself.

3.2. *Examples*

As discussed above, the new dynamical Chern-Simons densities

$$\tilde{\Omega}^{(n,D+1)}_{\mathrm{CS}}[A_\mu, \mathcal{F}]$$

¶There are, of course, Abelian CS densities in all odd spacetime dimensions but these do not concern us here since in all $D+1$ dimensions with $D = 2n \geq 4$, no regular solitons can be constructed.

are characterised by the dimensionality of the space D and the integer n specifying the dimension $2n$ of the bulk space from which the relevant residual system is arrived at.

The case $n = 2$ is empty, since according to (42) the largest spacetime in which a new CS density can be constructed is $2n - 2$, *i.e.*, in $1 + 1$ dimensional Minkowsky space which we ignore.

The case $n = 3$ is not empty, and affords two nontrivial examples. The largest spacetime $2n - 2$, in which a new CS density can be constructed in this case is $3 + 1$ and the next in $2 + 1$ Minkowski space. These, are, repectively,

$$\tilde{\Omega}_{\rm CS}^{(3,3+1)} = \varepsilon_{\mu\nu\rho\boldsymbol{\sigma}} \,\mathrm{Tr}\, F_{\mu\nu}\, F_{\rho\boldsymbol{\sigma}}\, \mathcal{F} \tag{43}$$

$$\tilde{\Omega}_{\rm CS}^{(3,2+1)} = \varepsilon_{\mu\nu\underset{\sim}{<}} \,\mathrm{Tr}\, \gamma_5 \left[-2\eta^2 A_{\underset{\sim}{<}} \left(F_{\mu\nu} - \frac{2}{3} A_\mu A_\nu\right) + \left(\mathcal{F}\, D_{\underset{\sim}{<}} \mathcal{F} - D_{\underset{\sim}{<}} \mathcal{F}\, \mathcal{F}\right) F_{\mu\nu}\right]. \tag{44}$$

The case $n = 4$ affords four nontrivial examples, those in $5 + 1$, $4 + 1$, $3 + 1$ and $2 + 1$ Minkowski space. These are, repectively,

$$\tilde{\Omega}_{\rm CS}^{(4,5+1)} = \varepsilon_{\mu\nu\rho\boldsymbol{\sigma}\tau\underset{\sim}{<}} \,\mathrm{Tr}\, F_{\mu\nu}\, F_{\rho\boldsymbol{\sigma}}\, F_{\tau\underset{\sim}{<}}\, \mathcal{F} \tag{45}$$

$$\tilde{\Omega}_{\rm CS}^{(4,4+1)} = \varepsilon_{\mu\nu\rho\boldsymbol{\sigma}\underset{\sim}{<}} \,\mathrm{Tr}\, \Gamma_7 \Big[A_{\underset{\sim}{<}} \left(F_{\mu\nu} F_{\rho\boldsymbol{\sigma}} - F_{\mu\nu} A_\rho A_{\boldsymbol{\sigma}} + \frac{2}{5} A_\mu A_\nu A_\rho A_{\boldsymbol{\sigma}}\right)$$
$$D_{\underset{\sim}{<}} \mathcal{F} \left(\mathcal{F} F_{\mu\nu} F_{\rho\boldsymbol{\sigma}} + F_{\mu\nu} \mathcal{F} F_{mn} + F_{\mu\nu} F_{mn} \mathcal{F}\right)\Big] \tag{46}$$

$$\tilde{\Omega}_{\rm CS}^{(4,3+1)} = \varepsilon_{\mu\nu\rho\boldsymbol{\sigma}} \,\mathrm{Tr}\Big[\mathcal{F}\left(\eta^2\, F_{\mu\nu} F_{\rho\boldsymbol{\sigma}} + \frac{2}{9} \mathcal{F}^2\, F_{\mu\nu} F_{\rho\boldsymbol{\sigma}} + \frac{1}{9} F_{\mu\nu} \mathcal{F}^2 F_{\rho\boldsymbol{\sigma}}\right)$$
$$-\frac{2}{9}\left(\mathcal{F} D_\mu \mathcal{F} D_\nu \mathcal{F} - D_\mu \mathcal{F}\mathcal{F} D_\nu \mathcal{F} + D_\mu \mathcal{F} D_\nu \mathcal{F}\mathcal{F}\right) F_{\rho\boldsymbol{\sigma}}\Big] \tag{47}$$

$$\tilde{\Omega}_{\rm CS}^{(4,2+1)} = \varepsilon_{\mu\nu\underset{\sim}{<}} \,\mathrm{Tr}\, \Gamma_5 \Big\{ 6\eta^4\, A_{\underset{\sim}{<}} \left(F_{\mu\nu} - \frac{2}{3} A_\mu\, A_\nu\right)$$
$$-6\,\eta^2 \left(\mathcal{F}\, D_{\underset{\sim}{<}} \mathcal{F} - D_{\underset{\sim}{<}} \mathcal{F}\, \mathcal{F}\right) F_{\mu\nu}$$
$$+ \left[\left(\mathcal{F}^2\, D_{\underset{\sim}{<}} \mathcal{F}\, \mathcal{F} - \mathcal{F}\, D_{\underset{\sim}{<}} \mathcal{F}\, \mathcal{F}^2\right) - 2\left(\mathcal{F}^3\, D_{\underset{\sim}{<}} \mathcal{F} - D_{\underset{\sim}{<}} \mathcal{F}\, \mathcal{F}^3\right)\right] F_{\mu\nu}\,.\Big\} \tag{48}$$

It is clear that in any $D + 1$ dimensional spacetime an infinite tower of CS densities $\tilde{\Omega}_{\rm CS}^{(n,D+1)}$ can be defined, for all positive integers n. Of these, those in even dimensional spacetimes are gauge invariant, *e.g.*, (43), (45) and (47), while those in odd dimensional spacetimes are gauge variant, *e.g.*, (44), (46) and (48), the gauge variations in these cases being given formally by (37) and (38), with g replaced by the appropriate gauge group here.

Static soliton solutions to models whose Lagrangians consist of the above introduced types of CS terms together with Yang-Mills–Higgs (YMH) terms are currently under construction. The only constraint in the choice of the detailed models employed is the requirement that the Derrick scaling requirement be satisfied. Such solutions are constructed numerically. In contrast to the monopole solutions, they are not endowed with topological stability because the gauge group must be larger than $SO(D)$, for which the solutions to the constituent YMH model is a stable monopole. Otherwise the CS term would vanish.

Acknowledgment

This work is carried out in the framework of Science Foundation Ireland (SFI) project RFP07-330PHY.

References

1. P.A.M. Dirac, *Proc. Roy. Soc. A* **133** (1931) 60.
2. C. N. Yang, *J. Math. Phys.* **19** (1978) 320.
3. T. Tchrakian, *Phys. Atom. Nucl.* **71** (2008) 1116.
4. A. S. Schwarz, *Commun. Math. Phys.* **56** (1977) 79.
5. V. N. Romanov, A. S. Schwarz and Yu. S. Tyupkin, *Nucl. Phys. B* **130** (1977) 209.
6. A. S. Schwarz and Yu. S. Tyupkin, *Nucl. Phys. B* **187** (1981) 321.
7. S. Deser, R. Jackiw and S. Templeton, *Phys. Rev. Lett.* **48** (1982) 975.
8. S. Deser, R. Jackiw and S. Templeton, Annals Phys. **140** (1982) 372 [Erratum-ibid. **185** (1988) 406] [Annals Phys. **185** (1988) 406] [Annals Phys. **281** (2000) 409].
9. R. Jackiw and S. Y. Pi, *Phys. Rev. D* **68** (2003) 104012 [arXiv:gr-qc/0308071].
10. R. Jackiw, "Chern-Simons terms and cocycles in physics and mathematics", in E.S. Fradkin *Festschrift*, Adam Hilger, Bristol (1985)
11. V. A. Rubakov and A. N. Tavkhelidze, *Phys. Lett. B* **165** (1985) 109.
12. Y. Brihaye, E. Radu and D. H. Tchrakian, *Phys. Rev. D* **81** (2010) 064005 [arXiv:0911.0153 [hep-th]].
13. J. Hong, Y. Kim and P. Y. Pac, *Phys. Rev. Lett.* **64** (1990) 2230.
14. R. Jackiw and E. J. Weinberg, *Phys. Rev. Lett.* **64** (1990) 2234.
15. F. Navarro-Lerida and D. H. Tchrakian, *Phys. Rev. D* **81** (2010) 127702 [arXiv:0909.4220 [hep-th]].

SPHERICAL SHELL COSMOLOGICAL MODEL AND UNIFORMITY OF COSMIC MICROWAVE BACKGROUND RADIATION

B. Vlahovic

Department of Physics, North Carolina Central University,
1801 Fayetteville Street, Durham, NC 27707 USA.
vlahovic@nccu.edu

Considered is spherical shell as a model for visible universe and parameters that such model must have to comply with the observable data. The topology of the model requires that motion of all galaxies and light must be confined inside a spherical shell. Consequently the observable universe cannot be defined as a sphere centered on the observer, rather it is an arc length within the volume of the spherical shell. The radius of the shell is 4.46 ± 0.06 Gpc, which is for factor π smaller than radius of a corresponding 3-sphere. However the event horizon, defined as the arc length inside the shell, has the size of 14.0 ± 0.2 Gpc, which is in agreement with the observable data. The model predicts, without inflation theory, the isotropy and uniformity of the CMB. It predicts the correct value for the Hubble constant $H_0 = 67.26 \pm 0.90$ km/s/Mpc, the cosmic expansion rate $H(z)$, and the speed of the event horizon in agreement with observations. The theoretical suport for shell model comes from general relativity, curvature of space by mass, and from holographic principle. The model explains the reason for the established discrepancy between the non-covariant version of the holographic principle and the calculated dimensionless entropy (S/k) for the visible universe, which exceeds the entropy of a black hole. The model is in accordance with the distribution of radio sources in space, type Ia data, and data from the Hubble Ultra Deep Field optical and near-infrared survey.

1. Introduction

The paradigm of ΛCDM cosmology works impressively well and with concept of inflation it explains universe after the time of decoupling. General relativity and standard model can predict with high accuracy decrease in orbital period of binary pulsar and angular power spectrum of CMB. Experimental data from BAO, CMB, SN Ia, and observations of large scale structures allow to put some constrains to cosmological models and parameters.

However there are still a few concerns. After all efforts there is no detection of dark matter and there are significant problems in theoretical description of dark energy. For that reason there are numerous attempts for alternative cosmologies that could modify GR theories (modify gravity $G_{\mu\nu}$, e.g. brane-worlds, modified action theories, f(R) gravity, tensor gravity, higher dimensional gravity)

$$G_{\mu\nu} = 8\pi G\tilde{T}_{\mu\nu} \quad where: \quad \tilde{T}_{\mu\nu} \equiv T_{\mu\nu} - \frac{\Lambda}{8\pi G} g_{\mu\nu} \tag{1}$$

or modify matter theories (modify $\tilde{T}_{\mu\nu}$, inhomogeneous universe, new matter, new interactions, quintessence, Chaplyging gas, k-essence).

The modification to GR will led to modified Friedmann and acceleration equations, e.g. modification of f(r) gravity will produce extra first four terms on the left side of equation:

$$-H^2 fr + \frac{a^2}{6} f + \frac{3}{2} H\dot{\mathrm{f}}r + \frac{1}{2}\ddot{\mathrm{f}}r + \frac{\ddot{\mathrm{a}}}{a} = -\frac{4\pi G}{3}(\rho + 3P) \tag{2}$$

and for instance higher dimensional gravity will add the first term on the left:

$$\frac{-\dot{\mathrm{H}}}{r_c} + \frac{\ddot{\mathrm{a}}}{a} = -\frac{4\pi G}{3}(\rho + 3P). \tag{3}$$

Because current data are not enough to discriminate between the GR and alternative theories there are ongoing and planned surveys to provide more accurate data and additional tests, such as: LSS observations (galaxy positions, weak lensing, redshift space distortions), dark energy survey, Euclid (an ESA mission to map the geometry of the dark Universe), evolutionary map of the universe, Westerbork observation of the deep aperitif northern sky. The goal of this paper is to consider a new cosmological model that will not require alternative GR and gravity or matter modification, but will allow to describe observable data without inflation.

It is well known that in the Big-Bang models homogeneity of space cannot be explained, it is simply assumed in initial conditions. The curvature and physical properties of the regions of the space which have never been in causal contact and should not be correlated are taken to be indistinguishable. In CMB spectrum points further apart than 2^0 should be not correlated, but correlations up to $\sim 60^0$ are observed. Homogeneity on the level of 10^{-5} is explained by inflation era. However, arguments in favor of inflation only exist if space was already homogeneous before inflation. If the pre inflationary universe was not already homogenous inflation will not lead to homogeneity.[1] So, the homogeneity problem is pushed only back in time, because Big-Bang itself is taken to be inherently free of correlations. In addition, no satisfactory model for inflation exist. We will show that proposed cosmological model could explain uniformity of CMB without inflation theory.

GR does not specify the topology of the space, Einstein's equation describe only local property of the spacetime, but do not fix the global structure, topology of spacetime. For the same matrix element different topologies can correspond leaving the possibility for a new universe models. Universe is correctly described by Friedmann equations but the values of the cosmic parameters are not known accurately to determine topology and curvature. We do not know the real value of the comoving space curvature radius or the present value of the scale factor $R(t_o) = R_o$, the only cosmological length that is directly observable is the Hubble length $L_{Hub} = cH_o^{-1}$. For non flat universe $R_o = L_{Hub}/\sqrt{|\Omega + \Lambda - 1|}$ and there is no scale for the flat case.

2. Spherical shell model

We will consider a model in which the universe is an expanding spherical shell with thickness much smaller than its radius. Such model can be justified by the short time interval of the Big-Bang universe creation and since then continuous Hubble expansion. We may assume, as it has been always emphasized, that galaxies do not move through space and that the universe is not expanding into empty space around it, for space does not exist apart from the universe.

There is no other space than that associated with the spherical shell. The motions of all galaxies and propagation of the light are confined to the volume of the shell, which expands with a radial velocity. As we will see this will have significant implication on the interpretation of the data. The light must follow geodesic lines, so like in torus models it is not traveling straight, it is bend. This will define observable universe for our model as the largest visible volume (from the point of the observer) inside the spherical shell that represents the universe, Fig. 1a). By this definition the cosmological horizon distance will be the largest possible distance inside the spherical shell. If for instance the universe is the same size as the observable universe and an observer is located at point A, the particle horizon for that observer will be the point B at the antipode and the observable universe will be the entire spherical shell. This is shown in Fig. 1b.

The dynamics of shell models has been investigated earlier. It was first introduced by Israel,[2] in the framework of the special-relativity by,[3] and a systematic study in the framework of general relativity is done for instance in[4] and.[5] However, our focus will be very different. We will present significant implications of the shell model when combined with a new proposed interpretation of experimental data.

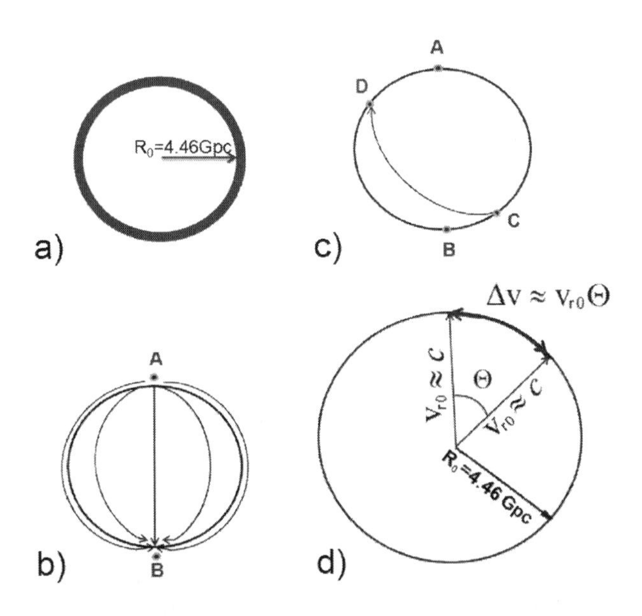

Fig. 1. a) The visible universe as an expanding shell with thickness much smaller than radius, b) Observable universe, as seen by an observer from the point A, is a volume of the shell, with event horizon located in the point B, c) CMB visible from Earth (by observer in point A) is originated in the antipodal point B and CMB visible from another place in the universe (point C) is emitted in the point D, d) The visible universe as a surface of the sphere with radius $R_0 = 4.46$ Gpc that expands with radial speed close to the speed of light.

The spherical shell model is in agreement with the main cosmological principles, isotropy and homogeneity of the space and as it will be shown it also satisfies Friedman-Lemaˆitre model. The model also must be in agreement with the observational data, the age of the universe and direct observations of matter and density. The dimension size of the shell (the arc length from pole to antipode) must correspond to the present size of the cosmological horizon, and the thickness of the shell must have the minimum size to explain present observation constrains: ghost images of sources; distribution and periodicity of clusters, super clusters, quasars, and gamma-ray bursts; statistical analysis of reciprocal distances between celestial objects;

and other limits obtained from CMB (uniformity and weak angular fluctuations of the CMB). As for example, from the statistical analysis of Abell catalog of spatial separation of clusters it appears that the shell thickness should be at least about 1 Gpc.[6]

In the current ΛCDM model the visible universe is defined as as a sphere centered on the observer and from our perspective it appears that the radius is $R_0 = 14.0 \pm 0.2$ Gpc (about 45.7 Gly). The value R_0 is the particle horizon and the quoted result corresponds to the direct WMAP7 measurements and the recombination redshift $z = 1090 \pm 1$.[7] In the standard FLRW model

$$R_o = a(t) \int_{t'=0}^{t} \frac{c}{a(t')} dt', \tag{4}$$

where

$$\frac{da}{dt} = \sqrt{\frac{\Omega_r}{a^2} + \frac{\Omega_m}{a} + \frac{\Omega_\Lambda}{a^{-2}}}. \tag{5}$$

The $R_0 = 14.0 \pm 0.2$ Gp corresponds to the following combinations of the parameters: $\Omega_m h^2 = 0.136 \pm 0.003$, $\Omega_r = \frac{8\pi G}{3H^2} \frac{\pi^2 k^4 T^4}{15 c^5 \hbar^4}$, $\Omega_\Lambda = 1 - \Omega_r - \Omega_m$. In our shell model we must obtain the same value for the particle horizon, which is now not the radius of the sphere, but an arc distance from pole to antipode. Therefore, the curvature radius of our spherical shell will be for factor π smaller than in the standard FLRW model. This will result in significantly different density, which will be at least for factor of π^3 (depending on the thickness of the shell) bigger in the spherical shell model.

3. Important implication of the model, uniformity of CMB

In addition to the significantly higher prediction for density, the most important implication of the shell model is prediction of the uniformity of the CMB without inflation. In the proposed spherical shell model, looking from the position of our galaxy (marked by A), the place of decoupling (the surface of last scattering, the source of CMB) is at the antipode (marked by B), Fig. 1c. Regardless of the direction we chose to measure CMB (for instance from point A looking in any arbitrary chosen direction), we will always measure CMB at the antipodal point (in this example point B). The reason for this is that the length of the arc on the sphere represents distance, which also represents past time. When the visible universe is of the same size as the whole universe, the point that is at the largest distance away from the point A (point B) represents the surface of last scatter, since we cannot see beyond that distance.

It is important to notice that measuring the same CMB by looking in the opposite directions of the universe does not represent or reflect the uniformity of the universe at the time of decoupling, because we always measure CMB originated from the same point regardless of the direction of observation. For that reason we always must obtain the same result. If from point A we observe to the right, left, backward or forward we will always measure CMB originated from the point B. Small variations for the CMB are possible and they are observed, but they are the result of the interaction between matter and light during its travel. For instance, depending on the direction we choose to measure CMB, light will travel from point B to A through different galaxies and will interact with different amounts of matter, which will result in the small observed variations of CMB. The observed fluctuations in the CMB are therefore created as the photons pass through nearby large scale structures by the integrated Sachs-Wolfe effect. The correlation between the fluctuations in the CMB and the matter distribution is well established.[8-11]

To establish a connection between the uniformity of the earlier universe at the time of decoupling and the CMB we will need to make a completely different kind of measurements of the CMB. We can see the CMB in any direction we can look in the sky. However, we must keep in mind that the CMB emitted by the matter that would ultimately form for instance the Milky Way is long gone. It left our part of the universe at the speed of light billions of years ago and now forms the CMB for observers in remote parts of the universe, actually exactly for an observer at the antipodal point B. For instance, if we perform measurement of the CMB at the point C, we will measure the CMB emitted by matter at the point D, Fig. 1c. To measure uniformity of the universe at the time of decoupling we will need to measure the CMB in at least two different points on the shell. If, for instance, the measurements from points A (CMB originated in B) and C (CMB originated at D) give the same result, then and only then may we speak about the uniformity of the CMB and uniformity of the universe at the time of decoupling. However, such measurements are not possible at the present time.

The strongest objection to this interpretation of the uniformity of CMB could be that observer located at the pole cannot see CMB originated at antipodal point, because the light originated at point B does not have enough time to reach observer at point A. In a matter dominated universe, light from an antipodal point on FLRW expanding sphere, can never reach an observer during the expanding phase $H > 0$. For closed universe in FRW metric

$$ds^2 = c^2dt^2 - a^2(t)[d\chi^2 + sin^2\chi(d\theta^2 + Sin^2\theta d\phi^2)], \tag{6}$$

where $\chi = \int cdt/a(t)$, which is by definition conformal time coordinate or arc parameter time η. Substituting for $ds^2 = 0$ and assuming that observer is located at $\chi = 0$ and CMB is originated at antipodal point $\chi_B = \pi$, the light will reach observer when $\eta = \pi$

$$\eta_0 - \eta = \pi = \int_0^{t_0} \frac{cdt}{a(t)} = \frac{c}{a_0} \int_0^z \frac{dz'}{H(z')}. \tag{7}$$

For matter dominated universe $a(\eta) = A(1 - cos\eta), ct = A(\eta - sin\eta)$. Therefore when antipodal light reach observer at $\eta = \pi$ the expansion will be at the maximum and $H = 0$. However this is not true for closed universe with a cosmological term. Because of the Λ term universe may not collapse and a light from antipodal point can reach observer during the expansion epoch. Also as we will show later the dynamics of the spherical shell is different from here considered dynamics of 3-sphere.

Another serious objection could be that the angular power spectrum of the CMB temperature fluctuations $\Delta T/T$ predicted by ΛCDM model agrees well with observations and that such agreement cannot be obtained for spherical shell model, because power spectrum coefficients a_{lm} are model dependent and will be different for 3-sphere and spherical shell. However, situation is not so simple because it is well known that the same CMB anisotropy spectra can be produced by two different models having different combinations of the parameters.[12,13] More importantly there is no reason to check if ΛCDM and spherical shell are degenerate models, which could produce the same anisotropy spectra, because the CMB power spectrum does not have any meaning for the spherical shell model.

The CMB temperature fluctuations $\Delta T/T$ are usually written in terms of a multipoles expansion on the celestial sphere:

$$\frac{\delta T}{T}(\theta, \phi) = \sum_{l=2}^{\infty} \sum_{m=-l}^{l} a_{lm} Y_l^m(\theta, \phi). \tag{8}$$

However, what is actually directly measured by observations is the angular correlation of the temperature anisotropy $\langle \frac{\delta T}{T}(\hat{n}_1) \frac{\delta T}{T}(\hat{n}_2) \rangle$ where $cos\theta = \hat{n}_1 \cdot \hat{n}_2$. This is expressed through power spectrum $C_l \equiv \langle |a_{lm}|^2 \rangle$, Legandre polinomials, and the filter function W_l as

$$C(\theta) = \frac{1}{4\pi} \sum_l \left[\frac{l + \frac{1}{2}}{l(l+1)} \right] C_l P_l(cos\theta) W_l. \tag{9}$$

The main contribution to C_l for $l > 60$ is from oscillations in the photon-baryon plasma before decoupling. However, in the spherical shell model we cannot see the imprint of these oscillations in the CMB a the time of last scattering, because we are always measuring CMB coming just from the single point, the antipodal point, of the surface of the last scatter. We cannot see the C_l contributions that are form the remaining part of the surface of the last scattering. For instance as we already mentioned, fluctuations from the surface of the last scattering that at the present corresponds to the Milky Way galaxy already left us bilions of years ago. Therefore, in principle, in shell model we cannot obtain the angular correlation of the CMB temperature anisotropy.

The contribution to C_l for low multipoles $l \leq 60$ is mainly from Sachs-Wolfe effect that relates temperature fluctuations to the integral of variations of the metric evaluated along the line of sight

$$\frac{\delta T}{T} = -\frac{1}{2} \int_{\eta_{rec}}^{\eta_0} \frac{\partial h_{\alpha\beta}}{\partial \eta} e^{\alpha} e^{\beta} d\eta. \tag{10}$$

One can argue that the line of sight is similar in 3-sphere and spherical shell model. For instance assume that we are on the surface of a 3-sphere and that propagation of the light is confined to its surface, then the observed distribution of the galaxies on surface of the 3-sphere and in sphere shell will be the same. Therefore for the spherical shell model we should obtain very similar spectrum for low multipoles C_l as in the standard ΛCDM model. However, that is only partially true, because curvature radius is different in these two models for factor π and density is different at least for factor π^3. However, if we assume that gravitational potential does not evolve with time, then equation (10) simplifies to

$$\frac{\delta T}{T} \simeq \frac{1}{3} \frac{\delta \phi}{c^2}. \tag{11}$$

Therefore the temperature asymmetry for $C_l \leq 60$ should be similar in both models and one can use this as the test of the spherical shell model. If one obtains with shell model parameters for the low multipoles part of the asymmetry spectrum as observed, then it may indicate that we are leaving in a shell model universe.

4. Justification for the spherical shell model

Dynamics of the shell model was considered in special relativity Newtonian (SRN) approach and in full relativistic approach. The similar results are obtained. In SRN approach the speed of the shell expansions depends on the total energy of the system and the maximum speed c is at the beginning of expansion. However, the system that has sufficient amount of energy can retain speed close to c for long time.[3]

$$E = \frac{M_0 c^2}{\sqrt{1-(v/c)^2}} - k\frac{M_0^2}{r}\frac{\delta\phi}{c^2}, \tag{12}$$

which has solutions

$$\frac{v}{c} = \left(1 - \frac{x^2}{(\frac{Ex}{M_0 c^2}+1)^2}\right)^{1/2} \tag{13}$$

and

$$x_{max} = \frac{1}{1-\frac{E}{M_0 c^2}}, \tag{14}$$

where $x = \frac{r}{R}$.

In GR approach at the beginning of the expansion system has infinite speed. However the total size of the expansion is at the end similar to that calculated by SRN approach. GR equation for spherical shell can be written[2] as:

$$\ddot{\mathrm{R}}\left[(1+\dot{\mathrm{R}}^2)^{1/2} + (1+\dot{\mathrm{R}}^2 - \frac{2m}{R})^{1/2}\right] = -\frac{m(1+\dot{\mathrm{R}}^2)^{1/2}}{R^2}, \tag{15}$$

where m is the gravitational mass of the shell, equivalent to the total energy E in SRN notation. In the units $x = v/R$ and $\dot{\mathrm{R}} = v/c$, the solution can be expressed as:

$$\frac{v}{c} = \left[\left(\frac{E}{M_0 c^2} + \frac{1}{2x}\right)^2 - 1\right]^{1/2} \quad and \tag{16}$$

$$x_{max} = \frac{1}{2}\frac{1}{1-\frac{E}{M_0 c^2}}. \tag{17}$$

This is qualitatively the same solution as in SRN case for $x \to \infty$ and $\frac{E}{M_0 c^2} > 1$, while for $\frac{E}{M_0 c^2} < 1$ it differs from SRN solution for factor 1/2, which could be due to a different scaling of the sphere radius applied in these two cases.

As mentioned earlier, propagation of light and galaxies is confined to the spherical shell; for that reason we cannot point to the center of the universe. As seen from our galaxy, all other galaxies are moving away from us (and from each other) with the speed $v = v_{r0}\Theta$ (where v_{r0} is the current radial speed of expansion and Θ is azimuthal angle, Fig. 1d), which is actually the Hubble law $v = H_0 \times distance$. So the spherical shell model predicts value for the Hubble constant that can be calculated from

$$v = v_{r0}\Theta = H_0 R_0 \Theta \tag{18}$$

by inserting for the radius of the shell $R_0 = 4.46 \pm 0.06$ Gpc (which in shell model corresponds to the particle horizon 14.0 ± 0.2 Gpc, obtained by direct WMAP7 measurement[7]), and for v_{r0} a value close to c, which is predicted by both GR and SRN models (if the energy of the system is high), gives

$$H_0 = 67.26 \pm 0.90\ km/s/Mpc, \tag{19}$$

which is in agreement with the observable data.

However, let us here rewrite equation (18) in the form

$$v_r = H(z)R(z) \tag{20}$$

to emphasize that the product HR is equal to the speed of radial expansion and that for models with high energy it is a constant close to c. From equation (20) one can also see that the cosmic expansion rate $H(z)$ changed with time and that it must have been larger for an earlier universe. If the universe expanded at an approximately constant speed equation (20) gives for $H(0.5)/H_0 = 1.6$, $H(1.0)/H_0 = 2.3$, and $H(1.4)/H_0 = 3.0$ which agrees with[14] within three standard deviations, Fig. 2. Equation (20) is basically the same as the first Friedmann equation

$$H^2 = \left(\frac{\dot{a}}{a}\right)^2 \tag{21}$$

where for a closed 3-sphere universe the scale factor a corresponds to the radius of curvature of the universe. As shown above, our model predicts proper value for $H(z)$. It is important to note that both the Hubble law and the first Friedmann equation directly follow from the model.

One can use GR to make arguments that will favor spherical shell model over ΛCDM, which defines the observable universe as a sphere, centered on the observer with the radius $R_0 = 14.0 \pm 0.2$ Gpc (about 45.7 Gly). To be consistent with general relativity we should require that the model must satisfy GR assumption that the presence of matter or energy causes warping or curvature of spacetime. The current definition is not consistent with this requirement, because it assumes that light is expanding straight, radially, in all directions for 14 Gpc.

As it is well known if a photon passes a massive object at an impact parameter b, the local curvature of space-time will cause the photon to be

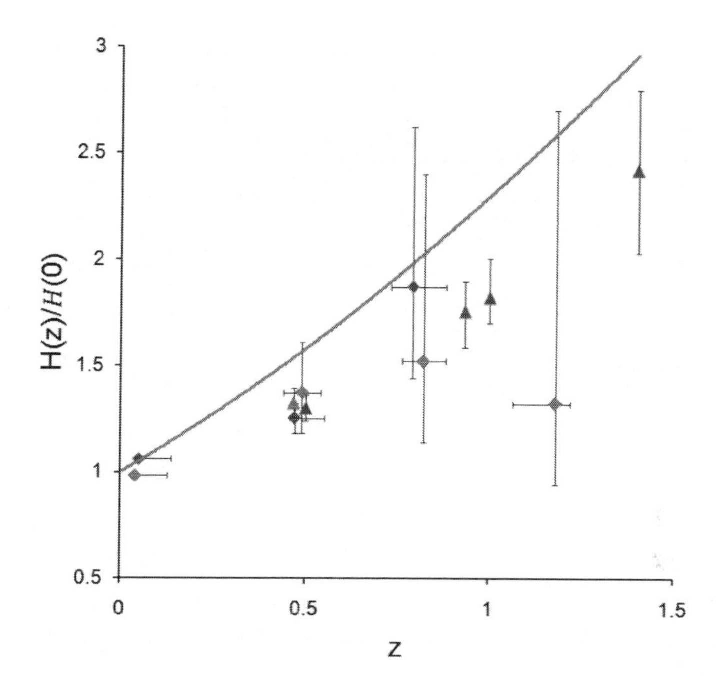

Fig. 2. The solid line are calculations for $H(z)/H_0$ at different z using equation (20), with distances corrected for a factor π, in accordance with our model. Data points are from.[14]

deflected by an angle

$$\alpha = \frac{4GM}{c^2 b}. \tag{22}$$

If photons are in a region of space where gravity is sufficiently strong, a *photon sphere* of radius

$$R_{ps} = \frac{3GM}{c^2}, \tag{23}$$

then the photons will be forced to travel in orbits. It is usually stated that the photon spheres can only exist in the space surrounding an extremely compact object, such as a black hole or a neutron star. However, as it will be shown the concept is also applicable to the visible universe.

The gravitational field $\mathbf{\Phi}$ on the boundary of the imaginary sphere that surrounds mass M is exactly the same as it would have been if all the mass had been concentrated at the center of the sphere.

By applying Birkhoff's theorem we can assume that the entire mass of the visible universe (considering the visible unverse as a sphere or spherical shell) is located in the center of that sphere. Using for $M = 10^{23} M_{\odot} = 2\text{x}10^{53}$ kg, it gives $R_{ps} = 14.3$ Gpc.

All photons that are inside sphere of 14.3 Gpc must follow circular orbits. Since the visible universe is smaller or about the calculated R_{ps} value all photons will be affected. Therefore, when we speak about the size and the radius of the visible universes we must take into account the bending of light. We cannot say that the visible universe is a sphere with a radius of 14 Gpc since photons cannot travel straight. The measured horizon distance of 14 Gpc is not the length of a straight line, because of the bending of light, it is an arc of a circle with a length of 14 Gpc. The radius of that circle is $14.0/\pi = 4.46$ Gpc, which suggest the proposed shell model.

Because the current observations are not enough accurate to directly test curvature of the space, can we in addition to the above example, which demonstrates that the current model is not taking into account bending of the light predicted by GR, find another test related to the curvature of space.

We will show that holographic principle also favors the shell model over the current ΛCDM. The entropy of the visible universe is calculated in[15] and it is shown that the dimensionless entropy S/k is 8.85 ± 0.37 times larger than allowed by a simplified and non-covariant version of the holographic principle, which requires that the entropy cannot exceed that of a black hole.

It was argued in[15] that by the holographic principle the entropy S/k has an upper limit equal to that of a black hole:

$$\left(\frac{S}{k}\right)_{Uni} \leq \left(\frac{S}{k}\right)_{BH} = \frac{4\pi R_S^2}{l_P^2}, \tag{24}$$

where $\left(\frac{S}{k}\right)_{Uni}$ is the entropy of the visible universe, $\left(\frac{S}{k}\right)_{BH}$ is the entropy of a black hole, l_P is the Planck length, and R_S is the Schwarzschild radius $R_S = 2\,GM$.

Equation (24) requires

$$\frac{\left(\frac{S}{k}\right)_{Uni}}{\left(\frac{S}{k}\right)_{BH}} = R_{BET}^4 \leq 1, \tag{25}$$

where R_{BET} is the Bond, Efstathiou, and Tegmark dimensionless shift parameter[16] defined as

$$R_{BET} = \frac{\sqrt{\Omega_m H_0^2}}{c} R_0. \tag{26}$$

Taking from[7] the size for the radius of the visible universe as $R_0 = 14.0 \pm 0.2$ Gpc gives the value $R_{BET}= 1.725 \pm 0.018$, and hence

$$R^4_{BET} = 8.85 \pm 0.1, \tag{27}$$

which, as pointed out in,[15] is in contradiction with equation (25) for 21σ. Therefore the current (larger) model for the visible universe violates the holographic principle. However, the author of[15] at this point speculates that equation (25) was fulfilled in the past when the radius of the universe was

$$R \leq 8.4 \pm 0.1\ Gpc \tag{28}$$

and further speculates that is when the cosmic deceleration ended and acceleration began.

Both of these assumptions are based on the evaluation of the relation (26) by using the present time Hubble constant H_0 instead of the cosmic expansion rate $H(z)$ that corresponds to the size of the universe at that time. Therefore the radius of the visible universe at past times cannot be obtained from relation (26) by using H_0.

According to our shell model $R_0 = 4.46 \pm 0.06\ Gpc$ and the ratio in equation (25) is satisfied and it has been always ≤ 1 as it is required by the holographic principle. Equation (26) is actually another expression for the equation (20), which can be seen by putting in equation (26), Ω_m=1. Therefore, because equation (26) and (20) are the same equations, and because in equation (20) speed $v_r \leq c$, the inequality (25) must be always ≤ 1 as it is required by the holographic principle.

The presented model has significant consequences for current cosmological theories. It explains uniformity in the CMB without inflation theory. The model also removes any requirements for superluminal speed expansion, since it can explain the size of the universe with speeds less than or equal to the speed of light. However, the size of the observable universe or the radius of the particle horizon is $\pi c t_0$. This is between the values $3cH_0^{-1}$ $= 3ct_0$, which corresponds to the particle horizon in Einstein-de Sitter universe ($\Omega_m = 1$ and $\Omega_\Lambda = 0$) and $3.4ct_0$, which corresponds to the currently favored cosmological model $\Omega_m = 0.3$ and $\Omega_\Lambda = 0.7$.[17] The velocity of the particle horizon of this model is $2c$ which is the same as in the Einstein-de Sitter model.

Our spherical shell model can be further tested. Assuming that matter is homogenously distributed in the universe, a simple experiment which will count the number of galaxies as function of redshift could provide a test for space curvature. If space is in form of a shell, the number of galaxies as function of altitude on the sphere, or function of redshift, should first increase and then decrease. This test is more complex than it appears, since it should take into account the expansion of the space with time and the detection limits of current instrumentation, but it is feasible at the present time. The Hubble Ultra Deep Field (HUDF) optical and near-infrared survey performed in 2004 covered only a tiny patch of the sky, just 3.5 arc minutes across, but due to the high sensitivity and long exposure time extends thousands of megaparsecs away. HUDF shows a uniform distribution of matter by distance. This is consistent with the model, since integration by longitude could result in different number of galaxies for different redshifts, but a survey that will confirm this needs to be performed.

It is important to note that a hollow shell model completely reproduces the distribution of the entire observed radio sources count for the flux density S from $S \approx 10$ μJy to $S \approx 10$ Jy.[18]

5. Conclusion

Considered is a model that interprets the visible universe as a spherical shell with radius 4.46 ± 0.06 Gpc and event horizon as the maximal length of the arc of the shell, which has the size of 14.0 ± 0.2 Gpc. Consistent with this model, the motion of the light and all galaxies is confined to the volume of the shell, which is radially expanding with the current speed v_{ro} close to c. The model predicts Hubble constant $H_0 = 67.26 \pm 0.90$ km/s/Mpc and values for the cosmic expansion rates $H(z)$ that are in agreement with observations. It explains uniformity of the CMB without the inflation theory, because by the model the entire observed CMB originates from a single antipodal point and for that reason the measured CMB must be exactly the same for all directions of observations, if corrected for Sachs-Wolfe fluctuations caused by large scale structures. The model predicts correct values for the particle horizon πct_0 and the velocity of the particle horizon $2c$. Justification for the shell model by GR is bending of the space by mass, which requires that the light trajectory should be an arc of circle rather than a straight line, if the mass of the universe is taken into account. The model is also favored by holographic principle and it allows to eliminate the established discrepancy between non-covariant version of the holographic principle and the calculated dimensionless entropy, S/k, for the visible universe that exceeds the entropy of a black hole, which is due to misinterpretation of the size of the visible universe. The model is in agreement with the distribution of radio sources in space, type SN Ia data, and with HUDF optical and near-infrared survey performed in 2004.

Acknowledgments

I would like to thank S. Matinyan and I. Filikhin for useful discussions. This work is supported by NSF award HRD-0833184 and NASA grant NNX09AV07A.

References

1. S.D. Goldwirth, T. Piran, in The 6th Marcel Grossmann meeting on general relativity, Sato and Nakamura ed., World Scientific, p. 1211 (1992).
2. W. Israel, Singular hypersurfaces and thin shells in general relativity, Nuovo Cimento **44B** (1966)1-14.
3. A. Czachor, Shell Model of the Big Bang in the special-relativity framework, Acta Physica Polonica B **38** (2007)2673.
4. V.A. Berezin and V. A. Kuzmin, Dynamics of bubbles in general relativity, Physical Review D **36** (1987)2919.
5. J.P. Krisch and E. N. Glass, Thin shell dynamics and equation of the state, Phys. Rev. D **78** (2008)044003.
6. R. Lehoucq, J.P. Luminet, M. Lachi'eze-Rej, preprint, 1994.
7. S. H. Suyu, P. J. Marshall, M. W. Auger, S. Hilbert, R. D. Blandford, L. V. E. Koopmans, C. D. Fassnacht and T. Treu, Dissecting the gravitational lens B1608+656. II. Precision measurements of the Hubble constant, spatial curvature, and the dark energy equation of state. Astrophysical Journal, **711** (2010) 201.
8. S. Ho, C. Hirata, N. Padmanabhan, U. Seljak, N. Bahcall, Correlation of CMB with large-scale structure. I. Integrated Sachs-Wolfe tomography and cosmological implications, Physical Review D **78** (4) (2008) 043519.
9. T. Giannantonio, R. Scranton, R.G. Crittenden, R. Nichol, S. Boughn, A.D. Myers, G.T. Richards, Combined analysis of the integrated Sachs-Wolfe effect and cosmological implications, Physical Review D **77** (12) (2008)123520.
10. A. Raccanelli, A. Bonaldi, M. Negrello, S. Matarrese, G. Tormen, G. De Zotti, A reassessment of the evidence of the Integrated Sachs-Wolfe effect through the WMAP-NVSS correlation, Monthly Notices of the Royal Astronomical Society **386** (2008) 2161-2166.
11. B. R. Granett, M.C. Neyrinck, I. Szapudi, , An imprint of superstructures on the microwave background due to the integrated Sachs-Wolfe effect, Astrophysical Journal **683** (2008) L99-L102.
12. G. Efstathiou and J.R. Bond, astro-ph/9807103 (1998).
13. G. Efstathiou et al. astro-ph/9812226 (1998).
14. Y. Wang and P. Mukherjee, Robust dark energy constrains from supernovae, galaxy clustering, and three-year Wilkinson microwave anisotropy probe observations, arXiv:astro-ph/0604051 v2 2006.
15. P. H. Frampton, Holographic principle and the surface of last scatter, arXiv:1005.2294v4.

16. J.R. Bond, G. Efstathiou and M. Tegmark, Forecasting cosmic parameter errors from microwave background anisotropy experiments, Monthly Notices of the Royal Astronomical Society **291** (1997) L33-L41.
17. T. M. Davis and C. H. Lineweaver, Expanding confusion: common misconceptions of cosmological horizons and the superluminal expansion of the universe, Publications of the Astronomical Society of Australia, **21** (2004) 97.
18. J. J. Condon, Radio sources and cosmology, published in "Galactic and Extragalactic Radio Astronomy", 2nd edition, eds. G.L. Verschuur and K. I. Kellermann, 1988.

PHOTON "MASS" AND ATOMIC LEVELS IN A SUPERSTRONG MAGNETIC FIELD

M. I. Vysotsky*

A.I. Alikhanov Institute of Theoretical and Experimental Physics,
Moscow 117218 Russia
**E-mail: vysotsky@itep.ru*

The structure of atomic levels originating from the lowest Landau level in a superstrong magnetic field is analyzed. The influence of the screening of the Coulomb potential on the values of critical nuclear charge is studied.

Keywords: Atomic levels, superstrong magnetic field, critical charge.

It is a great pleasure for me to present this paper to Sergei Gaikovich Matinyan on his 80th birthday.

1. Introduction

We will discuss the modification of the Coulomb law and atomic spectra in superstrong magnetic field. The talk is based on papers,[1–3] see also.[4]

2. $D = 2$ QED

Let us consider two dimensional QED with massive charged fermions. The electric potential of the external point-like charge equals:

$$\mathbf{\Phi}(k) = -\frac{4\pi g}{k^2 + \Pi(k^2)} , \tag{1}$$

where $\Pi(k^2)$ is the one-loop expression for the photon polarization operator:

$$\Pi(k^2) = 4g^2 \left[\frac{1}{\sqrt{t(1+t)}} \ln(\sqrt{1+t} + \sqrt{t}) - 1 \right] \equiv -4g^2 P(t) , \tag{2}$$

and $t \equiv -k^2/4m^2$, $[g] =$ mass.

In the coordinate representation for $k = (0, k_\parallel)$ we obtain:

$$\Phi(z) = 4\pi g \int\limits_{-\infty}^{\infty} \frac{e^{ik_\parallel z} dk_\parallel / 2\pi}{k_\parallel^2 + 4g^2 P(k_\parallel^2/4m^2)} \ . \tag{3}$$

With the help of the interpolating formula

$$\overline{P}(t) = \frac{2t}{3+2t} \tag{4}$$

the accuracy of which is better than 10% for $0 < t < \infty$ we obtain:

$$\Phi = 4\pi g \int\limits_{-\infty}^{\infty} \frac{e^{ik_\parallel z} dk_\parallel / 2\pi}{k_\parallel^2 + 4g^2 (k_\parallel^2/2m^2)/(3 + k_\parallel^2/2m^2)}$$

$$= \frac{4\pi g}{1 + 2g^2/3m^2} \left[-\frac{1}{2}|z| + \frac{g^2/3m^2}{\sqrt{6m^2+4g^2}} \exp(-\sqrt{6m^2+4g^2}|z|) \right] . \tag{5}$$

In the case of heavy fermions ($m \gg g$) the potential is given by the tree level expression; the corrections are suppressed as g^2/m^2.

In the case of light fermions ($m \ll g$):

$$\Phi(z) \Bigg|_{m \ll g} = \begin{cases} \pi e^{-2g|z|} & , z \ll \frac{1}{g} \ln\left(\frac{g}{m}\right) \\ -2\pi g \left(\frac{3m^2}{2g^2}\right) |z| & , z \gg \frac{1}{g} \ln\left(\frac{g}{m}\right) \end{cases} . \tag{6}$$

$m = 0$ corresponds to the Schwinger model; photon gets a mass due to a photon polarization operator with massless fermions.

3. Electric potential of the point-like charge in $D = 4$ in superstrong B

We need an expression for the polarization operator in the external magnetic field B. It simplifies greatly for $B \gg B_0 \equiv m_e^2/e$, where m_e is the electron mass and we use Gauss units, $e^2 = \alpha = 1/137....$ The following results were obtained in:[2]

$$\Phi(k) = \frac{4\pi e}{k_\parallel^2 + k_\perp^2 + \frac{2e^3 B}{\pi} \exp\left(-\frac{k_\perp^2}{2eB}\right) P\left(\frac{k_\parallel^2}{4m^2}\right)} \ , \tag{7}$$

$$\Phi(z) = 4\pi e \int \frac{e^{ik_\parallel z} dk_\parallel d^2 k_\perp / (2\pi)^3}{k_\parallel^2 + k_\perp^2 + \frac{2e^3 B}{\pi} \exp(-k_\perp^2/(2eB))(k_\parallel^2/2m_e^2)/(3 + k_\parallel^2/2m_e^2)}$$

$$= \frac{e}{|z|} \left[1 - e^{-\sqrt{6m_e^2}|z|} + e^{-\sqrt{(2/\pi)e^3 B + 6m_e^2}|z|} \right] . \tag{8}$$

For $B \ll 3\pi m^2/e^3$ the potential is Coulomb up to small corrections:

$$\left.\mathbf{\Phi}(z)\right|_{e^3 B \ll m_e^2} = \frac{e}{|z|}\left[1 + O\left(\frac{e^3 B}{m_e^2}\right)\right] , \tag{9}$$

analogously to $D = 2$ case with substitution $e^3 B \to g^2$.

For $B \gg 3\pi m_e^2/e^3$ we obtain:

$$\mathbf{\Phi}(z) = \begin{cases} \frac{e}{|z|} e^{(-\sqrt{(2/\pi)e^3 B}|z|)} , & \frac{1}{\sqrt{(2/\pi)e^3 B}} \ln\left(\sqrt{\frac{e^3 B}{3\pi m_e^2}}\right) > |z| > \frac{1}{\sqrt{eB}} \\ \frac{e}{|z|}(1 - e^{(-\sqrt{6m_e^2}|z|)}) , & \frac{1}{m_e} > |z| > \frac{1}{\sqrt{(2/\pi)e^3 B}} \ln\left(\sqrt{\frac{e^3 B}{3\pi m_e^2}}\right) , \\ \frac{e}{|z|} , & |z| > \frac{1}{m_e} \end{cases} \tag{10}$$

$$V(z) = -e\mathbf{\Phi}(z) . \tag{11}$$

The close relation of the radiative corrections at $B >> B_0$ in $D = 4$ to the radiative corrections in $D = 2$ QED allows to prove that just like in $D = 2$ case higher loops are not essential (see, for example,[5]).

4. Hydrogen atom in the magnetic field

For $B > B_0 = m_e^2/e$ the spectrum of Dirac equation consists of ultrarelativistic electrons with only one exception: the electrons from the lowest Landau level (LLL, $n = 0$, $\sigma_z = -1$) are nonrelativistic. So we will find the spectrum of electrons from LLL in the screened Coulomb field of the proton.

The wave function of electron from LLL is:

$$R_{0m}(\rho) = \left[\pi(2a_H^2)^{1+|m|}(|m|!)\right]^{-1/2} \rho^{|m|} e^{(im\varphi - \rho^2/(4a_H^2))} , \tag{12}$$

where $m = 0, -1, -2$ is the projection of the electron orbital momentum on the direction of the magnetic field.

For $a_H \equiv 1/\sqrt{eB} \ll a_B = 1/(m_e e^2)$ the adiabatic approximation is applicable and the wave function looks like:

$$\Psi_{n0m-1} = R_{0m}(\rho)\chi_n(z) , \tag{13}$$

where $\chi_n(z)$ satisfy the one-dimensional Schrödinger equation:

$$\left[-\frac{1}{2m_e}\frac{d^2}{dz^2} + U_{eff}(z)\right]\chi_n(z) = E_n\chi_n(z) . \tag{14}$$

Since screening occurs at very short distances it is not important for odd states, for which the effective potential is:

$$U_{eff}(z) = -e^2 \int \frac{|R_{0m}(\rho)|^2}{\sqrt{\rho^2 + z^2}} d^2\rho \ , \tag{15}$$

It equals the Coulomb potential for $|z| \gg a_H$ and is regular at $z = 0$.

Thus the energies of the odd states are:

$$E_{\text{odd}} = -\frac{m_e e^4}{2n^2} + O\left(\frac{m_e^2 e^3}{B}\right) \ , \quad n = 1, 2, \ldots \ , \tag{16}$$

and for the superstrong magnetic fields $B > m_e^2/e^3$ they coincide with the Balmer series with high accuracy.

For even states the effective potential looks like:

$$\tilde{U}_{eff}(z) = -e^2 \int \frac{|R_{0m}(\vec{\rho})|^2}{\sqrt{\rho^2 + z^2}} d^2\rho \left[1 - e^{-\sqrt{6m_e^2}\, z} + e^{-\sqrt{(2/\pi)e^3 B + 6m_e^2}\, z}\right] . \tag{17}$$

Integrating the Schrödinger equation with the effective potential from $x = 0$ till $x = z$, where $a_H \ll z \ll a_B$, and equating the obtained expression for $\chi'(z)$ to the logarithmic derivative of Whittaker function – the solution of the Schrödinger equation with Coulomb potential – we obtain the following equation for the energies of even states:

$$\ln\left(\frac{H}{1 + \frac{e^6}{3\pi} H}\right) = \lambda + 2\ln\lambda + 2\psi\left(1 - \frac{1}{\lambda}\right) + \ln 2 + 4\gamma + \psi(1 + |m|), \tag{18}$$

where $H \equiv B/(m_e^2 e^3)$, $\psi(x)$ is the logarithmic derivative of the gamma-function and

$$E = -(m_e e^4/2)\lambda^2 \ . \tag{19}$$

The spectrum of the hydrogen atom in the limit $B \gg m_e^2/e^3$ is shown in Fig. 1.

5. Screening versus critical nucleus charge

Hydrogen-like ion becomes critical at $Z \approx 170$: the ground level reaches lower continuum, $\varepsilon_0 = -m_e$, and two e^+e^- pairs are produced from vacuum. Electrons with the opposite spins occupy the ground level, while positrons are emitted to infinity.[6] According to[7] in the strong magnetic field

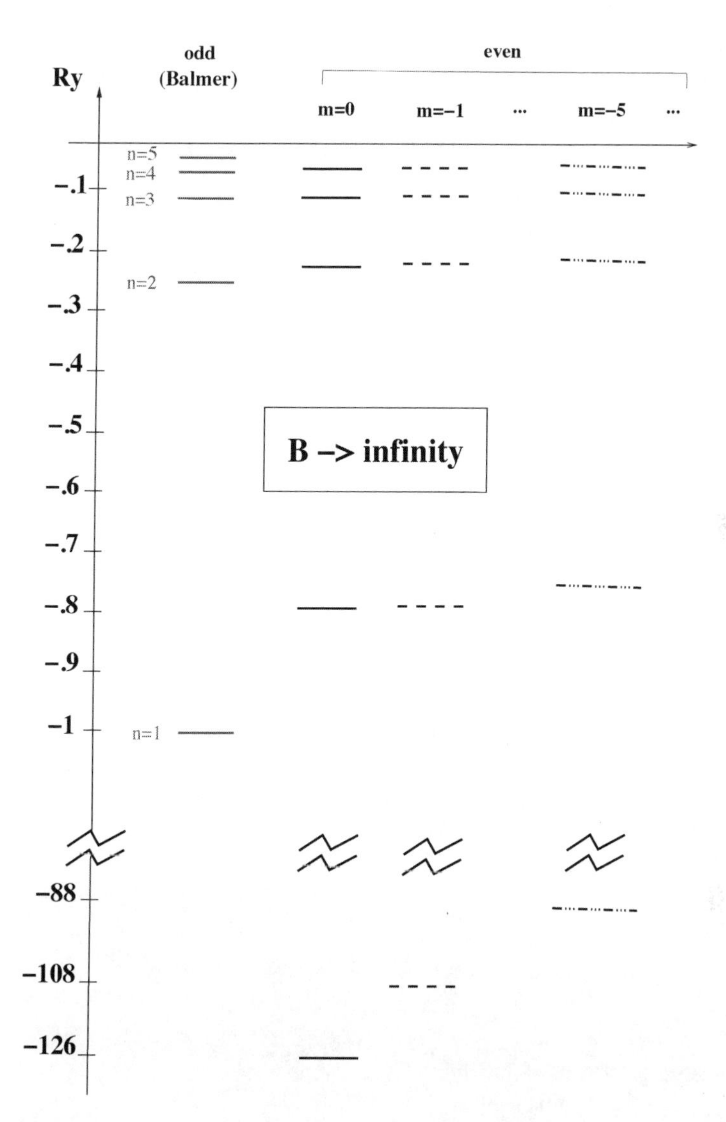

Fig. 1. Spectrum of the hydrogen atom in the limit of the infinite magnetic field. Energies are given in Rydberg units, $Ry \equiv 13.6\ eV$.

$Z_{\rm cr}$ diminishes: it equals approximately 90 at $B = 100B_0$; at $B = 3 \cdot 10^4 B_0$ it equals approximately 40. Screening of the Coulomb potential by the magnetic field acts in the opposite direction and with account of it larger magnetic fields are needed for a nucleus to become critical.

The bispinor which describes an electron on LLL is:

$$\psi_e = \begin{pmatrix} \varphi_e \\ \chi_e \end{pmatrix} ,$$

$$\varphi_e = \begin{pmatrix} 0 \\ g(z)\exp\left(-\rho^2/4a_H^2\right) \end{pmatrix} , \quad \chi_e = \begin{pmatrix} 0 \\ if(z)\exp\left(-\rho^2/4a_H^2\right) \end{pmatrix} . \qquad (20)$$

Dirac equations for functions $f(z)$ and $g(z)$ look like:

$$\begin{aligned} g_z - (\varepsilon + m_e - \bar{V})f = 0, \\ f_z + (\varepsilon - m_e - \bar{V})g = 0, \end{aligned} \qquad (21)$$

where $g_z \equiv dg/dz$, $f_z \equiv df/dz$. They describe the electron motion in the effective potential $\bar{V}(z)$:

$$\bar{V}(z) = -\frac{Ze^2}{a_H^2}\left[1 - e^{-\sqrt{6m_e^2}|z|} + e^{-\sqrt{(2/\pi)e^3B+6m_e^2}|z|}\right] \times \int_0^\infty \frac{e^{-\rho^2/2a_H^2}}{\sqrt{\rho^2+z^2}}\rho d\rho \; . \qquad (22)$$

Intergrating (21) numerically we find the dependence of $Z_{\rm cr}$ on the magnetic field with the account of screening. The results are shown in Fig. 2. For the given nucleus to become critical larger magnetic fields are needed and the nuclei with $Z < 52$ do not become critical.

Acknowledgments

I am grateful to the organizers for wonderful time in Nor-Amberd and Tbilisi and to my coauthors Sergei Godunov and Bruno Machet for helpful collaboration. I was partly supported by the grants RFBR 11-02-00441, 12-02-00193, by the grant of the Russian Federation government 11.G34.31.0047, and by the grant NSh-3172.2012.2.

References

1. M.I. Vysotsky, *JETP Lett.* **92**, 15(2010).
2. B. Machet and M.I. Vysotsky, *Phys. Rev. D* **83**, 025022 (2011).
3. S.I. Godunov, B. Machet, and M.I. Vysotsky, *Phys. Rev. D* **85**, 044058 (2012).
4. A.E. Shabad, V.V. Usov, *Phys. Rev. Lett.* **98**, 180403 (2009); *Phys. Rev. D* **77**, 025001 (2008).
5. V.B.Beresteckii, *Proceedings of LIYaF Winter School* **9, part 3**, 95 (1974).

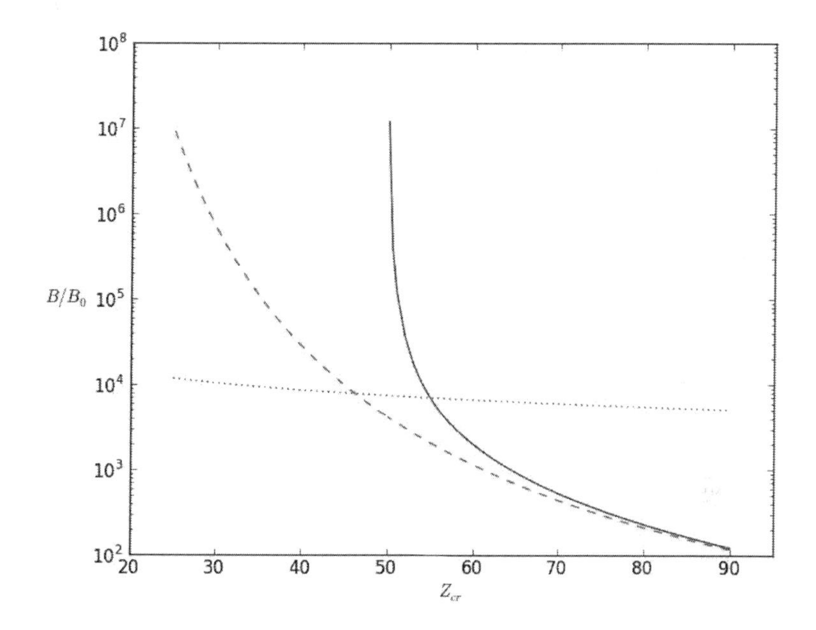

Fig. 2. The values of B_{cr}^{Z}: a) without screening according to,[7] dashed (green) line; b) numerical results with screening, solid (blue) line. The dotted (black) line corresponds to the field at which a_H becomes smaller than the size of the nucleus.

6. Ya.B. Zeldovich, V.S. Popov, *UFN* **105**, 403 (1971);
   W. Greiner, J. Reinhardt, *Quantum Electrodynamics* (Springer-Verlag, Berlin, Heidelberg, 1992);
   W. Greiner, B. Müller, and J. Rafelski, *Quantum Electrodynamics of Strong Fields* (Springer-Verlag, Berlin, Heidelberg,1985).
7. V.N. Oraevskii, A.I. Rez, and V.B. Semikoz, *Zh. Eksp. Teor. Fiz.* **72**, 820 (1977) [*Sov. Phys. JETP* **45**, 428 (1977)].

CONSTRAINTS ON PARAMETERS OF THE BLACK HOLE AT THE GALACTIC CENTER

A.F. Zakharov

Institute of Theoretical and Experimental Physics, 117218 Moscow, Russia
Bogoliubov Laboratory of Theoretical Physics, JINR, 141980 Dubna, Russia
zakharov@itep.ru

F. De Paolis, G. Ingrosso, A.A. Nucita

Dipartimento di Matematica e Fisica "Ennio De Giorgi", Università del Salento and INFN Sezione di Lecce, CP 193, I-73100 Lecce, Italy
depaolis,ingrosso,nucita@le.infn.it

We consider shadows around black holes as a tool to evaluate their parameters. Kerr and Reissner – Nordström cases are analyzed. Different ways to find signatures of extra dimensions are discussing in the literature. If the Randall–Sundrum II braneworld scenario is adopted then a metric of black holes may be different from a standard one and the Schwarzschild metric has to be changed with the Reissner – Nordström metric with a tidal charge. The model may be applied for the black hole at the Galactic Center. Since the Schwarzschild and Reissner – Nordström metrics are different for a significant charge one can expect that geodesics of bright stars near the black hole and their observed fluxes may be different for these metrics due to a difference in gravitational lensing. Therefore, a deviation from the Schwarzschild case flux may be measured for a significant charge and the signature of extra dimensions may found. However, as a theoretical analysis shows that a tidal charge has to be very close to zero and a suggested charge value which may lead to measurable flux deviations for S2 like stars (or their essential displacements or astrometrical lensing) is not consistent with observational constraints on a shadow size.

Keywords: Black hole physics; Kerr black holes; Reissner – Nordström black holes; Sgr A^*; The Galactic Center

1. Introduction

There are not too many observational signatures of black holes where we actually need a strong gravitational field approximation to explain observational data. For many years bright examples were shapes of relativistic lines near black holes[1] where sometimes we have to use a Kerr black hole metrics approximation to fit observations.[2] Numerical simulations of the line structure are be found in a number of papers.[3–5] In particular, results of simulations and corresponding approximation are given.[6–15] One could mention also that if the emitting region has a degenerate position with respect to the line of sight (for example, the inclination angle of an accretion disk is $\gtrsim 85^0$) strong bending effects found[17] and analyzed later[12] do appear. An influence of microlensing on spectral lines and spectra in different bands was analyzed[18,19] (an optical depth of microlensing for distant quasars was discussed for different locations of microlenses[20]). The investigations were inspired by discoveries of microlensing features in X-ray band for gravitational lensed systems. These results were obtained due to an excellent angular resolution in X-ray band of the Chandra satellite enabling us to resolve different images of gravitationally lensed systems and study their luminosities separately.

A couple of years ago, it was simulated a formation of images for supermassive black holes.[21,22] The authors concluded that a strong gravitational field is bent trajectories of photons emitted by accreting particles and an observer can see a dark spot (shadow) around a black hole position. For the black hole at the Galactic Center a size of shadow is around 50 μas. Based on results of simulations, the authors[21,22] concluded that the shadow may be detectable at sub-mm wavelength, however, interstellar scattering may be very significant at cm wavelength, so there are very small chances to observe the shadows at the cm band. Meanwhile, there is a tremendous progress a minimal size of spot for the Sgr A^*,[23] it was evaluated a shadow size as small as 37^{+16}_{-10} μas.[23] Practically, a minimal size of bright spot was evaluated, but a boundary of a dark spot (shadow) has to be bright, a size of bright boundary was measured.

2. Shadows for Kerr black holes

As usual, we use geometrical units with $G = c = 1$. It is convenient also to measure all distances in black hole masses, so we may set $M = 1$ (M is a black hole mass). Calculations of mirage forms are based on qualitative analysis of different types of photon geodesics in a Kerr metric.[24–28] In fact, we know that impact parameters of photons are very close to the critical ones (which correspond to parabolic orbits). This set (critical curve) of impact parameters separates escape and plunge orbits[24–28] or otherwise the critical curve separates scatter and capture regions for unbounded photon trajectories. Therefore the mirage shapes almost look like to critical curves but are just reflected with respect to z-axis. We assume that mirages of all orders almost coincide and form only one quasi-ring from the point of view of the observer. We know that the impact parameter corresponding to the π deflection is close to that corresponding to a $n\pi$ deflections (n is an odd number). For more details see paper[31] (astronomical applications of this idea was discussed[32] and its generalizations for Kerr black hole are considered[33]). We use prefix "quasi" since we consider a Kerr black hole case, so that mirage shapes are not circular rings but Kerr ones. Moreover, the side which is formed by co-moving (or co-rotating) photons is much brighter than the opposite side since rotation of a black hole squeeze deviations between geodesics because of Lense - Thirring effect. Otherwise, rotation stretches deviations between geodesics for counter-moving photons.

The full classification of geodesic types for Kerr metric is given.[26] As it was shown in this paper, there are three photon geodesic types: capture, scattering and critical curve which separates the first two sets. This classification fully depends only on two parameters $\xi = L_z/E$ and $\eta = Q/E^2$, which are known as Chandrasekhar's constants.[25] Here the Carter constant Q is given[29]

$$Q = p_\theta^2 + \cos^2\theta\left[a^2\left(m^2 - E^2\right) + L_z^2/\sin^2\theta\right], \tag{1}$$

where $E = p_t$ is the particle energy at infinity, $L_z = p_\phi$ is z-component of its angular momentum, $m = p_i p^i$ is the particle mass. Therefore, since photons have $m = 0$

$$\eta = p_\theta^2/E^2 + \cos^2\theta\left[-a^2 + \xi^2/\sin^2\theta\right]. \tag{2}$$

The first integral for the equation of photon motion (isotropic geodesics) for a radial coordinate in the Kerr metric is described by the following equation[25,26,29]

$$\rho^4(dr/d\lambda)^2 = R(r), \tag{3}$$

where $R(r) = r^4 + (a^2 - \xi^2 - \eta)r^2 + 2[\eta + (\xi - a)^2]r - a^2\eta$, and $\rho^2 = r^2 + a^2\cos^2\theta, \Delta = r^2 - 2r + a^2, a = S/M^2$. The constants M and S are the black hole mass and angular momentum, respectively. Eq. (3) is written in dimensionless variables (all lengths are expressed in black hole mass units M).

We will consider different types of geodesics on r - coordinate in spite of the fact that these type of geodesics were discussed in a number of papers and books, in particular in a classical monograph[25] (where the most suited analysis for our goals was given). However, our consideration is differed even Chandrasekhar's analysis in the following items.

i) Chandrasekhar[25] considered the set of critical geodesics separating capture and scatter regions as parametric functions $\eta(r), \eta(r)$, but not as the function $\eta(\xi)$ (as we do). However, we believe that a direct presentation of function $\eta(\xi)$ is much more clear and give a vivid illustration of different types of motion. Moreover, one could obtain directly form of mirages from the function $\eta(\xi)$ (as it will be explained below).

ii) Chandrasekhar[25] considered the function $\eta(r)$ also for $\eta < 0$ and that is not quit correct, because for $\eta < 0$ allowed constants of motion correspond only to capture (as it was mentioned in the book[25]). This point will be briefly discussed below.

We fix a black hole spin parameter a and consider a plane (ξ, η) and different types of photon trajectories corresponding to (ξ, η), namely, a capture region, a scatter region and the critical curve $\eta_{\text{crit}}(\xi)$ separating the scatter and capture regions. The critical curve is a set of (ξ, η) where the polynomial $R(r)$ has a multiple root (a double root for this case). Thus, the critical curve $\eta_{\text{crit}}(\xi)$ could be determined from the system[26]

$$R(r) = 0, \quad \frac{\partial R}{\partial r}(r) = 0, \tag{4}$$

for $\eta \geq 0, r \geq r_+ = 1 + \sqrt{1 - a^2}$, because by analyzing of trajectories along the θ coordinate we know that for $\eta < 0$ we have $M = \{(\xi, \eta) | \eta \geq -a^2 + 2a|\xi| - \xi^2, -a \leq \xi \leq a\}$ and for each point $(\xi, \eta) \in M$ photons will be captured. If instead $\eta < 0$ and $(\xi, \eta) \bar{\in} M$, photons cannot have such constants of motion, corresponding to the forbidden region.[25,26]

One can therefore calculate the critical curve $\eta(\xi)$ which separates the capture and the scattering regions.[26] We remind that the maximal value for $\eta_{\text{crit}}(\xi)$ is equal to 27 and is reached at $\xi = -2a$. Obviously, if $a \to 0$, the well-known critical value for Schwarzschild black hole (with $a = 0$) is obtained.

Thus, at first, we calculate the critical curves for chosen spin parameters a which are shown in Fig. 1. The shape of the critical curve for $a = 0$ (Schwarzschild black hole) is well-known because for this case we have $\eta_{\text{crit}}(\xi) = 27 - \xi^2$ for $|\xi| \leqslant 3\sqrt{3}$, but we show the critical curve to compare with the other cases.

By following this approach we can find the set of critical impact parameters (α, β), for the image (mirage or glory) around a rotating black hole. The sets of critical parameters form caustics around black holes and it is well-known that caustics are the brightest part of each image. We remind that (α, β) parameters could be evaluated in terms of $(\xi, \eta_{\text{crit}})$ by the

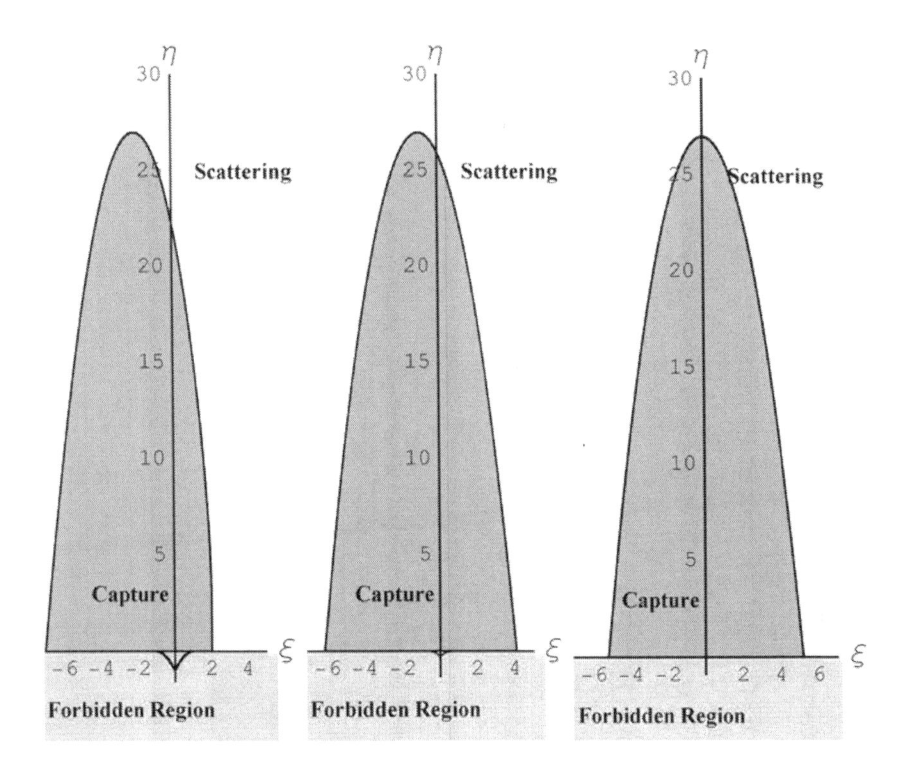

Fig. 1. Different types for photon trajectories and spin parameters ($a = 1., a = 0.5, a = 0.$). Critical curves separate capture and scatter regions. Here we show also the forbidden region corresponding to constants of motion $\eta < 0$ and $(\xi, \eta) \bar{\in} M$ as it was discussed in the text.

following way[25]

$$\alpha(\xi) = \xi / \sin\theta_0, \tag{5}$$

$$\begin{aligned}\beta(\xi) &= (\eta_{\rm crit}(\xi) + a^2\cos^2\theta_0 - \xi^2\cot^2\theta_0)^{1/2} \\ &= (\eta_{\rm crit}(\xi) + (a^2 - \alpha^2(\xi))\cos^2\theta_0)^{1/2}.\end{aligned}$$

Actually, the mirage shapes are boundaries for shadows considered in paper.[21]

We note that the precision we obtain by considering critical impact parameters instead of their exact values for photon trajectories reaching the observer is good enough. In particular, co-rotating photons form much brighter part of images with respect to retrograde photons. Of course, the larger is the black hole spin parameter the larger is this effect (i.e. the co-rotating part of the images become closest to the black hole horizon and brighter).

Let us assume that the observer is located in the equatorial plane ($\theta = \pi/2$). For this case we have from Eqs. (5) and (6)

$$\alpha(\xi) = \xi, \beta(\xi) = \sqrt{\eta_{\rm crit}(\xi)}. \tag{6}$$

As mentioned in section 2, the maximum impact value $\beta = 3\sqrt{3}$ corresponds to $\alpha = -2a$ and if we consider the extreme spin parameter $a = 1$ a segment of straight line $\alpha = 2, 0 < |\beta| < \sqrt{3}$ belongs to the mirage (see images in Fig. 2 for different spin parameters). It is clear that for this case one could easy evaluate the black hole spin parameter after the mirage shape reconstruction since we have a rather strong dependence of the shapes on spins. As it was explained earlier, the maximum absolute value for $|\beta| = \sqrt{27} \approx 5.196$ corresponds to $\alpha = -2a$ since the maximum value for $\eta(\xi)$ corresponds to $\eta(-2a) = 27$ as it was found.[26] Therefore, in principle it is possible to estimate the black hole spin parameter by measuring the position of the maximum value for β, but probably that part of the mirage could be too faint to be detected.

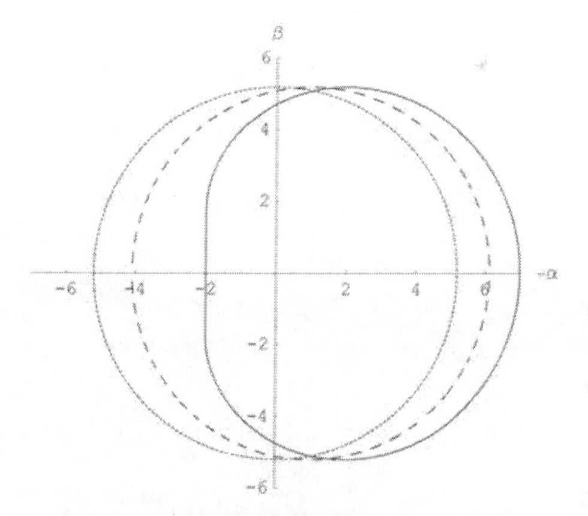

Fig. 2. Mirages around black hole for equatorial position of distant observer and different spin parameters. The solid line, the dashed line and the dotted line correspond to $a = 1, a = 0.5, a = 0$, respectively.

Let us consider different values for the angular positions of a distant observer $\theta = \pi/2, \pi/3, \pi/4$ and $\pi/6$ for $a = 1$. (Fig. 3). From these Figures one can see that angular positions of a distant observer could be evaluated from the mirage shapes only for rapidly rotating black holes ($a \sim 1$), but there are no chances to evaluate the angles for slowly rotating black holes, because even for $a = 0.5$ the mirage shape differences are too small to be distinguishable by observations. Indeed, mirage shapes weakly depend on the observer angle position for moderate black hole spin parameters.

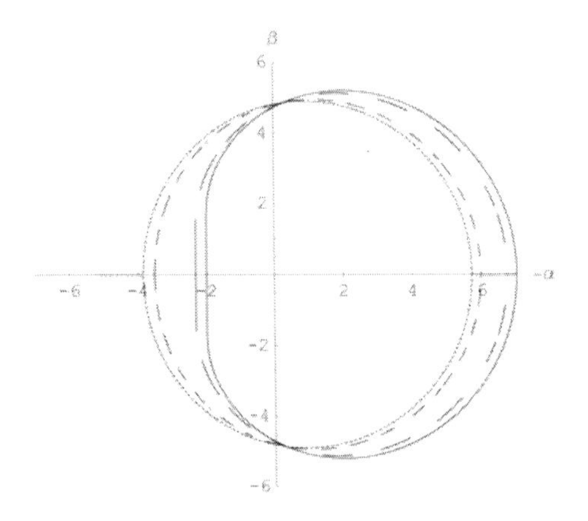

Fig. 3. Mirages around black hole for different angular positions of a distant observer and the spin $a = 1$. Solid, long dashed, short dashed and dotted lines correspond to $\theta_0 = \pi/2, \pi/3, \pi/6$ and $\pi/8$, respectively.

3. Shadows for Reissner-Nordström black holes

"Black holes have no hair" means that a black hole is characterized by only three parameters ("hairs"), its mass M, angular momentum J, and charge Q.[39] Therefore, in principle, charged black holes can form, although astrophysical conditions that lead to their formation may look rather problematic. Nevertheless, one can not claim that their existence is forbidden by theoretical or observational arguments.

Charged black holes are also objects of intensive studies, since they are described by Reissner-Nordström geometry which is a static, spherically symmetrical solution of Yang-Mills-Einstein equations with fairly natural requirements on asymptotic behavior of the solutions.[30] The Reissner - Nordström metric thus describes a spherically symmetric black hole with a color charge and (or) a magnetic monopole.

The formation of retro-lensing images (also known as mirage, shadows, or "faces" in the literature) due to the strong gravitational field effects near black holes has been investigated.[31–35] The question that naturally arises is whether these images are observable or not. It has been shown that the retro-lensing image around the black hole at the Galactic Center (Sgr A^*) due to S_2 star is observable in the K-band (peaked at 2.2 μm) by the next generation infra-red space-based missions. The effects of retro-lensing image shapes due to black hole spin has also been investigated.[33,34]

Here we focus on the possibility of measuring the black hole charge as well, and we present an analytical dependence of mirage size on the black hole charge. Indeed, a present space missions like RadioAstron in radio band have angular resolution close to the shadow size for massive black holes in the center of our own and nearby galaxies.

3.1. *Basic definitions and equations*

The expression for the Reissner - Nordström metric in natural units ($G = c = 1$) has the form

$$ds^2 = -\left(1 - \frac{2M}{r} + \frac{Q^2}{r^2}\right)dt^2 + \left(1 - \frac{2M}{r} + \frac{Q^2}{r^2}\right)^{-1} dr^2 + r^2(d\theta^2 + \sin^2\theta d\phi^2), \quad (7)$$

where M is the mass of the black hole and Q is its charge.

Applying the Hamilton-Jacobi method to the problem of geodesics in the Reissner - Nordström metric, the motion of a test particle in the r-coordinate can be described by following equation[39]

$$r^4(dr/d\lambda)^2 = R(r), \quad (8)$$

where $R(r) = P^2(r) - \Delta(\mu^2 r^2 + L^2), P(r) = Er^2 - eQr, \Delta = r^2 - 2Mr + Q^2$. Here, the constants $\mu, E, L,$ and e are associated with the particle; i.e. μ is its mass, E is energy at infinity, L is its angular momentum at infinity, and e is the particle's charge.

The photon capture cross section for an extremely charged black hole turns out to be considerably smaller than the capture cross section of a Schwarzschild black hole. The critical value of the impact parameter, characterizing the capture cross section for a Reissner - Nordström black hole, is determined by the equation[35,36,40,41]

$$l = \frac{b_1 + \sqrt{{b_1}^2 + 64q^3(1-q)}}{2(1-q)}, \quad (9)$$

where $b_1 = 8q^2 - 36q + 27$.

Substituting Eq.(9) into the expression for the coefficients of the polynomial $R(r)$, it is easy to calculate the radius of the unstable circular photon orbit, which is the same as the minimum periastron distance. The orbit of a photon moving from infinity with the critical impact parameter, determined in accordance with Eq.(9), spirals into circular orbit. As was explained[34-36] this leads to the formation of shadows described by the critical value of L_{cr}; or in other words, in the spherically symmetric case, shadows are circles with radii L_{cr}. Therefore, by measuring the shadow size, one could evaluate the black hole charge in black hole mass units M.

3.2. *Shadows for a Reissner–Nordström black holes with a tidal charge*

Theories with extra dimensions admit astrophysical objects (supermassive black holes in particular) which are rather different from standard ones. There were proposed tests which may help to discover signatures of extra dimensions in supermassive black holes since the gravitational field may be different from the standard one in the GR approach. So, gravitational lensing features are different for alternative gravity theories with extra dimensions and general relativity. Recently, Bin-Nun[42-44] discussed an opportunity that the black hole at the Galactic Center is described by a Reissner– Nordström metric with a tidal charge[45] which may be admitted by the Randall–Sundrum II braneworld scenario. Bin-Nun suggested an opportunity of evaluating the black hole metric analyzing (retro-)lensing of bright stars around the black hole in the Galactic Center.

Measurements of the shadow size around the black hole may help to evaluate parameters of black hole metric.[34,35] Another opportunity to evaluate parameters of the black hole is an analysis of trajectories bright stars near the Galactic Center.[37,38] We derive an analytic expression for the black hole shadow size as a function of charge for the tidal Reissner– Nordström metric. We conclude that observational data concerning shadow size measurements are not consistent with significant negative charges, in particular, the significant negative charge $Q/(4M^2) = -1.6$ (discussed in papers[42-44]) is practically ruled out with a very probability (the charge is roughly speaking is beyond 9σ confidence level, but a negative charge is beyond 3σ confidence level).

We could apply Eq. (9) for the black hole at the Galactic Center assuming that the black hole mass is about $4 * 10^6 M_\odot$ and a distance toward the Galactic Center is about 8 *kpc*. In this case a diameter of shadow is about 52 μas for the Schwarzschild metric and about 40 μas for the extreme Reissner–Nordström metric.

Doeleman et al.[23] evaluated a size of the smallest spot near the black hole at the Galactic Center such as 37^{+16}_{-10} microarcseconds at a wavelength of 1.3 mm with 3σ confidence level. Theoretical analysis and observations show that the size of shadow can not be smaller than a minimal spot size at the wavelength.[21,22,34,35] Roughly speaking, it means that a small positive q is consistent with observations but a significant negative q is not. For $q = -6.4$ (as it was suggested in papers[42–44]) we have a shadow size 84.38 μas.[36] It means that the shadow size is beyond of shadow size with a probability corresponding to a deviation about 9σ from an expected shadow size. Therefore, a probability to have so significant tidal charge for the black hole at the Galactic Center is negligible. So, we could claim that the tidal charge is ruled out with observations and corresponding theoretical analysis.

4. Conclusions

In spite of the difficulties of measuring the shapes of images near black holes is so attractive challenge to look at the "faces" of black holes because namely the mirages outline the "faces" and correspond to fully general relativistic description of a region near black hole horizon without any assumption about a specific model for astrophysical processes around black holes (of course we assume that there are sources illuminating black hole surroundings).

The angular resolution of the space RadioAstron interferometer will be high enough to resolve radio images around black holes. By measuring the mirage shapes one should be able to evaluate the black hole mass, inclination angle (e.g. the angle between the black hole spin axis and line of sight), and spin, if the black hole distance is known. For example, for the black hole at the Galactic Center, the mirage size is $\sim$ 52 μas for the Schwarzschild case. In the case of a Kerr black hole,[33–35] the mirage is deformed depending on the black hole spin a and on the angle of the line of sight, but its size is almost the same. In the case of a Reissner-Nordström black hole, its charge changes the size of the shadows up to 30 % for the extreme charge case. Therefore, the charge of the black hole can be measured by observing the shadow size, if the other black hole parameters are known with sufficient precision. In general, one could say that a measure of the mirage shape (in size) allows to evaluate all the black hole "hairs" to be evaluated.

However, there are two kinds of difficulties when measuring mirage shapes around black holes. First, the brightness of these images or their parts (arcs) may not be sufficient. But, numerical simulations[21,22] give hope that the shadow brightness could be high enough to be detectable. Second, turbulent plasma effects near the black hole horizon could give essential broadening of observed images[46] leading to a confusion of the shadow image.

The minimal fringe size of the RadioAstron space–ground interferometer is about 8 μas at the 1.3 cm and the size is smaller than the shadow size about 50 μas for the black hole at the Galactic Center, however probably scattering can cause blurring the shadow. Forthcoming observations with these facilities will give answer for the question. However, further observations (after observations presented in papers[23,47]) with VLBI facilities at mm wavelengths look more promising to improve a precision for evaluation of the shadow size near the black hole at the Galactic Center.

The Randall–Sundrum II braneworld approach can lead to a Reissner–Nordström metric with a tidal charge for the black hole at the Galactic Center,[42–44] moreover, a way to find an essential tidal charge (and signatures of extra dimensions) was proposed in these papers. However, as it was mentioned earlier, a size of shadow evaluated in paper[23] is not consistent with a significant tidal charge.[36] Therefore, there are no signatures of extra dimensions for the black hole at the Galactic Center at least for the tidal charge value considered by Bin-Nun.[42–44]

Acknowledgments

It is a pleasure to acknowledge W. Cash, P. Jovanović, A.M. Cherepashchuk, L.Č. Popović, M. V. Sazhin and Z. Shen for useful discussions and clarifications. The authors are grateful to V. Gurzadyan for his kind attention to this contribution.

References

1. A.C. Fabian, M.J. Rees, L. Stella, & N. E. White, *MNRAS*, **238**, 729 (1989).
2. Y. Tanaka et al., *Nature*, **375**, 659 (1995).
3. G. Bao, & Z. Stuchlik, *ApJ*, **400**, 163 (1992).
4. G. Bao, P. Hadrava, & E. Ostgaard, *ApJ*, **435**, 55 (1994).
5. B. C. Bromley, K. Chen & W. A. Miller, *ApJ*, **475**, 57 (1997).
6. A.F. Zakharov, *MNRAS*, **269**, 283 (1994).
7. A. F. Zakharov, *Intern. Journal Mod. Phys. A*, **20**, 2321 (2005).
8. A.F. Zakharov et al., *MNRAS*, **342**, 1325 (2003).

9. A. F. Zakharov, Z. Ma & Y. Bao, *New Astronomy*, **9**, 663 (2004).
10. A. F. Zakharov & S.V. Repin, *Astronomy Reports*, **43**, 705 (1999).
11. A. F. Zakharov & S. V. Repin, *Astronomy Reports*, **46**, 360 (2002).
12. A.F. Zakharov & S. V. Repin, *A &A*, **406**, 7 (2003).
13. A.F. Zakharov & S. V. Repin, *Astronomy Reports*, **47**, 733 (2003).
14. A.F. Zakharov & S. V. Repin, *Nuovo Cimento*, **118B**, 1193 (2003).
15. A.F. Zakharov & S. V. Repin, *Advances in Space Res.* **34**, 1837 (2004).
16. A.F. Zakharov & S. V. Repin, *New Astronomy*, **11**, 405 (2006).
17. G. Matt G., G. C. Perolla & L. Stella, *A & A*, **267**, 2, 643 (1993).
18. L. Č. Popović et al., *ApJ*, **637**, 620 (2006).
19. P. Jovanović, A. F. Zakharov, L.Č. Popović, T. Petrović, *MNRAS*, **386**, 397 (2008).
20. A.F. Zakharov, L. Č. Popović, P. Jovanović, *A & A*, 420, 881 (2004).
21. H. Falcke, F. Melia, E. Agol, *ApJ*, **528**, L13 (2000).
22. F. Melia, & H. Falcke, *Annual Rev. A&A*, **39**, 309 (2001).
23. S.S. Doeleman et al., *Nature*, **455**, 78 (2008).
24. P. Young, *Phys. Rev. D*, **14**, 3281 (1976).
25. S. Chandrasekhar, *Mathematical Theory of Black Holes,* (Clarendon Press, Oxford, 1983).
26. A.F. Zakharov, *Sov. Phys. - Journ. Experim. and Theor. Phys.*, **64**, 1 (1986).
27. A.F. Zakharov, *Sov. Astron.*, **32**, 456 (1988).
28. A.F. Zakharov, *Sov. Phys. – Journ. Experim. and Theor. Phys.*, **68**, 217 (1989).
29. B. Carter, *Phys. Rev.*, **174**, 1559 (1968).
30. D.V. Gal'tsov, A.A. Ershov, *Phys. Lett. A* **138**, 160 (1989).
31. D. Holz and J.A. Wheeler, *ApJ*, **578**, 330 (2002).
32. F. De Paolis et al. *A & A*, **409**, 804 (2003).
33. F. De Paolis et al. *A&A*, **415**, 1 (2003).
34. A.F. Zakharov et al. *New Astronomy*, **10**, 479 (2005).
35. A.F. Zakharov et al., *A & A*, **442**, 795 (2005).
36. A. F. Zakharov et al., *New Astron. Rev.*, **56**, 64 (2012).
37. A.F. Zakharov et al., *Phys. Rev. D* **76**, 062001 (2007).
38. D. Borka et al. *Phys. Rev. D*, **85**, 124004 (2012).
39. C.W. Misner, K.S. Thorne and J. A. Wheeler, 1973. Gravitation, (San Francisco, Freeman, 1973).
40. A.F. Zakharov A.F. *Sv A*, **35**, 147 (1991).
41. A.F. Zakharov, *Class. Quant. Grav.*, **11**, 1027 (1994).
42. A.Y. Bin-Nun, *Phys. Rev. D* **81**, 123011 (2010).
43. A.Y. Bin-Nun, *Phys. Rev. D* **82** 064009 (2010).
44. A.Y. Bin-Nun, *Class. Quant. Grav.* **28**, 114003 (2011).
45. N. Dadhich et al. *Phys. Lett. B* 487, 1 (2000).
46. F. C. Bower et al. *Sciencexpress*, www.sciencexpress.org/1 April 2004/.
47. A. E. Broderick et al. *ApJ*, **735**, 110 (2011).

DIFFUSION IN TWO-DIMENSIONAL DISORDERED SYSTEMS WITH PARTICLE-HOLE SYMMETRY

K. Ziegler

Institut für Physik, Universität Augsburg
D-86135 Augsburg, Germany
**E-mail: klaus.ziegler@physik.uni-augsburg.de*
www.physik.uni-augsburg.de/~klausz

We study the scattering dynamics of a spinor wavefunction in a random environment on a two-dimensional lattice. In the presence of particle-hole symmetry we find diffusion on large scales. The latter is described by a non-interacting Grassmann field, indicating a special kind of asymptotic freedom on large length scales in $d = 2$.

Keywords: Transport in two dimensions, disordered systems, diffusion, spontaneous symmetry breaking.

1. Introduction

Systems with particle-hole (PH) symmetry have become a subject of intense research after the discovery of the unusual electronic properties of graphene. The PH symmetry appears in the presence of two bands (valence and conduction band) when the Fermi energy is located between the bands. The system is always metallic when these bands touch each other at one or several separated points (subsequently called nodes). This situation is realized in many materials, such as graphene, semiconductors and topological insulators. Tuning the Fermi energy (e.g. by an external gate) from hole-like to particle-like behavior, the transport reveals a very robust minimal conductivity when the chemical potential passes through the node between the two bands.[1–5] The robustness of the minimal conductivity indicates a transport mechanism that is not affected by details of the sample. The latter rules out ballistic transport because this would be strongly affected by boundary conditions. Therefore, a natural candidate for transport at the node is diffusion. Diffusion as the fundamental transport mechanism in disordered system has been discussed in the literature,[6,7] whereas ballistic transport is ruled out[8] in two-dimensional systems.

The central idea for the following approach is that diffusion appears as a consequence of spontaneous symmetry breaking, where a massless mode is generated by a broken continuous symmetry. Since the PH symmetry is discrete, there must be an additional continuous symmetry. It is well known that this is not a symmetry of the Hamiltonian but of the dynamical quantities, which are given by a combination of advanced and retarded Green's functions.[9]

Subsequently we will consider a continuous symmetry between non-interacting fermions and bosons which are scattered by the same random potential. Averaging over the latter, we can describe the physics on large scales by an effective field theory which is controlled by a massless fermion mode of the spontaneously broken continuous symmetry.

2. Model

Now we introduce the concept of quantum diffusion, using the language of classical diffusion with a transition probability and mean-square displacement. This is applied to a two-band Hamiltonian with PH symmetry and with a random gap. The Hamiltonian is formulated in terms of a two-component spinor that represents the two bands.

2.1. *Quantum diffusion*

We consider a quantum particle on a lattice, where the lattice sites have coordinates $\mathbf{r}$. The motion of a quantum particle is characterized by the transition probability $P_{\mathbf{r},\mathbf{r}'}(i\epsilon)$ for a two-component spinor particle at site $\mathbf{r}'$ that moves to site $\mathbf{r}$ with frequency $i\epsilon$:

$$P_{\mathbf{r},\mathbf{r}'}(i\epsilon) = \frac{K_{\mathbf{r},\mathbf{r}'}(i\epsilon)}{\sum_{\mathbf{r}} K_{\mathbf{r},\mathbf{r}'}(i\epsilon)} \tag{1}$$

with

$$K_{\mathbf{r},\mathbf{r}'}(i\epsilon) = \langle \mathrm{Tr}_2 \left[G_{\mathbf{r},\mathbf{r}'}(i\epsilon) G^{\dagger}_{\mathbf{r}',\mathbf{r}}(i\epsilon) \right] \rangle_v = \langle \mathrm{Tr}_2 \left[G_{\mathbf{r},\mathbf{r}'}(i\epsilon) G_{\mathbf{r}',\mathbf{r}}(-i\epsilon) \right] \rangle_v \,, \tag{2}$$

where $G(i\epsilon) = (i\epsilon + H)^{-1}$ is the one-particle Green's function of the Hamiltonian H and $\langle ... \rangle_v$ is the average with respect to some random scatterers. $\mathrm{Tr}_2(...)$ is the trace with respect to the two spinor components. The last equation in Eq. (2) follows from the Hermitean Hamiltonian: $H^{\dagger} = H$.

After Fourier transforming the transition matrix (or two-particle Green's function) $K_{\mathbf{r},\mathbf{r}'}(i\epsilon) \to k_{\mathbf{r},\mathbf{r}'}(t)$ we study the motion of the quantum particle in time and calculate the mean-square displacement of the coordinate r_μ

$$\langle r_\mu^2 \rangle_t = \frac{\sum_{\mathbf{r}} r_\mu^2 k_{\mathbf{r},0}(t)}{\sum_{\mathbf{r}} k_{\mathbf{r},0}(t)} . \tag{3}$$

In the case of diffusion (ballistic motion) this expression grows linearly (quadratically) with time t. A direct physical interpretation of $k_{\mathbf{r},\mathbf{r}'}(t)$ can be given in terms of the (unnormalized) wavefunction $\Psi(\mathbf{r},t)$ as $k_{\mathbf{r},0}(t) = |\Psi(\mathbf{r},t)|^2$ which is generated from an initial state by the evolution from the initial state $\Psi(\mathbf{r},0)$ with H: $\Psi(\mathbf{r},t) = \exp(-iHt)\Psi(\mathbf{r},0)$.

The conductivity $\sigma_{\mu\mu}$ at frequency ω can be calculated from $K_{\mathbf{r},\mathbf{r}'}(i\epsilon)$ via the Kubo approach by an analytic continuation $\epsilon \to i\omega/2$ as[9,10]

$$\sigma_{\mu\mu} \sim -\frac{e^2}{2h}\omega^2 \sum_{\mathbf{r}} r_\mu^2 K_{\mathbf{r},0}(-\omega/2) . \tag{4}$$

2.2. *Hamiltonian*

In the following we assume that the Hamiltonian H satisfies the generalized PH symmetry under a unitary transformation

$$H \to -UH^*U^\dagger = H , \tag{5}$$

which belongs to class D according to Cartan's classification scheme.[11] In terms of the Green's function, this transformation provides a sign change of the energy:

$$G(i\epsilon) \to -UG^T(i\epsilon)U^\dagger = G(-i\epsilon) ,$$

since the Hermitean Hamiltonian ($H^* = H^T$, where ^{T} is the matrix transposition) implies from (5) the relation $H = -UH^TU^\dagger$.

In order to be more specific here, we consider a generalization of two-dimensional Weyl fermions with random gap m, where the Hamiltonian has the spinor form

$$H_m = \begin{pmatrix} m & (i\partial_1 + \partial_2)^n \\ (i\partial_1 - \partial_2)^n & -m \end{pmatrix} \tag{6}$$

with the spatial differential operator ∂_j in j–direction. Then the PH transformation of Eq. (5) becomes

$$H_m \to \sigma_{j_n} H_m^T \sigma_{j_n} = -H_m ,$$

where $j_n = 1$ ($j_n = 2$) for n odd (even) with Pauli matrices $\{\sigma_j\}_{j=0,\dots,3}$. For the random gap we assume an uncorrelated Gaussian distribution with zero mean and variance $\langle m_{\mathbf{r}} m_{\mathbf{r}'} \rangle = g\delta_{\mathbf{r}\mathbf{r}'}$.

A consequence of the PH symmetry is a symmetric spectrum of H_m with respect to the energy $E = 0$. The eigenstate Ψ_{-E} of energy $-E$ is obtained from the eigenstate Ψ_E by the transformation $\Psi_{-E} = \sigma_{j_n}\Psi^*_E$. For a homogeneous gap m the Hamiltonian can be diagonalized by a Fourier transformation and yields $E_{\mathbf{k}} = \pm\sqrt{m^2 + k^{2n}}$ (cf. plots in Fig. 1). Moreover, the density of states $\rho_n(E)$ for $m = 0$

$$\rho_n(E) = \frac{1}{2\pi^2} \lim_{\epsilon\to 0} Im \left[\int_0^\lambda \frac{E + i\epsilon}{-(E + i\epsilon)^2 + k^{2n}} kdk \right]$$

is singular at the PH symmetry point $E = 0$:

$$\rho_n(0) = const. \lim_{\epsilon\to 0} \epsilon^{2/n-1} \ .$$

Thus, the density of states at the PH symmetric point $E = 0$ is sensitive to the spectral power exponent n: It vanishes (diverges) for $n < 2$ ($n > 2$) and is finite for $n = 2$. Thus, $n = 2$ plays a marginal role that separates two different regimes. We will see later that this is also the case after averaging over the random gap.

It should be noticed that $n = 1$ ($n = 2$) represents the low-energy behavior of monolayer (bilayer) graphene. A possible realization of $n > 2$ was recently discussed in the literature as an ABC stack of n graphene layers.[12,13] Other realizations may exist in semiconducting materials with complex band structures and topological insulators.

3. Functional-integral representation

The Green's functions $G(i\epsilon)$ and the transposed Green's function $G^T(i\epsilon)$ can be expressed by a functional integral of a free complex (boson) field and a Grassmann (fermion) field, respectively.[14] Then the transition matrix $K_{\mathbf{rr}'}$ in Eq. (2) can be written in terms of a combination of a boson and a fermion functional integral. This approach has been described in detail in Refs. [10,15,16]. The corresponding boson-fermion functional integral is approximated by a saddle-point approximation, where the saddle-point is degenerated with respect to a fermion degree of freedom. The detailed derivation of the saddle-point approximation is given in Ref. [10] and can be directly extended to the new Hamiltonian in Eq. (6).

We start from the result of the saddle-point approximation, which takes into account only the integration with respect to the saddle-point manifold, and apply this approach to the Hamiltonian (6). In particular, the transition

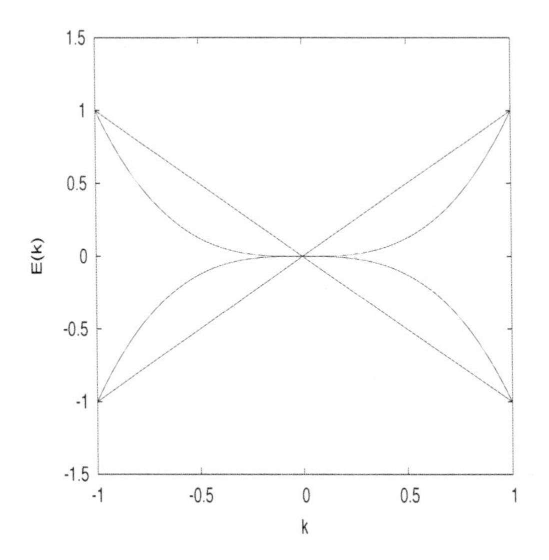

Fig. 1. Massless dispersion near the node for $n = 1$ (linear curves) and $n = 3$.

matrix then reads

$$K_{\mathbf{r},\mathbf{r}'} = \frac{4\eta^2}{g^2} \int \varphi_{\mathbf{r}} \varphi'_{\mathbf{r}'} \mathrm{detg} \left[\hat{H}_0 + i\epsilon + i\eta \hat{U}^2 \right]^{-1} \mathcal{D}[\hat{U}] \tag{7}$$

with block-diagonal Hamiltonian $\hat{H}_0 = diag(H_0, H_0^T)$ and

$$\hat{U}_{\mathbf{r}} = \exp(-\hat{S}_{\mathbf{r}}), \quad \hat{S}_{\mathbf{r}} = \begin{pmatrix} 0 & \varphi_{\mathbf{r}} \sigma_{j_n} \\ \varphi'_{\mathbf{r}} \sigma_{j_n} & 0 \end{pmatrix} \quad (\varphi_{\mathbf{r}}, \varphi'_{\mathbf{r}} \in \mathcal{G}) ,$$

where $\mathcal{G}$ is a Grassmann algebra and detg is the graded determinant. The scattering rate η is determined by a self-consistent (saddle-point) equation:

$$g \int_0^\infty \frac{k}{\eta^2 + k^{2n}} dk = 1 . \tag{8}$$

A continuous transformation characterizes (7) by the fact that $\hat{H}_0$ anticommutes with $\hat{S}$

$$(\hat{H}_0)_{\mathbf{r},\mathbf{r}'} S_{\mathbf{r}'} = -S_{\mathbf{r}'} (\hat{H}_0)_{\mathbf{r},\mathbf{r}'} .$$

This property reads for a global ($\mathbf{r}$ independent) $\hat{S}$

$$\hat{U} \hat{H}_0 \hat{U} = \hat{H}_0 .$$

$i\epsilon$ breaks this symmetry, since it commutes with $\hat{S}$. Thus, $\mathrm{detg}(\hat{U}) = 1$ implies that (7) is invariant under a global change of $\hat{U}$ if $\epsilon \to 0$. For the subsequent discussion it useful to express $\hat{U}$ as a polynomial in $\hat{S}$:

$$\hat{U}^2 = \mathbf{1} + 2\hat{S} + 2\hat{S}^2 \ . \tag{9}$$

Next we define the effective action

$$S' = \log \mathrm{detg}\left(\hat{H}_0 + i\epsilon + i\eta\hat{U}^2\right) \tag{10}$$

and write for (7)

$$K_{\mathbf{r},\mathbf{r}'} = \frac{4\eta^2}{g^2}\int \varphi_{\mathbf{r}}\varphi'_{\mathbf{r}'}e^{-S'}\mathcal{D}[\hat{U}] \ . \tag{11}$$

A gradient expansion up to second order is the conventional treatment of the action S' and leads to a nonlinear sigma model with action S'' for the nonlinear field[16] $\hat{U}^2$:

$$S' = S_0 + S'' + o(\eta^3), \quad S'' = i\eta \mathrm{Trg}[\hat{G}_0(\hat{U}^2 - \mathbf{1})] + \frac{\eta^2}{2}\mathrm{Trg}\left\{[\hat{G}_0(\hat{U}^2 - \mathbf{1})]^2\right\}$$

with $\hat{G}_0 = (\hat{H}_0 + i(\eta + \epsilon))^{-1}$ and the graded trace Trg. According to (9) the action of the nonlinear sigma model then reads

$$S'' = 2i\eta \mathrm{Trg}[\hat{G}_0(\hat{S} + \hat{S}^2)] + 2\eta^2 \mathrm{Trg}\left\{[\hat{G}_0(\hat{S} + \hat{S}^2)]^2\right\}$$

which, after a straightforward calculation and a Fourier transformation $\varphi_{\mathbf{r}} \to \varphi_{\mathbf{q}}$, reduces to a quadratic form in the Grassmann field $\varphi_{\mathbf{q}}$:

$$S'' = \int_{\mathbf{q}} \tilde{C}_{\mathbf{q}}\varphi_{\mathbf{q}}\varphi'_{\mathbf{q}} \ . \tag{12}$$

The coefficient of the quadratic form reads for $\mathbf{q} \sim 0$

$$\tilde{C}_{\mathbf{q}} \sim \frac{4\eta}{g}(\epsilon + Dq^2) \tag{13}$$

with the coefficient

$$D = -\frac{g\eta}{2}\frac{\partial^2}{\partial q_\mu^2}\int \mathrm{Tr}_2\left[G_{0,\mathbf{k}}(i\eta)G_{0,\mathbf{k}-\mathbf{q}}(-i\eta)\right]\frac{d^2k}{4\pi^2}\bigg|_{\mathbf{q}=0} \ .$$

This result gives us for the transition matrix eventually the diffusion propagator

$$\tilde{K}_{\mathbf{q}} = \frac{\eta/g}{\epsilon + Dq^2} \ , \tag{14}$$

where the diffusion coefficient reads

$$D = n\frac{g}{4\pi\eta} \ . \tag{15}$$

4. Discussion

For a two-dimensional system of non-interacting particles with PH symmetry and a random gap with mean zero our calculation has provided a diffusive behavior. The diffusion propagator in Eq. (14) allows us to evaluate the dynamics of the quantum walk and the corresponding conductivity in detail. For a quantitative result we must determine the scattering rate η from Eq. (8):

$$\eta = (gI_n)^{n/2(n-1)} , \quad I_n = \int_0^\infty \frac{x}{1+x^{2n}} dx \quad (n > 1) .$$

For $n = 1$ the scattering rate depends on the momentum cutoff λ:[10]

$$\eta = \frac{\lambda}{\sqrt{e^{2\pi/g} - 1}} .$$

This indicates a vanishing scattering rate for vanishing disorder strength g. From η we obtain directly the average density of states at $E = 0$ as

$$\rho_0 = \frac{\eta}{2\pi g} . \tag{16}$$

The scattering rate allows us also to determine the diffusion coefficient in Eq. (15) as

$$D = \frac{n}{4\pi} g^{(n-2)/2(n-1)} I_n^{-n/2(n-1)} \quad (n > 1) , \tag{17}$$

which is plotted in Fig. 2 for fixed g. There is a qualitative change in the behavior of the density of states and the diffusion coefficient at $n = 2$ when we increase the disorder strength g: For $n = 1$ ρ_0 (D) increases (decreases) with increasing g, whereas the behavior is the opposite for $n > 2$. At $n = 2$ both quantities are independent of g. The opposite behavior of ρ_0 and D for $n > 2$ can be interpreted as a suppression of diffusion by an increased density of states.

For fixed ϵ we can evaluate the corresponding localization length from $P_{\mathbf{r},\mathbf{r}'}(i\epsilon)$ in Eq. (2) together with (14):

$$\langle r_k^2 \rangle_\epsilon = \frac{\sum_{\mathbf{r}} r_\mu^2 K_{\mathbf{r}}(i\epsilon)}{\sum_{\mathbf{r}} K_{\mathbf{r}}(i\epsilon)} = \frac{2D}{\epsilon} = n\frac{g}{2\pi\eta\epsilon} . \tag{18}$$

Thus the localization length diverges with ϵ^{-1} for a vanishing symmetry-breaking term ϵ. The corresponding expression for the pure system $(m = 0)$ is obtained from Eq. (2) by replacing $G(i\epsilon)$ with $G_0(i\epsilon)$ as

$$\langle r_k^2 \rangle_\epsilon = \frac{n}{2\pi\epsilon^2} , \tag{19}$$

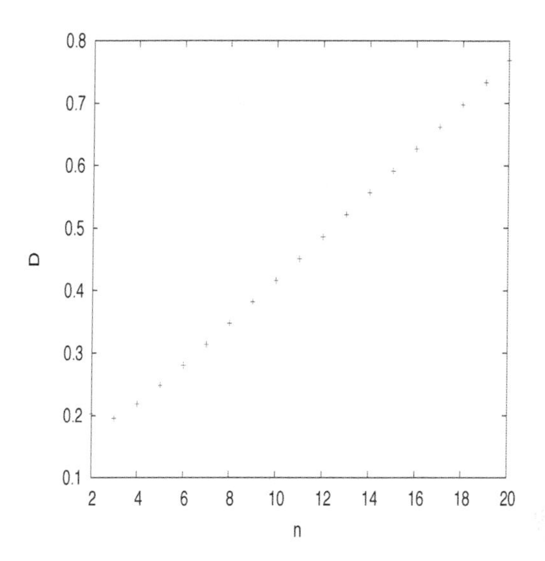

Fig. 2. Diffusion coefficient D as a function of the spectral exponent n for disorder strength $g = 0.1$ (from Eq. (17).

where the ϵ^{-2} divergence reflects the wave-like states in the absence of a random mass.

To study time-dependent quantities, we apply a Fourier transformation from frequency ϵ to time t and get from Eq. (14)

$$\tilde{K}_{\mathbf{q}}(i\epsilon) \to K_{\mathbf{q}}(t) = \frac{\eta}{g} e^{-Dq^2 t} ,$$

and a Fourier transformation from momentum $\mathbf{q}$ to real space coordinates $\mathbf{r}$ then yields

$$K_{\mathbf{q}}(t) \to k_{\mathbf{r}}(t) = \frac{\eta}{g} \frac{e^{-r^2/4Dt}}{\pi Dt} .$$

With this expression the mean-square displacement reads as a function of time

$$\langle r_\mu^2 \rangle_t = \frac{\sum_{\mathbf{r}} r_k^2 k_{\mathbf{r}}(t)}{\sum_{\mathbf{r}} k_{\mathbf{r}}(t)} \sim 2Dt . \tag{20}$$

Finally, the diffusion propagator in Eq. (14) allows us to evaluate the conductivity via the Kubo formula in Eq. (4) as

$$\bar{\sigma}_{\mu\mu} \sim \frac{n}{\pi} \frac{e^2}{h} , \tag{21}$$

which is the well-known minimal conductivity of graphene for $n = 1$ (except for an additional degeneracy factor 4 due to spin and valley degeneracy).[1] The disorder independent conductivity reflects the fact that the conductivity can not distinguish between ballistic and diffusive transport of Weyl fermions.[9] This is remarkable but may be valid only when we restrict our calculation of the functional integral to the saddle-point manifold. Including corrections to the saddle-point manifold would create an additional finite correction factor K_g to the conductivity as $\sigma_{\mu\mu} = K_g \bar{\sigma}_{\mu\mu}$ (cf. Ref. [17]).

It is important to notice that the saddle-point integration in Eq. (7) is restricted to the two-component Grassmann field $(\varphi_{\mathbf{r}}, \varphi'_{\mathbf{r}})$. This is crucial for the derivation of the main results. The integration would be over a larger manifold when the underlying Hamiltonian were replaced by $\hat{H}_0 = diag(H_0, H_0)$ instead of $\hat{H}_0 = diag(H_0, H_0^T)$ in Eq. (7). This would be the case in the conventional supersymmetric approach.[18] The complete equivalence of the fermionic and the bosonic sector in the latter case implies that the supersymmetric action has a higher dimensional symmetry space than our asymmetric action S' in Eq. (10). Then the corresponding effective action has an different symmetry which was identified by Bocquet et al. with the ortho-symplectic Lee group $OSp(2n|2n)/GL(n|n)$.[19] The saddle-point integration is more complex then. It was treated within a renormalization-group approach, which provides an ideal metallic fixed point with infinite conductivity $\sigma_{\mu\mu}$, in contrast to our finite conductivity in Eq. (21). It remains an open question how the infinite conductivity of the symmetric approach is connected to our finite conductivity.

Acknowledgment

Financial support by the DFG grant ZI 305/51 is gratefully acknowledged.

References

1. K.S. Novoselov, A.K. Geim, S.V. Morozov, D. Jiang, M.I. Katsnelson, I.V. Grigorieva, S.V. Dubonos, A.A. Firsov, Nature **438**, 197 (2005).
2. Y. Zhang, Y.-W. Tan, H.L. Stormer, P. Kim, Nature **438**, 201 (2005).
3. A.K. Geim and K.S. Novoselov, Nat. Mater. 6 (2007), p. 183.
4. K. S. Novoselov, E. McCann, S. V. Morozov, V. I. Fal'ko, M. I. Katsnelson, U. Zeitler, D. Jiang, F. Schedin, and A. K. Geim, Nat. Phys. **2**, 177 (2006).
5. M. I. Katsnelson, K. S. Novoselov, and A. K. Geim, Nat. Phys. **2**, 620 (2006).
6. E. Abrahams, P.W. Anderson, D.C. Licciardello and T.V. Ramakrishnan, Phys. Rev. Lett. **42**, 673 (1979).
7. M. Kaveh, Phil. Mag. B **51**, 453 (1985).
8. B. Simon, Comm. Math. Phys. **134**, 209 (1990).

9. D.S.L. Abergel, V. Apalkov, J. Berashevich, K. Ziegler and T. Chakraborty, Adv. Phys. **59**, 261 (2010).
10. K. Ziegler, Phys. Rev. Lett. **102**, 126802 (2009); Phys. Rev. B **79**, 195424 (2009).
11. M.R. Zirnbauer, J. Math. Phys. **37**, 4986 (1996).
12. Zhang F, Sahu B, Min H, MacDonald A H 2010 *Phys. Rev. B* **82** 035409
13. Min H, MacDonald A H 2008 *Phys. Rev. B* **77** 155416
14. J.W. Negele, H. Orland, *Quantum Many-particle Systems*, (Addison-Wesley, New York, 1988).
15. It should be noticed that this different from the conventional supersymmetric functional integral (cf. Ref. [18], where the same Green's functions are expressed with boson and fermion fields.
16. K. Ziegler, Phys. Rev. B 55, 10661 (1997); Phys. Rev. Lett. 80, 3113 (1998).
17. A. Sinner and K. Ziegler, Phys. Rev. B **84**, 233401 (2011).
18. K. Efetov, *Supersymmetry in Disorder and Chaos* (Cambridge University Press 1997).
19. M. Bocquet, D. Serban and M.R. Zirnbauer, Nucl. Phys. B **578**, 628 (2000).

AUTHOR INDEX